Date Due

Mathematical
Methods
of Optimal Control

Mathematical Methods of Optimal Control

V. G. Boltyanskii

Recipient of the 1962 Lenin Prize
for Science and Technology

Vladimir Grigor'evich Boltianskii

Authorized Translation from the Russian

Translator

K. N. Trirogoff

The Aeorspace Corporation

Editor

Ivin Tarnove

TRW Systems Group

Balskrishnan-Neustadt Series

Holt, Rinehart and Winston, Inc.

New York   Chicago   San Francisco   Atlanta   Dallas
Montreal   Toronto   London   Sydney

Preface to the
English Translation

This is an essentially faithful translation of the first Russian edition of Professor Boltyanskii's book enhanced by the inclusion of two additional sections (51 and 52 in this edition). The manuscript for these sections, representing significant new results of practical importance on the synthesis of optimal controls for two classes of nonlinear second-order systems, was provided to us by the author on a recent visit to the United States. The material contained in these additional sections has since been incorporated into the second Russian edition of the book.

References in the book to Russian works have been supplemented by information on English translations whenever these were known to exist. A few references to English books have been included.

To increase the usefulness of the book, the index has been expanded so as to make the contents more accessible. Wherever found, typographical errors have been corrected.

We are very grateful to Professor L. W. Neustadt of the University of Southern California who proposed the undertaking of the translation and provided guidance in the course of this effort. We are pleased to acknowledge the contribution of Dr. K. L. Miller of TRW Systems Group, whose careful reading of the manuscript led to many improvements. Finally, we wish to thank Mrs. Betty G. Griffie for expertly typing the manuscript.

I. Tarnove
K. N. Trirogoff

Los Angeles, California
July 1970

v

Preface to the
Russian Edition

This book is devoted to the study of controlled objects and to the finding of the best means of controlling them. Controlled objects have been firmly implanted in our daily life and, in fact, are beginning to become commonplace. We see them literally at every turn: automobiles, airplanes, various electrical devices equipped with regulators (for example, electrical refrigerators), and so on. Common to all these cases is our ability to "control" or influence to some extent the behavior of the object.

Usually, the transition of the controlled object from one state to another can be accomplished in many ways. This gives rise to the problem of choosing that way which will prove most advantageous from some (completely defined) point of view. This, in fact, is the (somewhat vaguely formulated) problem of optimal control.

It should be clearly stated at the outset that the reader will not find in this book any specific techniques for construction and operation of control systems. Rather, we consider the application of mathematical methods to the calculation of optimal controls. Mathematics does not deal with a real object, but instead, treats mathematical models thereof. The mathematical model of a controlled object is defined at the very beginning of this book. The task in practice is to decide whether the real object of interest can be "matched" to the mathematical framework considered here and to carry out those simplifications and idealizations which are deemed to be admissible. If the object falls into the mathematical framework considered here, then one can attempt to use the theory presented in this book.

The mathematical theory of optimal control is a recent development. The results obtained between 1956 and 1961 by a group of mathematicians led by Academician L. S. Pontryagin serve as a central core of this theory. Important results were also obtained in the U.S.A. by L. W. Neustadt, J. P. LaSalle, and by a group of mathematicians headed by R. Bellman. It is also worth mentioning the works of the Soviet mathematician N. N. Krasovskii, the Czechoslovakian mathematician Y. Kurzweil, and others. Finally, we must take note of the investigations of A. A. Fel'dbaum, one of the pioneers and enthusiasts of this new field.

A monograph containing the basic results of the theory of optimal control was published in 1962 [9]. It acquired considerable popularity, not only among mathematicians, but also among engineers. The basic results formulated in this monograph were comparatively simple and easily understood, but the proofs require a great deal of mathematical maturity. Suffice it to say that they contain the concept of the Lebesque integral, differential equations with measurable right-hand sides, the theorem on the weak compactness of a sphere in the space of linear potentials, and so on. In view of this, the author had long contemplated writing a simpler book covering the same problems. The plan became progressively clearer while the author was repeatedly teaching the course on optimal control theory. Finally, I succeeded in finding a proof of the existence theorem for optimal controls which does not use the concept of weak compactness of a sphere. This made it possible to eliminate the use of measurable functions and the Lebesque integral, and to return to my original proof of the maximum principle, using only piecewise-continuous functions. The presentation at once becomes noticeably simpler.

Because of the simplifications that were made, the book is accessible, for example, to a student proficient in engineering college mathematics. In addition to containing considerable simplifications, the book differs in content from the above-mentioned monograph. I have not included the problem of pursuit of a stochastically moving object and the problem involving constraints imposed on phase coordinates, since these problems are complicated (because of the nature of the results obtained) and are of more theoretical rather than practical value. In addition, some new results are included in the book, of which we must mention first the exceedingly interesting and elegant results obtained by Neustadt (and supplementary results due to Eaton) on the approximate computation of linear time-optimal controls; also included and due to the author are sufficient conditions for optimality, the foundations of the method of dynamic programming which were obtained by the author, a number of new examples, and so on. The appropriateness of this choice of material must be judged by the reader.

In conclusion, I would like to express my sincere graditude to my teacher and friend, L. S. Pontryagin: the continuous creative contact with him was the basic inspiration for my modest achievements. I also thank all those who, by their attention and assistance, helped in the publication of this book, especially my friends and colleagues E. F. Mishchenko and R. V. Gamkrelidze.

V. G. Boltyanskii

Moscow, Russia
December 1965

Contents

Mathematical
Methods
of Optimal Control

chapter 1 Introduction

§ 1. THE TIME-OPTIMAL PROBLEM

1. The Concept of Controlled Objects. Our basic aim throughout this book will be to study *controlled objects* and to find the best ways of *controlling* them. Accordingly, let us first explain the meaning of these concepts.

Of course, in carrying out the following mathematical investigation, we do not deal directly with real objects but with a mathematical model for these objects which we now describe.

Consider, for example, rectilinear motion of an automobile. At each instant of time the state of the automobile is characterized by two quantities: the distance travelled, s, and the velocity, v. These two quantities change with time but not in an arbitrary manner; rather, they are subject to the will of a driver who, at his discretion, can control the operation of the vehicle by increasing or decreasing the force F being developed by the engine. Thus, we have three interrelated parameters: s, v, and F, shown schematically in Figure 1. The quantities s and v which characterize the

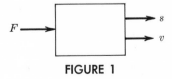

FIGURE 1

state of the automobile are called *phase coordinates*, and the quantity F is called a *control parameter*. If we were to consider the motion of an automobile on a plane (and not along a straight line), then there would be four phase coordinates (two "geographical" coordinates and two velocity components) and two control parameters (for example, the force of the engine and the angle of rotation of the steering wheel). In the case of an airplane in flight one may consider six phase coordinates (three spatial coordinates and three velocity components) and several control parameters (the thrust of the engine, quantities characterizing the position of the elevator, rudder, and ailerons, and so forth). For an electric iron with temperature control the current and temperature serve as phase coordinates, while the position of the controller is the control parameter.

The following mathematical description of controlled objects is a natural abstraction from the above discussion. The *state* of an object is specified (at each instant of time) by n numbers x^1, x^2, . . . , x^n, called *phase coordinates* of the object. The object's motion, from a mathematical point of view, consists in the state of the object changing with time, that is, x^1, x^2, . . . , x^n are variable quantities (functions of time). This motion of the object does not take place arbitrarily but can be *controlled;* for this purpose, the object is equipped with "controllers" whose positions are characterized (at each instant of time) by r numbers u^1, u^2, . . . , u^r; these numbers are called *control parameters*. The controllers can be "manipulated," that is, the control parameters u^1, u^2, . . . , u^r may be changed with time as desired. In other words, we can, at our discretion, choose the functions $u^1(t)$, $u^2(t)$, . . . , $u^r(t)$ which describe the change of the control parameters with time. On the other hand, the functions $x^1(t)$, $x^2(t)$, . . . , $x^n(t)$ are not completely at our disposal; we assume (as is usually the case) that knowing the state of the object at the initial time t_0 and having selected the control functions $u^1(t)$, $u^2(t)$, . . . , $u^r(t)$ (for $t > t_0$), we can mathematically calculate the exact behavior of the object for all $t > t_0$, that is, we can find the functions $x^1(t)$, $x^2(t)$, . . . , $x^n(t)$ characterizing the change of the phase coordinates with time. Thus, we can influence the change of the phase coordinates, to a certain extent, by appropriate choice of the control functions $u^1(t)$, $u^2(t)$, . . . , $u^r(t)$.

The object which we have discussed above is depicted in the theory of automatic control as shown in Figure 2. The quantities u^1, . . . , u^r (control parameters) are frequently also called "input variables," and the quantities x^1, . . . , x^n (phase coordinates) are termed "output variables." One also says that u^1, . . . , u^r are supplied "at the input" of the object, and x^1, . . . , x^n are obtained "at the output." Of course, Figure 2 shows only a *conventional diagram* for a controlled object but does not reflect its internal structure which must be known in order to clarify the *manner* in which knowledge of the control functions $u^1(t)$, . . . , $u^r(t)$ makes possible the calculation of the change in the phase coordinates $x^1(t)$, . . . , $x^n(t)$.

It is convenient to consider u^1, . . . , u^r as the components of a *vector* $u = (u^1, u^2, . . . , u^r)$ which is also called the *control parameter* (vector).

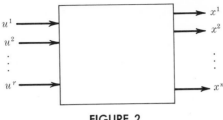

FIGURE 2

Similarly, it is convenient to consider $x^1, \ldots, x^n$ as the components of a vector (or point) $x = (x^1, \ldots, x^n)$; this point is called the state of the object. Every state $x = (x^1, \ldots, x^n)$ is a point of an n-dimensional space with coordinates $x^1, \ldots, x^n$. This n-dimensional space, in which the states of the object are denoted by points, is called the *phase space* of the object under consideration. If the object is such that its state is characterized by only two coordinates x^1, x^2 (compare Figure 1), then we speak of the *phase plane*, and not of the phase "space." In this case, the states of the object can be represented particularly clearly. For this reason and for the sake of clarity of exposition all the drawings and a number of proofs will be carried out for $n = 2$ (that is, in a phase plane); on the other hand, in formulating the theorems, we shall assume that n (the number of phase coordinates) is an arbitrary number.

Thus, using vector notation, the controlled object can be depicted as shown in Figure 3. The input $u = (u^1, u^2, \ldots, u^r)$ is the control parameter, and the output $x = (x^1, \ldots, x^n)$ is a point of the phase space (or, otherwise, the state of the object).

As mentioned previously, in order to completely define the motion of an object, it is necessary to specify its state at the initial time t_0 and to choose the control functions $u^1(t), u^2(t), \ldots, u^r(t)$ (for $t > t_0$), that is, to choose the vector function

$$u(t) = (u^1(t), u^2(t), \ldots, u^r(t)).$$

This function $u(t)$ will be called the *control*. The specification of the initial state x_0 and the control $u(t)$ uniquely defines the subsequent motion of the

FIGURE 3   **FIGURE 4**

object. This motion consists in the phase point

$$x(t) = (x^1(t), x^2(t), \ldots , x^n(t)),$$

which represents the state of the object, moving with time and describing a path in the phase space called the *phase trajectory* of the motion under consideration (Figure 4). The pair of vector functions $(u(t), x(t))$, that is, the control $u(t)$ and the corresponding phase trajectory $x(t)$, will be called in the following the *control process* or simply the *process*.

Thus, summarizing the above, we can say that the state of the *controlled object* at every instant of time is characterized by the *phase point* $x = (x^1, x^2, \ldots , x^n)$. The object's motion can be affected by the *control parameter*

$$u = (u^1, u^2, \ldots , u^r).$$

The evolution of u and x with time is called the *process;* the process $(u(t), x(t))$ consists of the *control* $u(t)$ and the *phase trajectory* $x(t)$. The process is completely determined by the control $u(t)$ (for $t > t_0$) and the initial state $x_0 = x(t_0)$.

2. The Control Problem.   One frequently encounters the following problem concerning controlled objects. At the initial time t_0, the object is in the state x_0; it is required to choose a control $u(t)$ such that the object will be transferred to a preassigned terminal state x_1 (different from x_0; Figure 5). Moreover, it is usually also required that the *transition process*

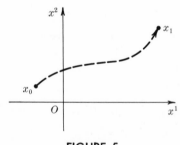

FIGURE 5

(that is, the process of transition from the initial state x_0 to the preassigned terminal state x_1) be "best" in a definite sense, for example, that the transition be accomplished in the shortest time or that minimum energy be expended during the transition process, and so forth. A "best" transition process is called an *optimal process*. We see that to make the term "optimal process" more precise it is necessary to clarify the sense in which optimality is to be understood. If we speak of the shortest transition time, then such processes are called *time-optimal processes*. In other words, a process resulting in the transfer of the object from the point x_0 to the point x_1 (Figure 5) is said to be time-optimal if there does not exist a process that transfers the object from x_0 to x_1 in a shorter time (it is assumed here and in the following that $x_1 \neq x_0$).

The study of optimal processes is the basic theme of this book. We shall indicate below a very general and precise definition of the term "optimal processes." However, the majority of results and examples will pertain to the case of time-optimality.

For the problem stated above (find a control which transfers the object from the initial state x_0 to the prescribed terminal state x_1), the initial state x_0 is frequently not known beforehand. Let us consider a typical example. An object is required to operate steadily in a given condition (that is, to remain in a state x_1). For various reasons, the object may depart from the operating state x_1 and be transferred to some other state x_0. In this case, the object must be controlled in such a manner as to return it from the state x_0 to the required operating state x_1. Moreover, the point x_0 to which the object may be transferred (as a result of an unexpected push or for some other reason) is not known beforehand, and we must be able to control the object so as to return it from any point x_0 to its operating state x_1 (Figure 6).

This type of control is frequently accomplished by a human (operator) who watches various instruments in trying to maintain the object in the required operating state. However, under modern conditions of highly developed technology, an operator cannot cope with this task because of the complexity of the object's behavior, the high speed of the processes, and so on. Therefore, it is extremely important to design devices that are capable of controlling the operation of an object (for example, to return the object to its operating state in the case of a departure from this state) without human participation. Such devices ("regulators," "automatic control systems," and so on) are widely used in technology; their study lies in the domain of the theory of automatic control. The first regulator of this kind

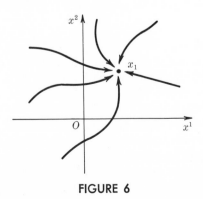

FIGURE 6   **FIGURE 7**

was Watt's centrifugal regulator designed to control the operation of a steam engine (Figure 7). Roughly speaking, this regulator operates in the following manner. A vertical bar connected to the shaft of the steam engine rotates at a certain angular velocity ω. Under the action of centrifugal force, the weights of the regulator move apart so that rods holding these weights deviate from the vertical bar and form a certain angle φ with the vertical bar. As the weights move apart, a sleeve M, fitted over the vertical bar, rises; the sleeve M, in turn, is connected to a valve in the steam pipe, so that as the sleeve M moves, the supply of steam to the engine's cylinders either decreases or increases. If the engine is operating under steady conditions, and its velocity *decreases* for some reason, then the weights come to a lower position and the sleeve moves down. This *opens* the valve and, as a result, the steam supply increases and the velocity of the engine begins to grow. Conversely, if the angular velocity *exceeds* the operating velocity, then the weights move apart and the sleeve rises. This *closes* the valve and, as a result, the steam supply decreases and the velocity of the engine begins to drop.

A schematic of this regulator is shown in Figure 8. The steam supply is the input quantity for the steam engine; it can be characterized by the angle φ (Figure 7) since the steam supply is determined by the valve's position and the valve is connected rigidly to the sleeve. The angular velocity ω serves as an output quantity for the steam engine. This angular velocity ω is transmitted to the input of the regulator which is designed in such a way as to "figure out" the quantity φ controlling the steam supply.

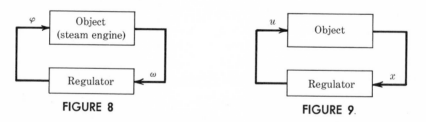

FIGURE 8   FIGURE 9.

In a general case (Figure 9), the phase coordinates are transmitted to the input of the regulator. The regulator is designed so that its output u, supplied at the input of the object, controls the operation of the object in an appropriate way (that is, returns the object to the operating state if, for some reason, the object departs from this state).

Let us recall that the regulator must not only return the object to its operating state, but must do so in the best possible way, for example, in the sense of rapidity of action (that is, it must return the object to the operating state in the shortest time). In this connection, regulators of various designs and degrees of sophistication are considered in the theory of automatic control. The study of regulators which have been used thus far leads, apparently, to the conclusion that reducing the transition time of a process can only be achieved by designing more complicated regulators; therefore, designing more sophisticated regulators makes it possible merely to approach the "ideal," "optimal" regulator which in all cases accomplishes the transition in the shortest time. Apparently, an "optimal" regulator cannot be realized precisely. However, this deduction is erroneous. One of the most important merits of the theory of optimal processes is that it establishes the existence of optimal regulators and creates the mathematical apparatus for designing such regulators. The optimal regulators differ substantially from those used thus far in the theory of automatic control. Presently, the first steps are being made in engineering and computing practice to create and implement such "optimal" regulators. It may be expected that optimal regulators will play an important role in future technology.

3. Equations of Motion of an Object.   We begin by considering a simple example. Let G be a particle in rectilinear motion (Figure 10). We shall assume that its mass is a constant equal to m and that its coordinate (measured from a certain point O on the straight line along which the particle moves) is denoted by x^1. During the motion of G, its coordinate x^1 changes with time. The derivative $\dot{x}^1$ is the velocity of G.

We shall assume that G is subjected to two external forces: the friction force $b\dot{x}^1$ and the elastic force kx^1; in addition, it is assumed that G is equipped with an engine. The force developed by the engine and acting on G will be denoted by u. Thus, according to the second law of Newton, the subsequent motion of G can be described by the differential equation

$$m\ddot{x}^1 = -b\dot{x}^1 - kx^1 + u.$$

Denoting the velocity by x^2 (that is, setting $\dot{x}^1 = x^2$), we can write this law of motion in the form of the following system of differential equations:

$$\dot{x}^1 = x^2,$$
$$\dot{x}^2 = -\frac{k}{m}x^1 - \frac{b}{m}x^2 + \frac{1}{m}u. \tag{1.1}$$

Here, the quantities x^1 and x^2 are the phase coordinates of G, and u is the control parameter, that is, we have the object represented schematically in Figure 11.

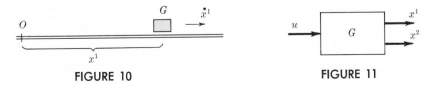

FIGURE 10   FIGURE 11

Equation (1.1) is the rule of change of the phase coordinates with time, that is, the rule of motion of the phase point in the phase plane.

We have considered only one specific case, but we could give a number of other examples in which the rule of motion of an object is described by differential equations. In general [compare (1.1)], these equations provide expressions for the derivatives of the phase coordinates in terms of the phase coordinates themselves and the control parameters, that is, they have the form

$$\begin{aligned}
\dot{x}^1 &= f^1(x^1, \ldots, x^n, u^1, \ldots, u^r), \\
\dot{x}^2 &= f^2(x^1, \ldots, x^n, u^1, \ldots, u^r), \\
&\quad\quad\quad \cdot \quad \cdot \quad \cdot \\
\dot{x}^n &= f^n(x^1, \ldots, x^n, u^1, \ldots, u^r),
\end{aligned} \tag{1.2}$$

where $f^1, f^2, \ldots, f^n$ are functions defined by the internal structure of the object. In the sequel we shall confine our attention to objects (Figure 2) whose rule of motion is described by a system of differential equations of the form (1.2). Using vector notation, (1.2) can be written in the form

$$\dot{x} = f(x, u), \tag{1.3}$$

where x is a vector with components $x^1, x^2, \ldots, x^n$, u is a vector with components $u^1, \ldots, u^r$, and finally, $f(x, u)$ is a vector whose components are the right-hand sides of (1.2).

Of course, it is not possible to solve the system of differential equations (1.2) (that is, to find the motion of an object) without knowing how the control parameters $u^1, \ldots, u^r$ change with time. Conversely, knowing the behavior of $u^1, \ldots, u^r$, that is, knowing the control functions $u^1(t)$, $u^2(t), \ldots, u^r(t)$ for $t > t_0$, we can, with the aid of the system of equations

$$\begin{aligned}
\dot{x}^1 &= f^1(x^1, \ldots, x^n, u^1(t), \ldots, u^r(t)), \\
\dot{x}^2 &= f^2(x^1, \ldots, x^n, u^1(t), \ldots, u^r(t)), \\
&\quad\quad\quad \cdot \quad \cdot \quad \cdot \\
\dot{x}^n &= f^n(x^1, \ldots, x^n, u^1(t), \ldots, u^r(t))
\end{aligned} \tag{1.4}$$

or, equivalently, with the aid of the vector equation

$$\dot{x} = f(x, u(t)), \tag{1.5}$$

uniquely determine the object's motion (for $t > t_0$) if we know the initial state of the object (at time $t = t_0$). In other words, specifying the control $u(t)$ and the initial state x_0 uniquely determines the phase trajectory $x(t)$ for $t > t_0$, which is in agreement with the assumption made previously on the properties of the object. The fact that giving the initial state (at time $t = t_0$) makes it possible to determine uniquely the phase trajectory $x(t)$, $t > t_0$, with the aid of (1.4) follows from the *theorem on the existence and uniqueness of solutions of systems of differential equations.* The formulation of this theorem will be given below (p. 46).

Suppose that, given an initial state x_0 and the control $u(t) = (u^1(t), \ldots, u^r(t))$, we have determined a trajectory $x(t)$ [with the aid of (1.4)]. If we change the control $u(t)$ (preserving the same initial state x_0), then we obtain some other trajectory emanating from the same point x_0; changing once again the control $u(t)$, we obtain still another trajectory, and so on. Thus, considering various controls $u(t)$, we obtain many trajectories emanating from the point x_0 (Figure 12). [Of course, this does not contradict

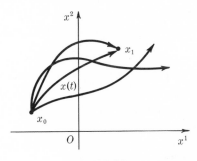

FIGURE 12

the uniqueness theorem in the theory of differential equations since, replacing the functions $u^1(t), \ldots, u^r(t)$ by other functions, we turn (1.4) into another system of differential equations in the phase coordinates $x^1, \ldots, x^n$.] Recall that the *time-optimal problem* consists in finding a control $u(t)$ for which the corresponding phase trajectory $x(t)$ passes through the point x_1 with the transition from x_0 to x_1 being accomplished in the shortest time. Such a control, $u(t)$, will be called an *optimal control;* similarly, the corresponding trajectory $x(t)$ along which the phase point is transferred in the shortest time from the state x_0 to the state x_1 will be called an *optimal trajectory.*

4. Admissible Controls. Usually the control parameters $u^1, \ldots, u^r$ cannot take on completely arbitrary values but are subject to certain constraints. Thus, for example, in the case of the object described on p. 6, it is natural to assume that the force u developed by the engine cannot be arbitrarily large in magnitude but is subject to the constraints

$$\alpha \leqslant u \leqslant \beta,$$

where α and β are constants characterizing the engine. In particular, for $\alpha = -1, \beta = 1$, we obtain

$$-1 \leqslant u \leqslant 1,$$

which means that the engine can develop a force acting in either direction, but not exceeding unity in magnitude.

The constraints have an analogous meaning in other cases: control parameters such as the amount of fuel supplied to an engine, temperature, current, voltage, and so on, cannot take on arbitrarily large values.

In applications, we frequently encounter the case of an object containing r control parameters $u^1, u^2, \ldots, u^r$ where these parameters may vary arbitrarily within the following bounds:

$$\alpha^1 \leqslant u^1 \leqslant \beta^1, \qquad \alpha^2 \leqslant u^2 \leqslant \beta^2, \ldots, \qquad \alpha^r \leqslant u^r \leqslant \beta^r.$$

In other words, each of the quantities $u^1, u^2, \ldots, u^r$ in (1.2) is an independent control parameter whose domain of values does not depend on the values of the other control parameters, and is given by the inequalities

$$\alpha^i \leqslant u^i \leqslant \beta^i, \qquad i = 1, \ldots, r. \tag{1.6}$$

It should be noted that for $r = 2$, the points $u = (u^1, u^2)$, whose coordinates are subject to the inequalities (1.6), fill out a rectangle (Figure 13); for $r = 3$, the inequalities (1.6) define a parallelepiped in u_1, u_2, u_3-space; in the case of an arbitrary r one says that the inequalities (1.6) define an *r-dimensional parallelepiped*.

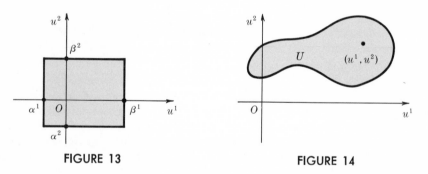

FIGURE 13   FIGURE 14

In the general case, it is assumed that, by virtue of the design of an object and the conditions of its operation, a certain set U (Figure 14) is given in $u^1, \ldots, u^r$-space, and that the control parameters $u^1, u^2, \ldots,$ u^r, at each instant of time, must take on only those values for which the point $u = (u^1, u^2, \ldots, u^r)$ belongs to U. In other words, one is permitted to consider only controls $u(t)$ for which $u(t) \in U$ for any t (the symbol $\in$ denotes that a point belongs to a set). In the sequel, U will be called the *control region*. The control region is not required to be a parallelepiped; geometrically it can have a more or less complicated character, since by virtue of the construction of the control portion of an object, there can exist relationships between the control parameters $u^1, u^2, \ldots, u^r$

which are expressed by equations of the type $\varphi(u^1, u^2, \ldots, u^r) = 0$ or by inequalities of the form $\psi(u^1, u^2, \ldots, u^r) \leqslant 0$. For example, if the parameters u^1, u^2 characterize a vector in a plane whose absolute value does not exceed unity but whose direction is arbitrary, then these parameters are subject only to the single condition

$$(u^1)^2 + (u^2)^2 - 1 \leqslant 0 \tag{1.7}$$

and the control region is a circle (Figure 15). In the following, we assume that the designation of the control region is included in the mathematical definition of the object. Thus the mathematical representation of an object entails specifying its rule of motion, (1.2), and the control region U. We note that in technical problems the case where U is a closed set, that is, the case where the point $u = (u^1, u^2, \ldots, u^r)$ may lie not only inside U but also on its boundary, is particularly important and characteristic [see inequalities (1.6) and (1.7)]. This condition means that the extreme positions of the "controllers" [that is, the values $u^i = \alpha^i$ or $u^i = \beta^i$ in inequalities (1.6), or boundary points of the circle (1.7)] are also admissible.

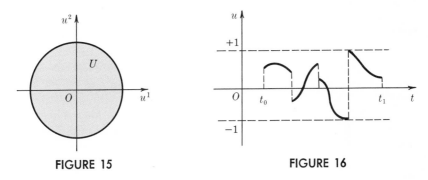

FIGURE 15   FIGURE 16

Finally, let us make one more assumption on the nature of the controls which will be of considerable importance in the sequel. Namely, we shall assume that the "controllers," whose positions are characterized by control parameters u^1, u^2, $\ldots$, u^r, are *inertialess*, so that we may, if needed, switch these "controllers" *instantaneously* from one position to another, that is, change (stepwise) the values of the control parameters u^1, u^2, $\ldots$, u^r. Accordingly, we shall consider not only continuous but also arbitrary *piecewise continuous* controls $u(t)$, that is, controls consisting of a finite number of continuous pieces (Figure 16). The class of piecewise continuous controls is apparently of great interest in practical applications of the theories presented here.

In order to avoid any misunderstanding, we specify that a function $u(t)$ (with values in the control region U) is called piecewise continuous if it is continuous for all t under consideration except for a finite number of points at which $u(t)$ may undergo discontinuities of the first kind; this means that finite limits from the left and the right are assumed to exist at every point of discontinuity τ:

$$u(\tau - 0) = \lim_{\substack{t \to \tau \\ t < \tau}} u(t), \qquad u(\tau + 0) = \lim_{\substack{t \to \tau \\ t > \tau}} u(t).$$

In the following, the value of a piecewise continuous control $u(t)$ at a point of discontinuity is of no particular importance. However, for the sake of definiteness, it is convenient to assume that the value of a control $u(t)$ at every point of discontinuity τ is equal to its limit from the right, namely,

$$u(\tau) = u(\tau + 0),$$

and that every control $u(t)$ under consideration is continuous at the endpoints of the interval $t_0 \leqslant t \leqslant t_1$ on which it is defined, that is, that all its points of discontinuity, if any, lie in the interval $t_0 < t < t_1$ (Figure 16).

For convenience, we shall agree to apply the term *admissible control* to any piecewise continuous function $u(t)$, $t_0 \leqslant t \leqslant t_1$, with values in the control region U, which is continuous at the endpoints of the interval $t_0 \leqslant t \leqslant t_1$ on which it is defined. The time-optimal problem can now be stated more precisely as follows:

Among all the admissible controls $u = u(t)$ which transfer an object from a given initial state x_0 to a prescribed terminal state x_1, find one for which this transfer is accomplished in the shortest time.

In conclusion, let us make several remarks concerning the assumption of "inertialess controllers." This assumption is sometimes met with objections. For example, letting u denote the angle of rotation of the rudder (of an airplane or a ship), it is noted that the "control parameter" u cannot change abruptly since the rudder may have to operate against the flow and hence the engine, which has finite power, can turn the rudder only with a limited velocity. However, this objection is evidently based on a misunderstanding. In fact, it suffices to write

$$\dot{u} = v,$$

where v is the velocity of the rudder's rotation, in order to realize that it is more appropriate to consider the quantity u not as a control parameter but as another phase coordinate, and to take the quantity v (which can vary within finite limits) as a control parameter. Of course, every real process has a certain "inertia," but in every real controlled object, it is always possible to find control parameters which can be considered inertialess to within reasonable accuracy.

Despite this explanation, a question naturally arises as to the need for piecewise continuous controls. Is it not possible to confine oneself to continuous controls? The answer to this question must be negative. As a matter of fact, as we shall see below in the simplest examples, the optimal control turns out, as a rule, to be discontinuous (that is, it contains instantaneous jumps, switchings). If, for example, the discontinuous function shown in Figure 17 by the solid line represents an optimal control, then "smoothing" this function (dotted line in Figure 17) we obtain a continuous

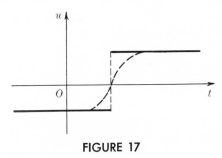

FIGURE 17

function which is close to the former. But no matter what continuous function we take, it is always possible to choose a steeper "smoothing" segment and obtain a continuous function which is even closer to the optimal control. Thus, the class of continuous functions simply does not contain the optimal control. Piecewise continuous controls are advantageous in that, in the first place, they make it possible to obtain the *exact mathematical solution* of the optimal problem for a sufficiently wide class of examples and, secondly, they are sufficiently intuitive and convenient for practical purposes.

§ 2. BASIC TRENDS IN THE THEORY OF OPTIMAL PROCESSES

5. The Method of Dynamic Programming. We now consider the problem of optimal transfer (in the sense of rapidity of action) from the state x to the state x_1 for the controlled object described in the previous section. Moreover, the terminal phase point x_1 will be considered to be fixed, and various points of the phase space will be taken as the initial point x. We shall assume here that the controlled object satisfies the following:

Hypothesis 1. *For any point x of the phase space (different from x_1) there exists an optimal process (in the sense of rapidity of action) transferring the point x to the point x_1 (Figure 6).*

The time during which the optimal transfer from the point x to the point x_1 is accomplished will be denoted by $T(x)$. In other words, for every point x there exists an admissible control $u = u_x(t)$ under whose action the object is transferred from the point x to the point x_1 during the time $T(x)$;

moreover, a transfer from the point x to the point x_1 in a time less than $T(x)$ is not possible.

Since every point x of the phase space has n components $x^1, x^2, \ldots, x^n$, $T(x)$ is a *function of n variables:*

$$T(x) = T(x^1, x^2, \ldots, x^n).$$

Therefore, it makes sense to speak of the continuity of this function (jointly in the variables $x^1, \ldots, x^n$) and the differentiability of this function with respect to each of the variables $x^1, \ldots, x^n$. We also assume in this section that the controlled object satisfies the following:

Hypothesis 2. The function $T(x)$ is continuous and has continuous partial derivatives $\partial T/\partial x^1$, $\partial T/\partial x^2$, $\ldots$, $\partial T/\partial x^n$ everywhere except at the point x_1.

It will be convenient for subsequent considerations to introduce the function $\omega(x)$ which differs from $T(x)$ only in sign:

$$\omega(x) = -T(x). \tag{1.8}$$

Since we assume that Hypotheses 1 and 2 are satisfied, the function $\omega(x)$ is defined and continuous in the entire phase space and has continuous partial derivatives $\partial\omega/\partial x^1$, $\partial\omega/\partial x^2$, $\ldots$, $\partial\omega/\partial x^n$ everywhere except at the point x_1.

Now let x_0 be an arbitrary point of the phase space different from x_1, and let u_0 be an arbitrary point of the control region U. Assume that the object, in the state x_0 at time t_0, moves under the action of a constant control $u = u_0$. The phase trajectory of the object during this motion will be denoted by

$$y(t) = (y^1(t), y^2(t), \ldots, y^n(t)).$$

Thus, the phase trajectory $y(t)$ for $t > t_0$ satisfies the equations

$$\dot{y}^i(t) = f^i(y(t), u_0), \qquad i = 1, 2, \ldots, n \tag{1.9}$$

[see (1.2) and (1.3)] and the initial condition

$$y(t_0) = x_0. \tag{1.10}$$

If we move from the point x_0 to the point $y(t)$ (along the phase trajectory under consideration), then the time spent on this motion will be $t - t_0$. Then, moving optimally from the point $y(t)$ (Figure 18), the time spent on the motion from the point $y(t)$ to the point x_1 will be $T(y(t))$. As a result, we accomplish the transfer from x_0 to x_1 in time $(t - t_0) + T(y(t))$. But since the optimal time of the motion from the point x_0 to x_1 is $T(x_0)$, that is, $T(y(t_0))$, we have

$$T(y(t_0)) \leqslant (t - t_0) + T(y(t)).$$

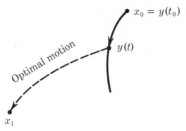

$x_0 = y(t_0)$

$y(t)$

Optimal motion

x_1

FIGURE 18

Replacing the function T by ω [see (1.8)] and dividing both sides of the inequality by the positive quantity $t - t_0$, we obtain

$$\frac{\omega(y(t)) - \omega(y(t_0))}{t - t_0} \leqslant 1,$$

and therefore, passing to the limit as $t \to t_0$, we find

$$\frac{d}{dt} \omega(y(t)) \Big|_{t=t_0} \leqslant 1. \qquad (1.11)$$

According to the formula for the total derivative, we have

$$\frac{d}{dt} \omega(y(t)) = \sum_{i=1}^{n} \frac{\partial \omega(y(t))}{\partial x^i} \dot{y}^i(t),$$

and therefore, from (1.9) and (1.10), the inequality (1.11) takes the form

$$\sum_{i=1}^{n} \frac{\partial \omega(x_0)}{\partial x^i} f^i(x_0, u_0) \leqslant 1.$$

The points x_0, u_0 were taken here arbitrarily.

Thus, *for any point x of the phase space (different from x_1) and for any point u of the control region U,*

$$\sum_{i=1}^{n} \frac{\partial \omega(x)}{\partial x^i} f^i(x, u) \leqslant 1. \qquad (1.12)$$

Now, let $(u(t), x(t))$ be an optimal process which transfers the object from the state x_0 to the state x_1, and let $t_0 \leqslant t \leqslant t_1$ be the time interval during which this optimal motion occurs, so that $x(t_0) = x_0$, $x(t_1) = x_1$, and $t_1 - t_0 = T(x_0)$. By virtue of the equations of motion of the object we have

$$\dot{x}^i(t) = f^i(x(t), u(t)), \qquad t_0 \leqslant t \leqslant t_1. \qquad (1.13)$$

The motion along the optimal trajectory under consideration from x_0 to $x(t)$ is accomplished during the time $t - t_0$, and the motion from $x(t)$ to x_1 requires the time $t_1 - t$, that is, the time $T(x_0) - (t - t_0)$. It is not possible to go from $x(t)$ to x_1 in a time less than $T(x_0) - (t - t_0)$; in fact, if such a more rapid motion were to exist (the dotted line in Figure 19) then moving

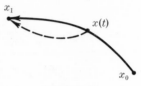

FIGURE 19

from x_0 to $x(t)$ during the time $t - t_0$, and then from $x(t)$ to x_1 during a time less than $T(x_0) - (t - t_0)$, we would accomplish the transfer from x_0 to x_1 during a time less than $T(x_0)$, which is not possible. Thus, $T(x_0) - (t - t_0)$ is the time of the optimal motion from $x(t)$ to x_1, namely,

$$T(x(t)) = T(x_0) - (t - t_0).$$

Replacing T by ω,

$$\omega(x(t)) = \omega(x_0) + t - t_0,$$

and differentiating with respect to t, we obtain

$$\sum_{i=1}^{n} \frac{\partial \omega(x(t))}{\partial x^i} \dot{x}^i(t) = 1,$$

that is [see (1.13)],

$$\sum_{i=1}^{n} \frac{\partial \omega(x(t))}{\partial x^i} f^i(x(t), u(t)) = 1, \qquad t_0 \leqslant t \leqslant t_1. \tag{1.14}$$

Thus, *for every optimal process*, (1.14) *is satisfied during the entire motion.* Introducing the function

$$B(x, u) = \sum_{i=1}^{n} \frac{\partial \omega(x)}{\partial x^i} f^i(x, u), \tag{1.15}$$

(1.12) and (1.14) may be written as follows:

$$B(x, u) \leqslant 1 \qquad \text{for all points } x \neq x_1 \text{ and } u; \tag{1.16}$$
$$B(x(t), u(t)) \equiv 1 \qquad \text{for any optional process } (x(t), u(t)). \tag{1.17}$$

We have thus proved the following:

Theorem 1.1.   *If Hypotheses* 1 *and* 2 *are satisfied for the controlled object under consideration and for the prescribed terminal state* x_1, *then* (1.16) *and* (1.17) *hold* (optimality is understood in the sense of rapidity of action).

This theorem constitutes the essence of the *method of dynamic programming* for the problem under consideration. It can be formulated somewhat differently. Writing (1.17) for $t = t_0$, we obtain

$$B(x_0,\ u(t_0)) = 1,$$

that is, *for any point* x_0 (different from x_1) *there is a point u in* U, *namely,* $u = u(t_0)$, *such that* $B(x_0, u) = 1$. In conjunction with inequality (1.16), we obtain the relation

$$\max_{u \in U} B(x,\ u) = 1 \qquad \text{for any point } x \neq x_1$$

or, equivalently,

$$\max_{u \in U} \sum_{i=1}^{n} \frac{\partial \omega(x)}{\partial x^i} f^i(x,\ u) = 1 \qquad \text{for any point } x \neq x_1. \qquad \textbf{(1.18)}$$

Thus, *if Hypotheses* 1 *and* 2 *hold, then the function* ω *satisfies* (1.18); *moreover, the maximum in* (1.18) *is attained for optimal processes* [see (1.17)]. This assertion represents an alternate formulation of the method of dynamic programming; (1.18) is called *Bellman's equation.*

The method of dynamic programming (1.16), (1.17) [or equivalently, (1.18), (1.17)] contains information on optimal processes, and therefore, may be used in finding them. However, it has a number of disadvantages. In the first place, the use of this method requires finding not only the optimal controls, but also the function $\omega(x)$, since this function enters into (1.16)–(1.18). Secondly, Bellman's equation [or (1.16) and (1.17)] is a partial differential equation for ω, made more complicated by the maximum sign. This fact hinders substantially the use of the method of dynamic programming for finding optimal processes. But the principal disadvantage of this method is the assumption that Hypotheses 1 and 2 be satisfied. In fact, the optimal controls and the function ω are not known beforehand, so that Hypotheses 1 and 2 contain assumptions on an unknown function and it is impossible to verify the validity of these hypotheses from the equations of motion of an object. This shortcoming could be considered of no particular importance if one could prove, after solving the optimal problem by the method of dynamic programming, that $\omega(x)$ is indeed continuously differentiable. But, as a matter of fact, even in the simplest linear problems of optimal control, the function ω, as we shall see below, is not, as a rule, differentiable everywhere, and the use of the above method is not justified.

Although the method of dynamic programming does not have a rigorous logical foundation in all cases, it nevertheless can frequently be valuable when used in a heuristic manner. Below (Chapter 4) we shall give more refined theorems which are very close in form to the principle of dynamic programming but have a much wider range of application. We shall note also that in the case of difference (not differential) equations the method of dynamic programming is fully justified and is the most efficient method.

6. The Maximum Principle. We now continue the discussion of the previous section assuming $\omega(x)$ to be twice continuously differentiable. Thus, we assume the validity of the following

Hypothesis 3. *The function $\omega(x)$ has continuous second derivatives*

$$\frac{\partial^2 \omega(x)}{\partial x^i \, \partial x^j}, \qquad i, j = 1, 2, \ldots, n,$$

and the functions $f^i(x, u)$ have continuous first derivatives

$$\frac{\partial f^i(x, u)}{\partial x^j}, \qquad i, j = 1, 2, \ldots, n.$$

Let $(u(t), x(t))$, $t_0 \leqslant t \leqslant t_1$, be an optimal process which transfers an object from the state x_0 to the state x_1. Let us fix the time t, $t_0 \leqslant t \leqslant t_1$, and consider the function $B(x, u(t))$ of the variable x. It follows from the expression for B [see (1.15)] and Hypothesis 3 that $B(x, u(t))$ has continuous derivatives with respect to the variables $x^1, x^2, \ldots, x^n$:

$$\frac{\partial B(x, u(t))}{\partial x^k} = \sum_{i=1}^{n} \frac{\partial^2 \omega(x)}{\partial x^i \, \partial x^k} f^i(x, u(t)) + \sum_{i=1}^{n} \frac{\partial \omega(x)}{\partial x^i} \cdot \frac{\partial f^i(x, u(t))}{\partial x^k}, \tag{1.19}$$

$$k = 1, \ldots, n.$$

Furthermore, by virtue of (1.16) and (1.17), we have

$$B(x, u(t)) \leqslant 1 \quad \text{for any } x \neq x_1;$$
$$B(x, u(t)) = 1 \quad \text{for } x = x(t);$$

consequently, $B(x, u(t))$ attains a maximum at $x = x(t)$ and therefore its partial derivatives with respect to $x^1, \ldots, x^n$ vanish at this point, namely,

$$\sum_{i=1}^{n} \frac{\partial^2 \omega(x(t))}{\partial x^i \, \partial x^k} f^i(x(t), u(t))$$

$$+ \sum_{i=1}^{n} \frac{\partial \omega(x(t))}{\partial x^i} \frac{\partial f^i(x(t), u(t))}{\partial x^k} = 0, \qquad k = 1, \ldots, n \tag{1.20}$$

[see (1.19)]. In addition, differentiating $\partial\omega(x(t))/\partial x^k$ with respect to t and taking account of (1.13), we find

$$\frac{d}{dt}\left(\frac{\partial\omega(x(t))}{\partial x^k}\right) = \sum_{i=1}^{n} \frac{\partial^2\omega(x(t))}{\partial x^k\,\partial x^i}\,\dot{x}^i(t) = \sum_{i=1}^{n} \frac{\partial^2\omega(x(t))}{\partial x^k\,\partial x^i}\,f^i(x(t),\,u(t)).$$

Therefore, (1.20) can be rewritten in the following form:

$$\frac{d}{dt}\left(\frac{\partial\omega(x(t))}{\partial x^k}\right) + \sum_{i=1}^{n} \frac{\partial\omega(x(t))}{\partial x^i}\,\frac{\partial f^i(x(t),\,u(t))}{\partial x^k} = 0, \qquad k = 1,\,\ldots\,,\,n$$

$$(1.21)$$

(it should be noted that

$$\frac{\partial^2\omega}{\partial x^i\,\partial x^k} = \frac{\partial^2\omega}{\partial x^k\,\partial x^i}$$

by virtue of the assumption on the continuity of second derivatives).

We note that (1.16), (1.17), (1.18), and (1.21) involve only the partial derivatives $\partial\omega/\partial x^1,\,\ldots\,,\,\partial\omega/\partial x^n$ but not the function ω itself. Therefore, for convenience, we introduce the following notation:

$$\frac{\partial\omega(x(t))}{\partial x^1} = \psi_1(t), \quad \frac{\partial\omega(x(t))}{\partial x^2} = \psi_2(t),\,\ldots\,, \quad \frac{\partial\omega(x(t))}{\partial x^n} = \psi_n(t). \quad (1.22)$$

Then B [see (1.15)] can be written as follows:

$$B(x(t),\,u(t)) = \sum_{i=1}^{n} \psi_i(t)f^i(x(t),\,u(t)),$$

and (1.17) takes the form

$$\sum_{i=1}^{n} \psi_i(t)f^i(x(t),\,u(t)) \equiv 1 \qquad \text{for the optional process } (x(t),\,u(t)),$$

$$(1.23)$$

$$t_0 \leqslant t \leqslant t_1,$$

where $\psi_1(t),\,\ldots\,,\,\psi_n(t)$ are defined by (1.22). Furthermore, according to (1.16),

$$\sum_{i=1}^{n} \psi_i(t)f^i(x(t),\,u) \leqslant 1 \qquad \text{for any point } u \in U \ (t_0 \leqslant t \leqslant t_1). \quad (1.24)$$

Finally, (1.21) can be written in the form

$$\dot{\psi}_k(t) + \sum_{i=1}^{n} \psi_i(t)\,\frac{\partial f^i(x(t),\,u(t))}{\partial x^k} = 0, \qquad k = 1,\,\ldots\,,\,n. \quad (1.25)$$

Thus, *if* $(x(t), u(t))$, $t_0 \leqslant t \leqslant t_1$, *is an optimal process, then there exist functions* $\psi_1(t)$, $\psi_2(t)$, . . . , $\psi_n(t)$ [*they are defined by* (1.22)] *such that* (1.23), (1.24), *and* (1.25) *hold.*

An examination of the left-hand sides of (1.23) and (1.24) suggests that it is appropriate to introduce the function

$$H(\psi, x, u) = \sum_{i=1}^{n} \psi_i f^i(x, u) = \psi_1 f^1(x, u) + \psi_2 f^2(x, u) + \cdots + \psi_n f^n(x, u)$$

$$(1.26)$$

depending on the $2n + r$ arguments ψ_1, ψ_2, . . . , ψ_n, x^1, . . . , x^n, u^1, . . . , u^r. With the aid of this function, (1.23) and (1.24) can be written as follows:

$$H(\psi(t), x(t), u(t)) \equiv 1 \qquad \text{for the optimal process } (x(t), u(t)), \ t_0 \leqslant t \leqslant t_1,$$

$$(1.27)$$

where $\psi(t) = (\psi_1(t), \ . \ . \ . \ , \psi_n(t))$ is defined by (1.22) and

$$H(\psi(t), x(t), u) \leqslant 1 \qquad \text{for any point } u \in U \qquad (t_0 \leqslant t \leqslant t_1). \quad (1.28)$$

By virtue of (1.27), we can write the following relation in place of inequality (1.28):

$$\max_{u \in U} H(\psi(t), x(t), u) = H(\psi(t), x(t), u(t)), \qquad t_0 \leqslant t \leqslant t_1. \quad (1.29)$$

Finally, (1.25) can obviously be rewritten as follows:

$$\dot{\psi}_k(t) = -\frac{\partial H(\psi(t), x(t), u(t))}{\partial x^k}, \qquad k = 1, \ . \ . \ . \ , n. \qquad (1.30)$$

Thus, *if* $(x(t), u(t))$ *is an optimal process, then there exists a function* $\psi(t) = (\psi_1(t), \ . \ . \ . \ , \psi_n(t))$, *such that* (1.27), (1.29), *and* (1.30) *hold, where* H *is defined by* (1.26). Since $\omega(x)$ does not appear explicitly in (1.26), (1.27), (1.29), (1.30), Equations (1.22), expressing the functions $\psi_1(t)$, . . . , $\psi_n(t)$ in terms of ω, provide no additional information and can be disregarded, and one can confine oneself to the statement that there exist functions $\psi_1(t)$, . . . , $\psi_n(t)$, satisfying (1.27), (1.29), and (1.30). The relation (1.30) is a system of equations which these functions satisfy. We note that $\psi_1(t)$, . . . , $\psi_n(t)$ constitute a nontrivial solution of this system (that is, these functions do not vanish simultaneously at any time t); in fact, if we were to have $\psi_1(t) = \psi_2(t) = \cdots = \psi_n(t) = 0$, then, by virtue of (1.26), we would obtain $H(\psi(t), x(t), u(t)) = 0$, which contradicts (1.27). Thus, we have the following theorem which is called the *maximum principle*.

Theorem 1.2.   *Suppose that Hypotheses* 1, 2, *and* 3 *are satisfied for a controlled object described by the equation* (*in vector form*)

$$\dot{x} = f(x, u), \qquad u \in U \tag{A}$$

and for the prescribed terminal state x_1. *Let* $(u(t), x(t))$, $t_0 \leqslant t \leqslant t_1$, *be any process which transfers the object from* x_0 *to* x_1. *We introduce the function* H *which depends on the variables* $x^1, \ldots, x^n, u^1, \ldots, u^r$ *and on the auxiliary variables* $\psi_1, \ldots, \psi_n$ [*compare* (1.26)]:

$$H(\psi, x, u) = \sum_{i=1}^{n} \psi_i f^i(x, u). \tag{B}$$

With the aid of this function H, *we write the following system of differential equations for the auxiliary variables* $\psi_1, \ldots, \psi_n$:

$$\dot{\psi}_k = - \frac{\partial H(\psi, x(t), u(t))}{\partial x^k}, \qquad k = 1, \ldots, n, \tag{C}$$

where $(u(t), x(t))$ *is the process under consideration* [*compare* (1.30)]. *Then, if the process* $(u(t), x(t))$ *is optimal, there exists a nontrivial solution* $\psi(t) = (\psi_1(t), \ldots, \psi_n(t))$, $t_0 \leqslant t \leqslant t_1$, *of* (C), *such that for any time* t, $t_0 \leqslant t \leqslant t_1$, *the maximum condition*

$$H(\psi(t), x(t), u(t)) = \max_{u \in U} H(\psi(t), x(t), u) \tag{D}$$

[compare (1.29)] *and the condition*

$$H(\psi(t), x(t), u(t)) = 1$$

are satisfied.

This theorem is much more convenient for finding optimal processes than the method of dynamic programming, as we shall see in the examples below. However, in the form given here, the maximum principle has the same shortcomings as the method of dynamic programming: it is derived under the assumption that $\omega(x)$ is differentiable (and twice at that), but as we mentioned previously, this function is actually not differentiable everywhere (in cases usually encountered). Due to the assumption on the validity of the hypotheses stated above [concerning $\omega(x)$], the method of dynamic programming and the maximum principle as formulated above are not convenient conditions for optimality. They are derived in the form of necessary conditions for optimality; if the process is optimal, then (1.18), respectively (D), are satisfied, that is, the validity of these conditions is necessary for optimality. However, these conditions are derived under the assumption that Hypotheses 1, 2, and 3 are satisfied, but these are by no means necessary for optimality. For this reason the theorems formulated above cannot be considered as necessary conditions for optimality.

However, it is a remarkable fact that *if we replace the last condition* $H(\psi(t), x(t), u(t)) = 1$ *by the weaker requirement*

$$H(\psi(t_1), x(t_1), u(t_1)) \geqslant 0, \qquad \text{(E)}$$

then in this form, the maximum principle holds without any assumptions on ω, *that is, the maximum principle becomes a very convenient and broadly applicable* **necessary condition for optimality.** The maximum principle will be proved in this form in § 5. Its proof is entirely different from the arguments presented above; it is rather complicated but does not even require the assumption of the existence of the function $\omega(x)$. This complexity is unavoidable and fully justified: whereas Theorems 1.1 and 1.2 proved above have almost no applicability (by virtue of the unnecessarily rigid requirements contained in Hypotheses 1, 2, and 3), the theorem proved in § 5 (the maximum principle) has a wide range of applications. We shall see this below in numerous examples.

7. Discussion of the Maximum Principle.   As we have seen, the formulation of the maximum principle is rather complicated. In addition to the basic variables (that is, phase coordinates) $x^1, x^2, \ldots, x^n$ and the control parameters $u^1, u^2, \ldots, u^r$, the formulation of the maximum principle also contains "auxiliary" variables $\psi_1, \psi_2, \ldots, \psi_n$ which have no relation to the statement of the optimal control problem.

However, as we shall show in this section, the maximum principle provides "sufficient" information for solving the problem of optimal control. The arguments of this section make no pretense of being rigorous and are used nowhere in the following. The aim of these arguments is to show that *of all the trajectories emanating from* x_0 *and passing through* x_1, *the maximum principle allows us to single out those generally speaking, separate, isolated trajectories which satisfy the above-formulated necessary conditions* (A)–(E). In fact, only these separate, isolated trajectories may prove to be optimal (since the maximum principle provides necessary conditions for optimality). The situation here is analogous, to a certain extent, to that observed in the search for maxima and minima of functions with the aid of the first derivative: the necessary condition that a function have a maximum or minimum (the vanishing of the first derivative, provided the point is not an endpoint) is satisfied, generally speaking, only at separate, isolated points, and only at these points may the function attain a maximum or minimum.

Thus, let us consider the relations (A)–(E) in the maximum principle. Altogether, there are $2n + r$ unknown functions $x^1, \ldots, x^n, \psi_1, \ldots, \psi_n, u^1, \ldots, u^r$ in the formulation of the maximum principle. How many relations are there for determining these unknown functions? Let us consider, first of all, the relation (D). It is not difficult to convince oneself that this equation yields r relations among the unknown functions. In fact, if $u(t)$

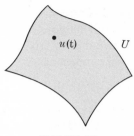

FIGURE 20

is an interior point of the control region U (Figure 20), then, to satisfy the maximum condition (D), it is necessary that

$$\frac{\partial H\left(\psi(t),\, x(t),\, u\right)}{\partial u^i}\Bigg|_{u\,=\,u(t)} = 0, \qquad i = 1,\, \ldots,\, r, \qquad (1.31)$$

yielding r relations among the unknown functions. If, however, $u(t)$ lies, for example, on an $(r-1)$-dimensional "face" of the control region U, then the condition that $u(t)$ belongs to this face must be satisfied (this yields one relation); in order to satisfy the maximum condition (D), the partial derivatives of the function $H(\psi(t),\, x(t),\, u)$ must vanish in all directions on this face which yields $r-1$ additional relations. A similar situation exists also in the case of a lower dimensional face (or on the curved portions of the boundary of the control region U, see Figure 21). Thus, it can be assumed

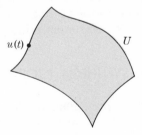

FIGURE 21

in all cases that if the control region U is r-dimensional, then the maximum condition (D) yields r relations among the unknown functions. These relations are finite [that is, they do not contain derivatives of the unknown functions, compare (1.31)].

In addition to the relation (D) already considered, the maximum principle contains also the relations (A) and (C) which constitute a system of $2n$ differential equations. Thus, we have $2n + r$ relations (A), (C), and (D) for finding $2n + r$ unknown functions $x^1,\, \ldots,\, x^n,\, \psi_1,\, \ldots,\, \psi_n,\, u^1,\, \ldots,\, u^r$, that is, the number of relations equals the number of unknown functions.

Moreover, the r relations (D) are finite, and the $2n$ relations (A) and (C) are ordinary differential equations. Therefore, it can be expected that all the unknown functions can be found from the relations (A), (C), and (D), provided one knows the initial conditions

$$x^1(t_0), \ldots, x^n(t_0), \psi_1(t_0), \ldots, \psi_n(t_0) \qquad (1.32)$$

for the differential equations (A) and (C). Thus, all solutions $x(t)$, $\psi(t)$, $u(t)$ of the system of equations (A), (C), (D) depend on $2n$ numerical parameters (1.32).

It is not difficult to see that one of these parameters is redundant. As a matter of fact, since H is a linear homogeneous function of the variables $\psi_1, \psi_2, \ldots, \psi_n$ [see (B)], the validity of (C), (D), (E) will not be violated if all the quantities $\psi_1, \psi_2, \ldots, \psi_n$ are multiplied by the same positive constant. In other words, $\psi_1(t_0), \ldots, \psi_n(t_0)$ are determined only to within a common, constant multiple, and therefore, one of the parameters (1.32) is superfluous.

Thus, the entire manifold of solutions of systems (A), (C), (D) depends on $2n - 1$ numerical parameters. These $2n - 1$ parameters must be used in such a way that the trajectory $x(t)$ passes through x_0 at a given $t = t_0$, and through x_1 at some $t_1 > t_0$. The number $t_1 - t_0$ (not known beforehand) is also a parameter, so that altogether there are $2n$ essential parameters. The condition that the trajectory passes through x_0 and x_1 provides $2n$ relations. Thus, altogether, there are $2n$ relations from which to obtain the $2n$ arbitrary parameters. Therefore, it may be expected that there exist only separate, isolated trajectories passing through x_0 and x_1 satisfying the conditions contained in the maximum principle.

§ 3. AN EXAMPLE. THE SYNTHESIS PROBLEM

8. An Example of the Application of the Maximum Principle. In this section, we analyze an example of the calculation of optimal processes. Namely, we consider the controlled object described in § 1.3 [see (1.1)], under the condition that the friction and elastic forces are absent (that is, $b = 0$, $k = 0$), the mass m is unity ($m = 1$), and the control parameter is subject to the constraint $|u| \leqslant 1$. In other words, we consider a particle G of mass $m = 1$ (see Figure 10), moving freely and without friction along a horizontal line, and equipped with a motor which develops a force u, where $|u| \leqslant 1$. According to (1.1), the equations of motion of this object have the form

$$\begin{aligned} \dot{x}^1 &= x^2, \\ \dot{x}^2 &= u; \end{aligned} \qquad (1.33)$$

$$-1 \leqslant u \leqslant 1. \qquad (1.34)$$

We consider the problem of the time-optimal arrival of this object at the origin $(0, 0)$ from a given initial state x_0. In other words, we consider the time-optimal problem for the case in which the origin $x_1 = (0, 0)$ serves as the terminal state. Physically, this means that we want to transfer a particle from a given initial position and with given initial velocity so that it arrives at the origin with zero velocity and further, so that this process is carried out in minimum time.

The function H in this case has the form

$$H = \psi_1 x^2 + \psi_2 u \tag{1.35}$$

[compare (1.33) and (B)]. Furthermore, for the auxiliary variables ψ_1, ψ_2, we obtain the system of equations

$$\dot{\psi}_1 = 0, \qquad \dot{\psi}_2 = -\psi_1.$$

From this system of equations, we readily find that $\psi_1 = d_1$ and $\psi_2 = -d_1 t + d_2$, where d_1 and d_2 are integration constants. Further, by virtue of the maximum relation (D) we find, considering (1.35) and (1.34),

$$\begin{aligned} u(t) &= +1 \quad \text{if} \quad \psi_2(t) > 0; \\ u(t) &= -1 \quad \text{if} \quad \psi_2(t) < 0. \end{aligned}$$

In other words, $u(t) = \text{sign } \psi_2(t) = \text{sign } (-d_1 t + d_2)$. Hence, it follows that *every optimal control $u(t)$, $t_0 \leqslant t \leqslant t_1$, is a piecewise-constant function which takes on the values ± 1 and has at most two intervals of constancy* (since the linear function $-d_1 t + d_2$ does not change sign more than once on the interval $t_0 \leqslant t \leqslant t_1$).

For the time interval on which $u \equiv 1$, we have [by virtue of (1.33)]

$$x^2(t) = t + c^2, \qquad x^1(t) = \int x^2(t)\, dt = \tfrac{1}{2}(t + c^2)^2 + c^1,$$

where c^1 and c^2 are integration constants, from which we find

$$x^1 = \tfrac{1}{2}(x^2)^2 + c^1. \tag{1.36}$$

Thus, the portion of a phase trajectory for which $u \equiv 1$ is an arc of the parabola (1.36). The family of parabolas (1.36) is shown in Figure 22 (each may be obtained from another by a displacement along the x^1 axis). The phase points move upward along these parabolas (since $\dot{x}^2 = u \equiv 1$, that is, $\dot{x}^2 > 0$).

Similarly, for the time interval on which $u \equiv -1$, we have

$$x^2(t) = -t + c'^2, \qquad x^1(t) = \int x^2(t)\, dt = -\tfrac{1}{2}(-t + c'^2)^2 + c'^1,$$

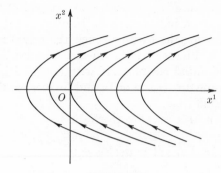

FIGURE 22

from which we find

$$x^1 = -\tfrac{1}{2}(x^2)^2 + c'^1. \tag{1.37}$$

The family of parabolas (1.37) (again each may be obtained from another by a displacement along the x^1 axis) is shown in Figure 23. The phase points move downward along the parabolas (1.37) (since $\dot{x}^2 = u \equiv -1$, that is, $\dot{x}^2 < 0$).

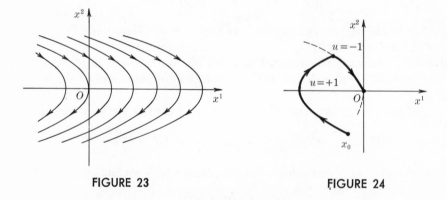

FIGURE 23   **FIGURE 24**

As indicated above, every optimal control $u(t)$ is a piecewise constant function taking on the values ± 1 and having not more than two intervals of constancy. If the control $u(t)$ starts out equal to $+1$ for some time, and then equals -1, the phase trajectory consists of two portions of parabolas (Figure 24) which abut one another; moreover, the second of these portions lies on that parabola (1.37) which passes through the origin (since the trajectory being sought must lead to the origin). If, however, $u = -1$ from the beginning, and then $u = +1$, the phase trajectory, shown in Figure 24, is replaced by a centrally symmetric trajectory (Figure 25). The values of

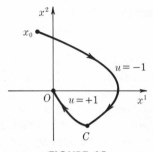

FIGURE 25

the control parameter u are marked in Figures 24 and 25 next to the corresponding arcs of the parabolas. The entire family of phase trajectories thus obtained is shown in Figure 26 (where AO is the arc of the parabola $x^1 = \frac{1}{2}(x^2)^2$ lying in the lower half-plane and BO is the arc of the parabola $x^1 = -\frac{1}{2}(x^2)^2$ lying in the upper half-plane). If the initial point x_0 lies above the curve AOB, then the phase point moves along the arc of the parabola (1.37) which passes through x_0; if, however, the point x_0 is below AOB, then the phase point moves along the parabola (1.36) passing through x_0. In other words, if the initial state x_0 lies above AOB, then the phase point must move under the action of the control $u = -1$ until it reaches the arc AO; at the instant the phase point reaches the arc AO, the value of u switches to $+1$ until the phase point reaches the origin. If, however, the initial state lies below AOB, then u must be equal to $+1$ until the phase point reaches the arc BO, and at this instant the value of u switches to -1.

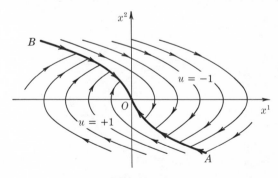

FIGURE 26

Thus, according to the maximum principle, *only the trajectories depicted in Figure 26 can be optimal;* moreover, as seen from the above investigation, from each point of the phase plane there emanates only one trajectory leading to the origin which can be optimal (that is, assigning the initial point x_0 uniquely determines the corresponding trajectory).

9. Proof of Optimality of the Trajectories.

The above discussion still does not prove conclusively that the trajectories shown in Figure 26 are indeed optimal (since the maximum principle is only a necessary condition for optimality). As a matter of fact, it could happen that there do not exist any optimal trajectories at all, that is, even the trajectories depicted in Figure 26 could turn out not to be optimal. (For example, from the fact that no one but student A can solve a difficult problem, it does not follow that student A can necessarily solve it!)

In fact, *all the trajectories depicted in Figure 26 are optimal*, as we shall now prove. Let us consider the process shown in Figure 25. Denote by $t_0 \leqslant t \leqslant t_1$ the time during which this process takes place, and let α be the instant of switching. Thus, the control parameter u for the process under consideration takes on the following values:

$$
\begin{aligned}
u(t) &= -1 && \text{for } t_0 \leqslant t < \alpha; \\
u(t) &= +1 && \text{for } \alpha \leqslant t \leqslant t_1.
\end{aligned}
\tag{1.38}
$$

Suppose that this process is not optimal. Then, there exists a control $\tilde{u}(t)$ satisfying the conditions (1.34) under the action of which the phase point, proceeding from x_0 at time t_0, reaches the origin at the time $\theta < t_1$ (that is, sooner than for the trajectory shown in Figure 25). The phase trajectory depicted in Figure 25 will be denoted by $x(t) = (x^1(t), x^2(t))$, and the phase trajectory emanating from the same point x_0 and corresponding to the control $\tilde{u}(t)$ will be denoted by $\tilde{x}(t) = (\tilde{x}^1(t), \tilde{x}^2(t))$ (Figure 27). According to the

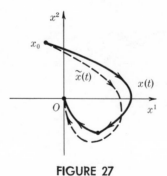

FIGURE 27

assumption, the trajectory $\tilde{x}(t)$ arrives at the origin at time θ, that is, $\tilde{x}^1(\theta) = 0$, $\tilde{x}^2(\theta) = 0$. The trajectory $x(t)$, however, arrives at the origin at time t_1, that is, $x^1(t_1) = 0$, $x^2(t_1) = 0$. In addition, both trajectories are subject to (1.33), namely,

$$\frac{d}{dt} x^1(t) = x^2(t),$$

$$\frac{d}{dt} x^2(t) = u(t) \qquad (t_0 \leqslant t \leqslant t_1);$$

$$\frac{d}{dt} \tilde{x}^1(t) = \tilde{x}^2(t),$$

$$\frac{d}{dt} \tilde{x}^2(t) = \tilde{u}(t) \qquad (t_0 \leqslant t \leqslant \theta).$$

(1.39)

Consider now the following two functions:

$$\Phi(t) = -x^1(t) + x^2(t)(t - \alpha), \qquad \Psi(t) = -\tilde{x}^1(t) + \tilde{x}^2(t)(t - \alpha).$$

Since both trajectories $x(t)$ and $\tilde{x}(t)$ emanate at time t_0 from the same point x_0 (that is, $x(t_0) = \tilde{x}(t_0) = x_0$),

$$\Phi(t_0) = \Psi(t_0).$$

(1.40)

In addition, it is obvious that

$$\Phi(t_1) = 0, \qquad \Psi(\theta) = 0.$$

(1.41)

Calculating the derivatives, we obtain by virtue of (1.39),

$$\dot{\Phi}(t) = u(t)(t - \alpha), \qquad \dot{\Psi}(t) = \tilde{u}(t)(t - \alpha).$$

According to (1.38), the first of these equations can be rewritten in the form

$$\dot{\Phi}(t) = |t - \alpha|$$

and, therefore [considering that $|\tilde{u}(t)| \leqslant 1$, see (1.34)], we have

$$\dot{\Phi}(t) \geqslant |\dot{\Psi}(t)| \geqslant \dot{\Psi}(t).$$

Integrating this inequality from t_0 to θ, we find

$$\int_{t_0}^{\theta} \dot{\Phi}(t) \, dt \geqslant \int_{t_0}^{\theta} \dot{\Psi}(t) \, dt,$$

or

$$\Phi(\theta) - \Phi(t_0) \geqslant \Psi(\theta) - \Psi(t_0).$$

By virtue of (1.40) and (1.41), the latter inequality takes the form $\Phi(\theta) \geqslant 0$.

On the other hand, we have

$$-\Phi(\theta) = \Phi(t_1) - \Phi(\theta) = \int_{\theta}^{t_1} \dot{\Phi}(t) \, dt = \int_{\theta}^{t_1} |t - \alpha| \, dt > 0$$

(since $t_1 > \theta$, and the integrand is positive). Thus, $\Phi(\theta) < 0$, which contradicts the inequality $\Phi(\theta) \geqslant 0$ obtained previously. This contradiction shows that the inequality $\theta < t_1$ cannot be satisfied, that is, departing from the point x_0 at time t_0, it is impossible to arrive at the origin sooner than at time t_1. In other words, the process $(u(t), x(t))$, $t_0 \leqslant t \leqslant t_1$, is optimal.

Thus, *all the trajectories shown in Figure 26 are optimal*. It should be noted that this fact is proved here without resorting to any assumptions on $\omega(x)$. In fact, it is completely immaterial how we were led to the trajectories shown in Figure 26. We have arrived at these trajectories with the aid of the maximum principle. But there is nothing to prevent us from assuming that these trajectories were taken so-to-say "out of thin air," and only then did we apply to them the arguments given in the last two pages; moreover, neither the maximum principle, nor the method of dynamic programming, nor the function $\omega(x)$ were mentioned in these arguments. On the other hand, knowing that all the trajectories shown in Figure 26 are optimal, we can now calculate $\omega(x)$, which we shall do at once.

10. On the Differentiability of Bellman's Function.   Let the point x_0 lie above the curve AOB (as in Figure 25). We denote the coordinates of this point by $(a; b)$. In order for the parabola (1.37) to pass through x_0, it is necessary that the coordinates of this point satisfy (1.37):

$$a = -\tfrac{1}{2}b^2 + c'^1.$$

Hence, we find $c'^1 = a + \tfrac{1}{2}b^2$. Thus, the parabola (1.37) passing through x_0 has the equation

$$x^1 = -\tfrac{1}{2}(x^2)^2 + a + \tfrac{1}{2}b^2. \tag{1.42}$$

The switching point C, shown in Figure 25, can be found as the point of intersection of the parabola (1.42) with the curve AO (see Figure 26) whose equation has the form

$$x^1 = \tfrac{1}{2}(x^2)^2. \tag{1.43}$$

In order to find the point of intersection, it is necessary to solve (1.42) and (1.43) simultaneously as a system of equations. Subtracting (1.43) from (1.42), we find $(x^2)^2 = a + \tfrac{1}{2}b^2$, hence $x^2 = \pm \sqrt{a + \tfrac{1}{2}b^2}$. The minus sign must be taken for the point C since it lies on AO, that is, below the axis of abscissas. Thus, we have found the ordinate of the point C:

$$x_C^2 = - \sqrt{a + \tfrac{1}{2}b^2}$$

(the abscissa of C will not be needed).

Since $u \equiv -1$ in moving from x_0 to C, the second of Equations (1.33) has the form $\dot{x}^2 \equiv -1$, and, integrating, we obtain (as before, α denotes the instant of switching, that is, the time at which the trajectory passes through C)

$$x_C^2 - b = \int_{t_0}^{\alpha} \dot{x}^2 \, dt = \int_{t_0}^{\alpha} (-1) \, dt = t_0 - \alpha.$$

Similarly, during the motion from C to the origin, we have $u \equiv 1$, that is, $\dot{x}^2 \equiv 1$, and therefore,

$$-x_C^2 = 0 - x_C^2 = \int_\alpha^{t_1} \dot{x}^2 \, dt = \int_\alpha^{t_1} dt = t_1 - \alpha.$$

Subtracting the first relation from the second, we find

$$b - 2x_C^2 = t_1 - t_0.$$

But $t_1 - t_0$ is the duration of the motion along the optimal trajectory from x_0 to the origin, that is, the time $T(x_0)$ of the optimal motion. Thus,

$$T(x_0) = b - 2x_C^2 = b + 2\sqrt{a + \tfrac{1}{2}b^2} \qquad (1.44)$$

(if $x_0 = (a; b)$ lies on or above AOB). If x_0 lies below AOB, then the duration of the optimal motion $T(x_0)$ can be calculated in a similar manner. However, it is simpler to notice that if the point $x_0 = (a; b)$ lies below AOB (Figure 28), then the point x_0' with the coordinates $(-a; -b)$ is symmetric to the

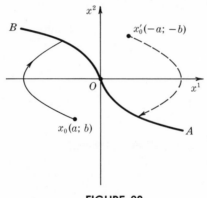

FIGURE 28

point x_0 with respect to the origin, lies above AOB, and the duration of the optimal motion for x_0 and x_0' are the same: $T(x_0) = T(x_0')$. Therefore, replacing a and b in (1.44) by $-a$ and $-b$, we obtain the function $T(x_0)$ for the points x_0 which lie below AOB:

$$T(x_0) = -b + 2\sqrt{-a + \tfrac{1}{2}b^2} \qquad (1.45)$$

(if $x_0 = (a; b)$ lies on or below AOB). Thus,

$$T(x_0) = \begin{cases} b + 2\sqrt{a + \tfrac{1}{2}b^2}, & \text{if the point } x_0 = (a; b) \text{ lies on or above } AOB; \\ -b + 2\sqrt{-a + \tfrac{1}{2}b^2}, & \text{if the point } x_0 = (a; b) \text{ lies on or below } AOB. \end{cases}$$

We note that if $x_0 = (a; b)$ lies on the arc AO (that is, $a = \tfrac{1}{2}b^2$, with $b < 0$), then (1.44) takes the form $T(x_0) = b + 2\sqrt{b^2} = b + 2|b| = b +$

$2(-b) = -b$, and (1.45) clearly also yields $T(x_0) = -b$. In other words, on AO both formulas (1.44) and (1.45) yield identical values for $T(x_0)$. This also holds on BO. This shows that although the function $T(x_0)$ is defined by different formulas above and below AOB, (1.44) and (1.45), these formulas coincide on AOB, and therefore $T(x_0)$ is continuous on the entire plane.

Having the expression for the function $T(x_0)$, we can now resolve the question of its differentiability. It is clear, first of all, that exterior to AOB, $T(x_0)$ has continuous derivatives with respect to a and b, since near any point not lying on AOB, $T(x_0)$ is defined by one of the formulas (1.44) or (1.45), and can be readily differentiated. But what will happen at points on AOB itself? We shall see immediately that the *function $T(x_0)$ does not have continuous derivatives with respect to a and b at any point of this curve.* In fact, let C be any point on AO and let $(a_0; b_0)$ be its coordinates, so that $a_0 = \frac{1}{2}(b_0)^2$, with $b_0 < 0$. At this point

$$\sqrt{a_0 + \tfrac{1}{2}(b_0)^2} = \sqrt{(b_0)^2} = |b_0| = -b_0.$$

Now, it is easy to find the derivatives of the function (1.44) at C:

$$\frac{\partial T}{\partial a}\bigg|_C = \frac{1}{\sqrt{a + \frac{1}{2}b^2}}\bigg|_C = -\frac{1}{b_0},$$

$$\frac{\partial T}{\partial b}\bigg|_C = 1 + \frac{b}{\sqrt{a + \frac{1}{2}b^2}}\bigg|_C = 1 + \frac{b_0}{-b_0} = 0,$$

and the derivatives of the function (1.45) at C:

$$\frac{\partial T}{\partial a}\bigg|_C = \frac{-1}{\sqrt{-a + \frac{1}{2}b^2}}\bigg|_C = -\infty,$$

$$\frac{\partial T}{\partial b}\bigg|_C = -1 + \frac{b}{\sqrt{-a + \frac{1}{2}b^2}}\bigg|_C = -\infty.$$

Thus (Figure 29), shifting upward from the point C we find $\partial T/\partial b = 0$, and shifting downward, we find $\partial T/\partial b = -\infty$, that is, the derivative $\partial T/\partial b$ does not exist at the point C. In exactly the same way the derivative $\partial T/\partial a$ does not exist at the point C. An analogous calculation can be carried out also for the points lying on BO.

Thus, the function $T(x_0)$ [and consequently, $\omega(x_0) = -T(x_0)$] does not have derivatives with respect to the coordinates at points on AOB.

In spite of the fact that $\omega(x)$ is not differentiable only at points on AOB, and has derivatives at the remaining points of the plane, all the arguments of § 2.5 immediately lose their validity. As a matter of fact, every optimal trajectory (Figure 26) moves along the curve AOB during some

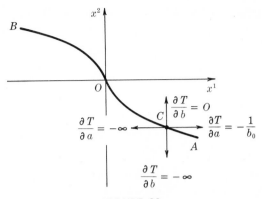

FIGURE 29

time interval, and therefore, the assumption of the differentiability of $\omega(x)$ (Hypothesis 2) does not hold on any of these trajectories. It is not merely that the proof given in § 2.5 does not work: it is not even possible to write Bellman's equation (1.18), since the derivatives $\partial\omega/\partial x^i$, appearing in this equation, do not exist on any of the optimal trajectories over an entire time interval.

Of course, this applies (to an even greater extent) to the arguments presented in § 2.6. However, the great merit of the maximum principle lies in the fact that its formulation does not contain the derivatives $\partial\omega/\partial x^i$ (nor the function ω itself), so that the formulation of the maximum principle does not become pointless even if the function ω is not differentiable. Of course, the proof given in § 2.6 is completely destroyed, but, as mentioned previously, there is another proof, based on entirely different principles, which will make it possible to establish the validity of the maximum principle in all cases. We note also that a modified formulation of the method of dynamic programming makes it widely applicable, not as a necessary condition but as a sufficient condition for optimality. This will be discussed in § 11.

11. The Problem of the Synthesis of Optimal Controls. Let us consider the example analyzed in the previous sections from a somewhat different viewpoint. The solution of the optimal problem found above can be interpreted as follows. Let $v(x) = v(x^1, x^2)$ denote a function given on the phase plane in the following manner:

$$v(x) = \begin{cases} +1 & \text{below the curve } AOB \text{ and on the arc } AO; \\ -1 & \text{above the curve } AOB \text{ and on the arc } BO. \end{cases}$$

Then (see Figure 26), on every optimal trajectory the value $u(t)$ of the control parameter (at an arbitrary time t) equals $v(x(t))$, that is, equals the value of the function v at the position occupied by the moving phase point at time t as it traverses the optimal trajectory:

$$u(t) = v(x(t)).$$

This means that, replacing the quantity u in (1.33) by the function $v(x)$, we obtain the system

$$\dot{x}^1 = x^2,$$
$$\dot{x}^2 = v(x^1, x^2), \tag{1.46}$$

whose solution (for an arbitrary initial state x_0) yields the optimal phase trajectory leading to the origin. In other words, (1.46) is a system of differential equations (with a discontinuous right-hand side) for finding optimal trajectories leading to the origin.

This can also be expressed in the following manner. Suppose, for example, that the initial state x_0 lies above the curve AOB. Then we set $u = -1$, and carry out the motion with this value of the control parameter; as soon as the moving phase point reaches AO (compare Figures 25 and 26), we switch the control parameter u to $+1$ until the end of the motion (that is, up to the time that the point reaches the origin). Thus, it is only necessary not to "miss" the time at which the moving phase point reaches the switching curve AOB. The above function $v(x)$ makes it possible to obtain a good idea of a device which automatically realizes the required switchings. This device, which we shall call a nonlinear relay, depicted as shown in Figure 30, must have the following property. It measures the state x of the object (that is, its coordinate x^1 and velocity x^2); in other words, the input to the nonlinear relay is the state x of the object, and the function $u = v(x)$ is the output. This means that if the state x lies below AOB or on AO, then we must obtain $+1$ at the output of the relay. If, however, the phase point x lies above AOB or on BO then we must have -1 at the relay's output. If such a relay is connected to the object whose behavior is described by (1.33), then we obtain a closed loop system (Figure 31) which, as can be readily understood, will operate automatically under optimal conditions. In fact, if, for example, the initial state lies above AOB, then the relay will supply the value $u = -1$ to the object's input and will maintain this value of the control parameter until the object's state reaches AOB. From this time until the end of the motion, the relay will supply the value $u = +1$ to the object's input. As a result, the object will move along the trajectory shown in Figure 25, that is, the relay will automatically realize the optimal conditions. In other words, the above nonlinear relay (Figure 30) is (for the object under consideration), in fact, the optimal regulator we have discussed in § 1.2. It may also be noted that since the object is described by (1.33), and the nonlinear relay (Figure 30) has the characteristic $u = v(x)$, the operation of the system depicted in Figure 31 will be described by the

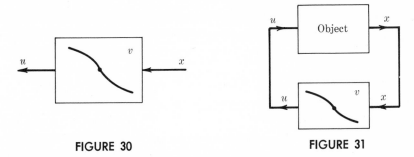

FIGURE 30   FIGURE 31

system of equations (1.46). But this means that the system shown in Figure 31 will automatically accomplish the optimal motion.

The term "nonlinear relay" can be explained as follows. In engineering practice, "ordinary" relays (Figure 32) are used rather widely. The char-

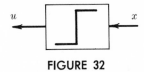

FIGURE 32

acteristic of these relays (in a mathematically ideal case) can be represented by the relationship $u = \text{sign } x$, where x is a scalar supplied to the relay's input. For example, in case the sum $x^1 + x^2$ of two phase coordinates (Figure 33) is supplied to the relay's input, the quantity u equals $+1$ if the phase point lies to the right of the bisector of the second and the fourth

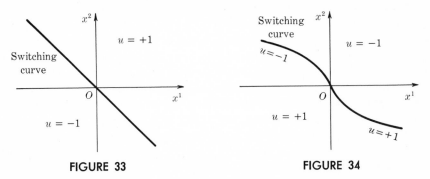

FIGURE 33   FIGURE 34

quadrants, and is equal to -1 if the phase point lies to the left of this bisector. In other words, the switching curve in this case is the straight line in the x^1, x^2 phase plane shown in Figure 33. A similar situation exists also in the case of the device shown schematically in Figure 30, but here the switching curve is not a straight line, as in Figure 33, but a curve consisting of two semiparabolas (Figure 34; compare Figure 26).

The considerations presented here are of a general nature. The above example shows that the solution of an optimal control problem is naturally expected to be obtained in the following form. We solve the optimal problem in the general form

$$\dot{x} = f(x, u)$$

(see § 1.3) considering all possible initial states x_0, each time prescribing the origin O of the phase space as the terminal state. Then (as far as can be judged from the example analyzed above), *there exists a function $v(x)$, defined in the phase space and taking on values in the control region U, such that the equation*

$$\dot{x} = f(x, v(x)) \tag{1.47}$$

[see (1.46)] *determines all the optimal trajectories leading to the origin.* In other words, it is natural to seek an optimal control in the form $u = v(x)$ rather than in the form $u = u(t)$, that is, the desired optimal control depends only on the location of the phase point at each instant of time. This is understandable: in fact, if we have reached the phase point x, the subsequent motion (from the point x to O) must also be optimal (see Figure 19). Therefore, the value of the optimal control $u(t)$ at the instant the phase point passes through the position x depends only on x and not on the point at which the motion started or on how long the phase point was in motion prior to reaching the position x.

The function $v(x)$ which yields the equation of optimal trajectories in the form (1.47) is called the *synthesizing function,* and the problem of finding the synthesizing function is called the *synthesis problem* of optimal controls. In the example we have discussed above, the synthesizing function was piecewise continuous.

Knowledge of the synthesizing function $v(x)$ makes it possible to assert that the problem of reaching the origin optimally is mathematically completely solved. In fact, if the controlled object is equipped with a nonlinear relay having the characteristic $v(x)$ [that is, with a measuring device that measures the states, and a servomechanism that puts the controllers into the position $u = v(x)$], then the object will move optimally. In other words, the above nonlinear relay is the required optimal controller (Figure 31).

The synthesis of time-optimal controls for linear second-order systems will be treated in § 10. In the general case, the synthesis problem is very complicated (compare § 9).

chapter 2 The Maximum Principle—
A Necessary Condition
for Optimality

§ 4. INFORMATION FROM GEOMETRY AND THE THEORY OF ORDINARY DIFFERENTIAL EQUATIONS

12. Elementary Concepts of n-Dimensional Geometry. In this section we consider the concepts of a segment, a ray, and a hyperplane, which are needed in the proofs given in the subsequent sections. All considerations will be carried out in an n-dimensional space with coordinates $x^1, \ldots, x^n$ (that is, in the phase space of the controlled object); this space will be denoted by X. The coordinate system $x^1, \ldots, x^n$ will be fixed throughout the discussion.

We shall make use of vector notation. Recall that the vector $\overrightarrow{PQ}$ with origin at $P(x^1, \ldots, x^n)$ and endpoint at $Q(y^1, \ldots, y^n)$ has the components

$$y^1 - x^1, \qquad y^2 - x^2, \ldots, \qquad y^n - x^n.$$

In particular, a vector from the origin $O = (0, 0, \ldots, 0)$ to the point M with coordinates $(x^1, x^2, \ldots, x^n)$ has the same components $(x^1, x^2, \ldots, x^n)$. In this connection, no distinction will generally be made between points and vectors, the point M being regarded as equivalent to the vector $x = \overrightarrow{OM}$.

Furthermore, we recall that vectors may be added to one another, and multiplied by real numbers: if

$$x = (x^1, \ldots, x^n), \qquad y = (y^1, \ldots, y^n)$$

are two vectors and λ is a real number, then the sum $x + y$ and the product

λx are defined by the formulas

$$x + y = (x^1 + y^1, x^2 + y^2, \ldots, x^n + y^n),$$
$$\lambda x = (\lambda x^1, \lambda x^2, \ldots, \lambda x^n).$$

For any three points M, N, and P of X

$$\overrightarrow{MN} + \overrightarrow{NP} = \overrightarrow{MP}.$$

In addition, for any two points M and N,

$$\overrightarrow{MN} = -\overrightarrow{NM}.$$

We now introduce the concept of a *segment*. Let A and B be two distinct points in X. One says that the *point C lies on the segment AB if the*

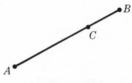

FIGURE 35

vectors $\overrightarrow{AB}$ and $\overrightarrow{AC}$ are related by (Figure 35)

$$\overrightarrow{AC} = \lambda \overrightarrow{AB}, \tag{2.1}$$

where λ is a real number satisfying $0 \leqslant \lambda \leqslant 1$. If, in addition to the points A, B, and C, we take an arbitrary point Q of X, then we can write

$$\overrightarrow{AC} = \overrightarrow{QC} - \overrightarrow{QA}, \qquad \overrightarrow{AB} = \overrightarrow{QB} - \overrightarrow{QA},$$

and therefore (2.1) takes the form

$$\overrightarrow{QC} - \overrightarrow{QA} = \lambda(\overrightarrow{QB} - \overrightarrow{QA}),$$

or

$$\overrightarrow{QC} = (1 - \lambda)\overrightarrow{QA} + \lambda\overrightarrow{QB}. \tag{2.2}$$

Reversing the order of the calculation, we obtain (2.1) from (2.2). Thus, (2.1) and (2.2) are equivalent, and therefore the following statement is valid:

The point C lies on the segment AB if and only if (2.2) holds, where λ is a real number satisfying $0 \leqslant \lambda \leqslant 1$ (moreover, the point Q is arbitrary).

Let us now consider the concept of a *ray*. Let Q and A be two distinct points in X. The *ray* emanating from Q and passing through A is defined as the set of all points B for which

$$\overrightarrow{QB} = \lambda\overrightarrow{QA},$$

where $\lambda \geqslant 0$ (Figure 36).

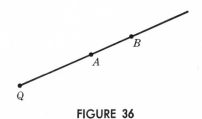

FIGURE 36

The *scalar product* xy of two vectors

$$x = (x^1, \ldots, x^n),$$
$$y = (y^1, \ldots, y^n)$$

is the real number

$$xy = x^1y^1 + x^2y^2 + \cdots + x^ny^n.$$

It can be directly verified that the scalar product has the properties of *commutativity* and *distributivity*:

$$xy = yx, \qquad x(y + z) = xy + xz$$

for any three vectors x, y, and z. The scalar product of the vector x with itself is

$$x^2 = xx = (x^1)^2 + (x^2)^2 + \cdots + (x^n)^2,$$

which for any vector is a non-negative number. The (positive) square root of this number is denoted by $|x|$ and is called the *length of the vector x:*

$$|x| = \sqrt{x^2} = \sqrt{(x^1)^2 + (x^2)^2 + \cdots + (x^n)^2}.$$

The length of any nonzero vector is a positive number. It is proved in geometry (see Shilov [10], p. 139) that the absolute value of the scalar product of two vectors does not exceed the product of their lengths:

$$|xy| \leqslant |x|\,|y|.$$

Consequently, for any two nonzero vectors x and y,

$$-1 \leqslant \frac{xy}{|x|\,|y|} \leqslant 1,$$

and therefore, there exists an angle φ, contained in the interval $0 \leqslant \varphi \leqslant \pi$, such that

$$\frac{xy}{|x|\,|y|} = \cos \varphi.$$

This angle φ is called the *angle between the vectors x and y*. Thus,

$$xy = |x|\,|y|\cos \varphi,$$

which agrees with the usual definition of the scalar product for vectors in a plane or in space. It follows from the last relation that the angle between two nonzero vectors is

$$\begin{array}{ll} \text{acute} & \text{if } xy > 0, \\ \text{right} & \text{if } xy = 0, \\ \text{obtuse} & \text{if } xy < 0. \end{array}$$

If this angle is a right angle, that is, if $xy = 0$, then the vectors x and y are said to be *orthogonal*.

The set of all points $x = (x^1, \ldots, x^n)$ which satisfy the linear equation

$$\alpha_1 x^1 + \alpha_2 x^2 + \cdots + \alpha_n x^n + \beta = 0, \tag{2.3}$$

in which at least one of the coefficients $\alpha_1, \alpha_2, \ldots, \alpha_n$ differs from zero, is called a *hyperplane* of the space X. The relation (2.3) is called the *equation* of this hyperplane. For $n = 2$ (that is, in the plane of the variables x^1, x^2), (2.3) takes the form

$$\alpha_1 x^1 + \alpha_2 x^2 + \beta = 0,$$

which is the equation of a straight line; thus, for $n = 2$ (in a plane), ordinary straight lines serve as hyperplanes. Furthermore, for $n = 3$, (2.3) takes the form

$$\alpha_1 x^1 + \alpha_2 x^2 + \alpha_3 x^3 + \beta = 0,$$

which is the equation of a plane in three-dimensional space; thus, for $n = 3$ (in three-dimensional space), the hyperplanes are ordinary planes.

Just as a straight line divides a plane into two *half-planes* or as a plane divides a three-dimensional space into two *half-spaces*, for an arbitrary n, any hyperplane divides the space X into two *half-spaces*. Namely, the hyperplane (2.3) divides X into two half-spaces. One of these half-spaces consists of the points satisfying the inequality

$$\alpha_1 x^1 + \alpha_2 x^2 + \cdots + \alpha_n x^n + \beta \geqslant 0, \tag{2.4}$$

and the other consists of the points satisfying

$$\alpha_1 x^1 + \alpha_2 x^2 + \cdots + \alpha_n x^n + \beta \leqslant 0. \tag{2.5}$$

The half-space (2.4) is called the *positive half-space*, and the half-space (2.5) is called the *negative half-space*. This distinction between positive and negative half-spaces is, of course, a convention: in fact, changing the signs of all the coefficients α_1, α_2, . . . , α^n, β in (2.3) does not change the hyperplane itself, but interchanges the roles of the positive and negative half-spaces. This remark makes it possible to consider either of the two half-spaces into which the hyperplane (2.3) divides the space X as negative.

Let us now indicate the vector notation for (2.3), (2.4), and (2.5). Let Q be an arbitrary point of the hyperplane (2.3) and let a^1, a^2, . . . , a^n be the coordinates of this point; then,

$$\alpha_1 a^1 + \alpha_2 a^2 + \cdots + \alpha_n a^n + \beta = 0. \tag{2.6}$$

Furthermore, let $M(x^1, \ldots, x^n)$ be an arbitrary point of X. Let us consider the sum

$$\alpha_1 x^1 + \alpha_2 x^2 + \cdots + \alpha_n x^n + \beta$$

on the left-hand sides of (2.3), (2.4), and (2.5). Considering (2.6), we may write

$$\begin{aligned}
\alpha_1 x^1 + \alpha_2 x^2 + \cdots + \alpha_n x^n + \beta &= \alpha_1 x^1 + \alpha_2 x^2 + \cdots + \alpha_n x^n + \beta \\
- (\alpha_1 a^1 + \cdots + \alpha_n a^n + \beta) &= \alpha_1(x^1 - a^1) + \alpha_2(x^2 - a^2) \\
&\quad + \cdots + \alpha_n(x^n - a^n).
\end{aligned} \tag{2.7}$$

The numbers $x^1 - a^1$, $x^2 - a^2$, . . . , $x^n - a^n$ are the components of the vector $\overrightarrow{QM}$. Now, let n denote the vector with components α_1, α_2, . . . , α_n. Then it is obvious that the right-hand side of (2.7) is the scalar product of the vectors n and $\overrightarrow{QM}$ and therefore,

$$\alpha_1 x^1 + \alpha_2 x^2 + \cdots + \alpha_n x^n + \beta = n\overrightarrow{QM}.$$

Now (2.3), (2.4), and (2.5) can be rewritten in the form

$$n\overrightarrow{QM} = 0, \qquad n\overrightarrow{QM} \geqslant 0, \qquad n\overrightarrow{QM} \leqslant 0. \tag{2.8}$$

In other words, the hyperplane (2.3) consists of those points $M(x^1, \ldots, x^n)$ for which $n\overrightarrow{QM} = 0$, that is, of the points M for which the vectors n and $\overrightarrow{QM}$ are orthogonal (Figure 37). In this connection, the vector n with components α_1, . . . , α_n is called the *vector orthogonal to the hyperplane* (2.3), or *the vector normal to the hyperplane* (2.3). The second and third relations in (2.8) indicate that the *positive half-space* (2.4) *consists of the points* $M(x^1, \ldots, x^n)$ *for which the scalar product* $n\overrightarrow{QM}$ *is non-negative, and the negative half-space* (2.5) *consists of the points for which the scalar product is*

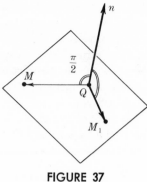

FIGURE 37

nonpositive. Here Q is an arbitrary point of the hyperplane (2.3), and n is a vector normal to this hyperplane.

13. Some Properties of Convex Sets. Let M be any set of points in the space X. The set M is called *convex* if for any two points A and B of this set, the segment AB belongs entirely to the set M (Figure 38). For $n = 2$

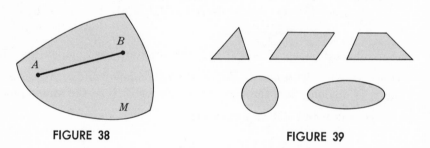

FIGURE 38   FIGURE 39

(that is, for the case in which X is a plane), the following are examples of convex sets: triangle, parallelogram, trapezoid, circle, and ellipse (Figure 39). The configuration shown in Figure 40 is not convex.

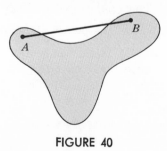

FIGURE 40

A set M in X is called a *cone with vertex at the point* Q if, together with every point A from Q, the set M contains also the entire ray emanating from Q and passing through A. If the set M is convex and, in addition, is a cone with the vertex at Q, then it is called a *convex cone* (with vertex Q).

It is easily seen that the only convex cones in a plane are a single ray, a straight line, an angle not exceeding π, a half-plane, and the entire plane (Figure 41). (An angle greater than π is not convex, Figure 42.) Examining

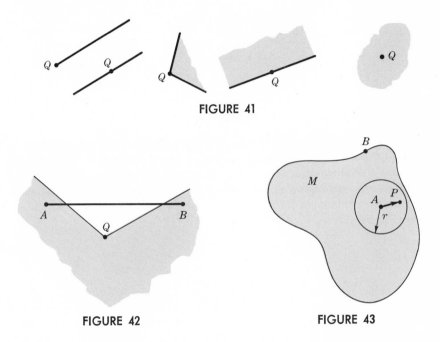

FIGURE 41

FIGURE 42   **FIGURE 43**

Figure 41, it is readily seen that either the convex cone M with vertex Q coincides with the entire plane, or there exists a straight line passing through Q such that the entire cone M lies completely in one of the two half-planes into which this straight line divides the plane. In spaces of dimension greater than two (even for $n = 3$), convex cones may have a more complicated structure; however, in this case also, as we shall see, every convex cone either coincides with the entire space X or lies completely in a "half-space."

We remind the reader that a point A is called an *interior* point of the set M if all points sufficiently close to A belong to M (that is, if there exists an $\epsilon > 0$ such that every point P, for which the length of the vector $\overrightarrow{AP}$ is less than ϵ, belongs to the set M, Figure 43). A point B is called a *boundary* point of the set M if there exist points which belong to the set M and points which do not belong to the set M lying as close as desired to B.

Let M be a convex set of the space X and let Q be one of its boundary points. A hyperplane passing through Q is called a *hyperplane of support* of the set M if the entire set M lies completely in one of the two half-spaces into which this hyperplane divides the space X. Hyperplanes of support for various convex sets (in the case $n = 2$) are shown in Figure 44.

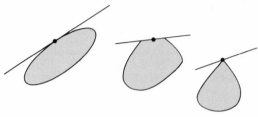

FIGURE 44

It is proved in geometry that *it is possible to draw a hyperplane of support to a convex set M through each of its boundary points*. It should be noted that this theorem asserts the existence of at least one hyperplane of support passing through a given boundary point; it can happen that more than one hyperplane of support can be drawn through some boundary points (Figure 45).

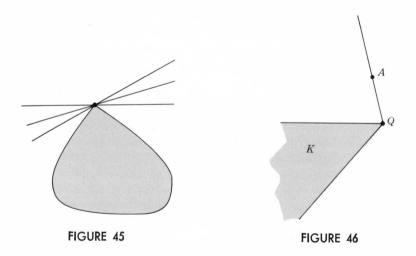

FIGURE 45   FIGURE 46

Now, let K be a convex cone of the space X with vertex Q. If the cone K does not coincide with the entire space X, then there exists a point A not belonging to this cone, and therefore, no point of the ray emanating from Q and passing through A belongs to the cone K (Figure 46); consequently, there are points as close as desired to Q which do not belong to K, that is, Q is a boundary point of the convex cone K.

Thus, if the cone K does not coincide with the entire space X, then the vertex Q is a boundary point of this cone, and therefore, it is possible to draw a hyperplane of support Γ through the point Q, that is, a hyperplane such that (Figure 47) *the entire cone K lies completely in one of the two half-*

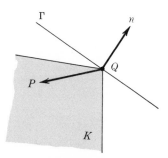

FIGURE 47

spaces defined by this hyperplane (compare p. 43). We shall assume that the cone K lies in the negative half-space, and denote by n the vector normal to the hyperplane Γ. Since for any point P lying in the negative half-space the scalar product $n\overrightarrow{QP}$ is nonpositive, we arrive at the following theorem:

Let K be a convex cone in the space X with vertex Q. If the cone K does not coincide with the entire space X, then there exists a nonzero vector n such that

$$n\overrightarrow{QP} \leqslant 0$$

for any point P of the cone K.

14. An Existence and Uniqueness Theorem. We have agreed above to consider objects whose behavior is described by systems of ordinary differential equations [see (1.2)]. Accordingly, we will have to deal with systems of differential equations and their solutions. For the convenience of the reader, we have compiled here information concerning solutions of systems of differential equations which will be used in the sequel.

In this section, we formulate the *theorem on the existence and uniqueness of solutions* in a form convenient for later use. Proofs of Theorems 2.1 and 2.3 stated below can be found in the book by Pontryagin ([8], pp. 159–169).

We consider the system of differential equations

$$\begin{aligned}
\dot{x}^1 &= \varphi^1(x^1, \ldots, x^n, t), \\
\dot{x}^2 &= \varphi^2(x^1, \ldots, x^n, t), \\
&\quad\cdots \\
\dot{x}^n &= \varphi^n(x^1, \ldots, x^n, t),
\end{aligned}$$

(2.9)

assuming that the right-hand sides φ^i, $i = 1, \ldots, n$, are defined on an open set Γ of $x^1, \ldots, x^n$, t-space. Recall that a set Γ is called *open* if every one of its points is an interior point, that is, for any point $(x_0^1, \ldots, x_0^n, t_0)$ $\in \Gamma$, it is possible to find a number $\epsilon > 0$ such that every point $(x^1, \ldots, x^n, t)$, whose coordinates satisfy $|x^1 - x_0^1| < \epsilon, \ldots, |x^n - x_0^n| < \epsilon$, $|t - t_0| < \epsilon$, belongs to Γ. Geometrically, an open set can be visualized as a region without boundaries (a region which does not include boundary points, that is, points of a line or of a surface bounding this region). For example, the relation

$$(x^1)^2 + (x^2)^2 < 1$$

defines an open set (a circle without boundaries, or, as one says, an open circle) in the x^1, x^2-plane; the relations

$$\alpha^1 < x^1 < \beta^1, \qquad \alpha^2 < x^2 < \beta^2$$

also define an open set (an open rectangle) in the x^1, x^2-plane. In general, any set defined in $x^1, \ldots, x^n$, t-space by a finite system of inequalities

$$g_1(x^1, \ldots, x^n, t) < 0, \ldots, \qquad g_s(x^1, \ldots, x^n, t) < 0,$$

where $g_1, \ldots, g_s$ are given continuous functions of the variables $x^1, \ldots, x^n$, t, is an open set.

Recall that a system of functions $x^1(t), \ldots, x^n(t)$ given on an interval $\theta_0 < t < \theta_1$ is called a *solution* of (2.9) if these functions are differentiable and if each equation in (2.9) becomes an identity (for any t, $\theta_0 < t < \theta_1$) when the above functions are substituted therein. This statement means, in particular, that for any θ, $\theta_0 < \theta < \theta_1$, the point $(x^1(\theta), x^2(\theta), \ldots, x^n(\theta), \theta)$ belongs to the set Γ. In other words, the curve, defined in $x^1, x^2, \ldots, x^n$, t-space by the parametric equations

$$x^1 = x^1(\theta), \ldots, x^n = x^n(\theta), t = \theta \qquad (\theta_0 < \theta < \theta_1),$$

lies entirely in the open set Γ. This curve is called an *integral curve* of the system (2.9). It is clear that considering the solution $x^1(t), \ldots, x^n(t)$ of system (2.9) is equivalent to considering the corresponding integral curve. Let γ_0 be a point of the set Γ and let $(x_0^1, x_0^2, \ldots, x_0^n, t_0)$ be its coordinates. It is obvious that the integral curve passes through the point γ_0 if and only if the corresponding solution $x^1(t), \ldots, x^n(t)$ of system (2.9) satisfies the conditions

$$x^1(t_0) = x_0^1, x^2(t_0) = x_0^2, \ldots, x^n(t_0) = x_0^n. \tag{2.10}$$

These relations are frequently called *initial conditions*.

Theorem 2.1. If the right-hand sides $\varphi^i(x^1, \ldots, x^n, t)$ of (2.9) and their partial derivatives $\dfrac{\partial \varphi^i(x^1, \ldots, x^n, t)}{\partial x^j}$, $i, j = 1, \ldots, n$, *exist and are* *continuous (jointly in the variables x^1, . . . , x^n, t) on the set Γ, then there is one and only one integral curve of (2.9) passing through each point $\gamma_0 = (x_0^1, \ldots, x_0^n, t_0)$ of the set Γ (Figure 48). In other words, there exists a solution $x^1(t)$,*

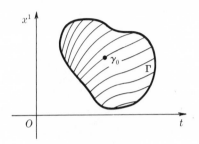

FIGURE 48

. . . , $x^n(t)$ defined on an interval $\theta_0 < t < \theta_1$, which contains the point t_0, such that the initial conditions (2.10) are satisfied; this solution is uniquely defined in the sense that any two solutions satisfying the same initial conditions (2.10) coincide on the common portion of the intervals on which they are defined.

Corollary 2.2.   Let the right-hand sides $\varphi^i(x^1, \ldots, x^n, t)$ be defined and continuous and have continuous derivatives $\partial \varphi^i / \partial x^j$, $i, j = 1, \ldots, n$, for all t satisfying $a \leqslant t < b$ (where a and b are arbitrary numbers). Then for any numbers $x_0^1, x_0^2, \ldots, x_0^n$ there exist continuous functions $x^1(t), x^2(t), \ldots, x^n(t)$ defined on an interval $a \leqslant t < \theta$ (where $\theta \leqslant b$) such that, considered on the interval $a < t < \theta$, these functions constitute a solution of (2.9), and in addition, satisfy the conditions

$$x^1(a) = x_0^1, \ldots, \qquad x^n(a) = x_0^n. \tag{2.11}$$

These functions $x^1(t), \ldots, x^n(t)$ are uniquely defined by the given numbers $x_0^1, \ldots, x_0^n$.

For the proof, it suffices to extend the functions φ^i to all values $t < b$ by setting

$$\varphi^i(x^1, \ldots, x^n, t) = \varphi^i(x^1, \ldots, x^n, a) \qquad \text{for } t < a.$$

Then the functions φ^i and their derivatives $\partial \varphi^i / \partial x^j$ turn out to be continuous on the entire open half-space $t < b$ of the variables $x^1, \ldots, x^n, t$, and therefore Theorem 2.1 is applicable. Determining, by virtue of Theorem 2.1, the solution which satisfies the initial conditions (2.11), we obtain the desired functions $x^1(t), \ldots, x^n(t)$. The uniqueness of these functions is proved verbatim as the uniqueness of the solution in Theorem 2.1 (see [8], pp. 166, 167).

Linear systems of differential equations, that is, systems of the form

$$
\begin{aligned}
\dot{x}^1 &= a_1^1(t)x^1 + a_2^1(t)x^2 + \cdots + a_n^1(t)x^n + b^1(t), \\
\dot{x}^2 &= a_1^2(t)x^1 + a_2^2(t)x^2 + \cdots + a_n^2(t)x^n + b^2(t), \\
&\quad \cdots \\
\dot{x}^n &= a_1^n(t)x^1 + a_2^n(t)x^2 + \cdots + a_n^n(t)x^n + b^n(t),
\end{aligned}
\tag{2.12}
$$

will play a very important role in the sequel. For such systems, the existence and uniqueness theorem can be substantially strengthened. Namely, if a solution, whose existence is assured by Theorem 2.1, is defined on *some* (possibly small) interval $\theta_0 < t < \theta_1$, then, for linear systems, it can be asserted that the solution is defined on the *entire* interval on which the coefficients $a_j^i(t)$ and $b^i(t)$ are defined. More precisely, we have the following:

Theorem 2.3.   If the functions $a_j^i(t)$ and $b^i(t)$, $i, j = 1, \ldots, n$, are defined and continuous on the interval $a < t < b$, then for any number t_0 satisfying $a < t_0 < b$ and for any numbers $x_0^1, \ldots, x_0^n$, there exist functions $x^1(t), \ldots, x^n(t)$, defined on the entire interval $a < t < b$, which constitute a solution of (2.12) and which satisfy the initial conditions (2.10) [these functions are uniquely defined by the initial conditions (2.10) by virtue of Theorem 2.1].

The following proposition, proved in the same way as Corollary 2.2, is a consequence of this theorem.

Corollary 2.4.   If the functions $a_j^i(t)$ and $b^i(t)$, $i, j = 1, \ldots, n$, are defined and continuous on an interval $a \leqslant t \leqslant b$, then, for any numbers $x_0^1, x_0^2, \ldots, x_0^n$, there exist continuous functions $x^1(t), \ldots, x^n(t)$, defined on the entire interval $a \leqslant t \leqslant b$, which constitute a solution of (2.12) on $a < t < b$, and in addition, satisfy (2.11). The functions $x^1(t), \ldots, x^n(t)$ are uniquely defined (by the numbers $x_0^1, \ldots, x_0^n$).

Let us now apply the propositions stated above to the consideration of control processes. Namely, we suppose that the law of motion of an object is given by (1.2) and that the right-hand sides of these equations f^i and their partial derivatives $\partial f^i / \partial x^j$ exist and are continuous jointly in the arguments $x^1, \ldots, x^n, u^1, \ldots, u^r$. In the following, these assumptions on the functions f^i will always be considered as satisfied (without necessarily mentioning this fact every time). Now, let $u(t) = (u^1(t), \ldots, u^r(t))$ be an arbitrary admissible control given on some interval $t_0 \leqslant t \leqslant t_1$, and let $x_0 = (x_0^1, \ldots, x_0^n)$ be any point of the phase space. Let $\theta_1, \theta_2, \ldots, \theta_k$ be all the points at which at least one of the functions $u^1(t), \ldots, u^r(t)$ has a discontinuity (recall that every admissible control is piecewise continuous); moreover, let $t_0 < \theta_1 < \theta_2 < \cdots < \theta_k < t_1$. Substituting the functions $u^1(t), \ldots, u^r(t)$ into the right-hand sides of (1.2), we arrive at the system of equations (1.4). We first consider (1.4) for values of t satisfying $t_0 \leqslant t < \theta_1$. Since the functions $u^1(t), \ldots, u^r(t)$ are continuous for $t_0 \leqslant t < \theta_1$, it follows that for these values (and for arbitrary $x^1, \ldots, x^n$) the right-hand sides of (1.4) and their partial derivatives with respect to x^j are continuous jointly in the variables $x^1, x^2, \ldots, x^n, t$; therefore Corollary 2.2 is applicable to (1.4) for $t_0 \leqslant t < \theta_1$. Thus, the solution $x(t) = (x^1(t), \ldots, x^n(t))$ of (1.4) with the initial condition (2.10), that is, $x(t_0) = x_0$, is (uniquely) defined on some interval $t_0 \leqslant t < \theta$ (where $\theta \leqslant \theta_1$). It is not always possible to extend the solution $x(t)$ to the entire interval $t_0 \leqslant t < \theta_1$: for example, the solution may become infinite as t tends to a value less than θ_1 (Figure 49).

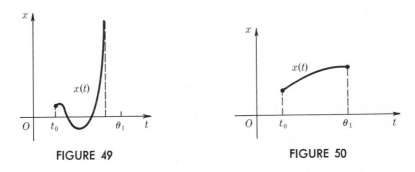

FIGURE 49   FIGURE 50

Let us suppose, however, that this solution is defined on the entire interval $t_0 \leqslant t < \theta_1$ and that

$$\lim_{t \to \theta_1} x(t)$$

exists and is finite. Then, denoting this limit by $x(\theta_1)$ we see that the function $x(t)$ is now defined on the entire interval $t_0 \leqslant t \leqslant \theta_1$, that it is, moreover, continuous on this interval, and satisfies (1.4) on the interval $t_0 < t < \theta_1$ (Figure 50).

We can now consider (1.4) on the interval $\theta_1 \leqslant t < \theta_2$ using the point $x(\theta_1)$ as the initial value. Here (that is, for $\theta_1 \leqslant t < \theta_2$), Corollary 2.2 is again applicable, since for $\theta_1 \leqslant t < \theta_2$, the functions $u^1(t), \ldots, u^r(t)$ are continuous [by virtue of the relation $u(\theta_1) = u(\theta_1 + 0)$, see p. 12]. Consequently, the solution with initial value $x(\theta_1)$ is defined on some interval $\theta_1 \leqslant t < \theta'$. This solution is again denoted by $x(t)$; thus, the solution $x(t)$, $t_0 \leqslant t < \theta'$, is continuous at all points for which it is defined, in particular, at the "junction point" θ_1 (Figure 51). Now, if the solution $x(t)$ is defined on the entire interval $t_0 \leqslant t < \theta_2$ and has a definite (finite) limit as $t \to \theta_2$, then, denoting this limit by $x(\theta_2)$, the function $x(t)$ will be defined and continuous on the entire interval $t_0 \leqslant t \leqslant \theta_2$. We can now consider (1.4) on the interval $\theta_2 \leqslant t < \theta_3$ using the point $x(\theta_2)$ as the initial value (since for $\theta_2 \leqslant t < \theta_3$, Corollary 2.2 is once again applicable) which will permit us to extend the solution $x(t)$ beyond the point θ_2 (Figure 52), and so on. The function $x(t)$ thus obtained is continuous and *piecewise differentiable;*

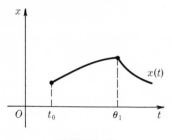

FIGURE 51

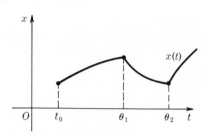

FIGURE 52

namely, the function $x(t)$ is continuously differentiable wherever it is defined, except at the points θ_1, θ_2, . . . , θ_n. The function $x(t)$ which we have constructed will be called the *solution* of system (1.2) *corresponding* to the control $u(t)$ for the initial condition $x(t_0) = x_0$. This solution may not be defined on the entire interval $t_0 \leqslant t \leqslant t_1$ on which the control $u(t)$ is given (it may become infinite).

Finally, we shall say that the admissible control $u(t)$, $t_0 \leqslant t \leqslant t_1$, *transfers* the phase point from the state x_0 to the state x_1 according to the law of motion (1.2) if the corresponding solution $x(t)$ of (1.2) satisfying the initial condition $x(t_0) = x_0$ is defined on the entire interval $t_0 \leqslant t \leqslant t_1$ and passes through the point x_1 at time t_1, that is, satisfies the terminal condition $x(t_1) = x_1$.

Recall that in formulating the time-optimal problem on p. 12 we introduced the idea of a control $u(t)$ which "transfers" an object from a state x_0 to a state x_1. Here this term has acquired a formal definition.

15. The System of Variational Equations.   We consider once again the system of equations (2.9). We assume that the right-hand sides of these equations φ^i and their partial derivatives $\partial\varphi^i/\partial x^j$ are defined and continuous (jointly in the variables x^1, x^2, . . . , x^n, t) on an open set Γ in x^1, x^2, . . . , x^n, t-space. Furthermore, let $x^1(t)$, $x^2(t)$, . . . , $x^n(t)$ be the solution of (2.9) satisfying the initial condition (2.10), which we suppose defined on some interval containing the interval $t_0 \leqslant t \leqslant t_1$. Let $y_0 = (y_0^1, \ldots, y_0^n)$ be a point such that $(y_0^1, \ldots, y_0^n, t_0) \in \Gamma$. We denote by

$$y(t, y_0) = (y^1(t, y_0), y^2(t, y_0), \ldots, y^n(t, y_0))$$

the solution of system (2.9) with initial condition

$$y(t_0, y_0) = y_0$$

[that is, $y(t, y_0)$ is the integral curve passing through the point y_0 at time $t = t_0$]. Under these conditions we have the following:

Theorem 2.5.   If the point y_0 is sufficiently close to the point x_0, that is, if the relations

$$|x_0^1 - y_0^1| < \delta, \; |x_0^2 - y_0^2| < \delta, \; \ldots, \; |x_0^n - y_0^n| < \delta$$

are satisfied (where δ is a sufficiently small positive number), then the solution $y(t, y_0)$ is defined on the entire interval $t_0 \leqslant t \leqslant t_1$, is continuous jointly in the variables $t, y_0^1, y_0^2, \ldots, y_0^n$, and has continuous partial derivatives with respect to the variables $y_0^1, y_0^2, \ldots, y_0^n$. We denote by $\xi_j^i(t)$ the value of the derivative $\partial y^i(t, y_0)/\partial y_0^j$ for $y_0 = x_0$. The functions $\xi_j^i(t)$ satisfy the following linear system of equations (called the system of variational equations):

$$\xi_j^i = \sum_{\alpha = 1}^n \frac{\partial \varphi^i(x^1(t), \ldots, x^n(t), t)}{\partial x^\alpha} \xi_j^\alpha, \tag{2.13}$$

$$i, j = 1, 2, \ldots, n.$$

A proof of this theorem may be found, for example, in Pontryagin ([8], pp. 198, 199).

We shall derive a corollary from the above theorem which will be used in the sequel. We consider an infinitesimal parameter ϵ and let the point y_0 depend on ϵ as follows:

$$y_0 = y_0(\epsilon) = x_0 + \epsilon h + o(\epsilon),$$

where h is a vector (in $x^1, \ldots, x^n$-space) and $o(\epsilon)$ is an infinitesimal of order higher than ϵ [that is,

$$\lim_{\epsilon \to 0} \frac{|o(\epsilon)|}{\epsilon} = 0,$$

where $|o(\epsilon)|$ is the length of the vector $o(\epsilon)$]. In other words,

$$y_0^i = y_0^i(\epsilon) = x_0^i + \epsilon h^i + o(\epsilon), \qquad i = 1, 2, \ldots, n, \tag{2.14}$$

where $o(\epsilon)$ is an infinitesimal of order higher than ϵ. [In the following, we shall use the same symbol $o(\epsilon)$ in denoting various infinitesimals of order higher than ϵ, so that, for example, $o(\epsilon) + o(\epsilon) = o(\epsilon)$.] Then, by virtue of Theorem 2.5, the solution $y(t, y_0) = y(t, y_0(\epsilon))$ will depend also on ϵ; moreover, this solution will be defined on the interval $t_0 \leqslant t \leqslant t_1$ for all sufficiently small ϵ [since for sufficiently small ϵ, $y_0(\epsilon)$ is close to x_0]. It is easily seen that

$$\left. \frac{dy_0^i(\epsilon)}{d\epsilon} \right|_{\epsilon = 0} = h^i$$

[see (2.14)], and therefore the solution $y(t, y_0(\epsilon))$ has the following derivative with respect to ϵ:

$$\frac{dy^i(t, y_0(\epsilon))}{d\epsilon}\bigg|_{\epsilon=0} = \sum_{\beta=1}^{n} \left(\frac{\partial y^i(t, y_0(\epsilon))}{\partial y_0^\beta}\bigg|_{\epsilon=0}\right)\left(\frac{dy_0^\beta(\epsilon)}{d\epsilon}\bigg|_{\epsilon=0}\right)$$

$$= \sum_{\beta=1}^{n} \left(\frac{\partial y^i(t, y_0)}{\partial y_0^\beta}\bigg|_{y_0=x_0}\right) h^\beta = \sum_{\beta=1}^{n} \xi_\beta^i(t) h^\beta.$$

We denote this derivative by $\delta x^i(t)$:

$$\delta x^i(t) = \frac{dy^i(t, y_0(\epsilon))}{d\epsilon}\bigg|_{\epsilon=0} = \sum_{\beta=1}^{n} \xi_\beta^i(t) h^\beta, \qquad i = 1, \ldots, n. \quad (2.15)$$

It follows directly from (2.13) that the functions $\delta x^i(t)$ satisfy the following linear system of differential equations:

$$\frac{d(\delta x^i(t))}{dt} = \sum_{\alpha=1}^{n} \frac{\partial \varphi^i(x^1(t), \ldots, x^n(t), t)}{\partial x^\alpha} \delta x^\alpha(t), \qquad i = 1, \ldots, n.$$

Furthermore, it follows directly from (2.15) (from the meaning of the derivative) that

$$\lim_{\epsilon \to 0} \frac{y^i(t, y_0(\epsilon)) - x^i(t)}{\epsilon} = \delta x^i(t),$$

or,

$$\lim_{\epsilon \to 0} \frac{y^i(t, y_0(\epsilon)) - x^i(t) - \epsilon \delta x^i(t)}{\epsilon} = 0.$$

But this means that the expression in the numerator is $o(\epsilon)$, that is, we have

$$y^i(t, y_0(\epsilon)) = x^i(t) + \epsilon \delta x^i(t) + o(\epsilon), \qquad i = 1, \ldots, n,$$

or, in vector form,

$$y(t, y_0(\epsilon)) = x(t) + \epsilon \delta x(t) + o(\epsilon).$$

Thus, we have arrived at the following proposition:

Corollary 2.6. Let $x(t) = (x^1(t), \ldots, x^n(t))$ be the solution of (2.9) satisfying the initial conditions (2.10), and defined on the interval $t_0 \leqslant t \leqslant t_1$. Furthermore, let $y(t)$ be the solution of this system with the initial condition

$$y(t_0) = y_0 = x_0 + \epsilon h + o(\epsilon), \qquad (2.16)$$

where $h = (h^1, \ldots, h^n)$ *is a vector. Then, for sufficiently small* ϵ*, the solution* $y(t)$ *is defined on the interval* $t_0 \leqslant t \leqslant t_1$ *and has the form*

$$y(t) = x(t) + \epsilon \delta x(t) + o(\epsilon), \qquad (2.17)$$

where $\delta x(t) = (\delta x^1(t), \ldots, \delta x^n(t))$ *is the solution of the linear system of equations*

$$\frac{d(\delta x^i)}{dt} = \sum_{\alpha=1}^{n} \frac{\partial \varphi^i(x^1(t), \ldots, x^n(t), t)}{\partial x^\alpha} \delta x^\alpha, \qquad i = 1, \ldots, n, \quad (2.18)$$

with the initial condition $\delta x(t_0) = h$.

We note that the quantity $o(\epsilon)$ in (2.17) depends, of course, on t, that is, it has the form $o_t(\epsilon)$. Moreover, it is an infinitesimal of order higher than ϵ uniformly in t, that is, $o_t(\epsilon)/\epsilon$ tends to zero uniformly in t, $t_0 \leqslant t \leqslant t_1$, as $\epsilon \to 0$. (In other words, if δ is an arbitrary positive number, then there exists an $\epsilon_0 > 0$, such that for $\epsilon < \epsilon_0$ and for any t, $t_0 \leqslant t \leqslant t_1$, $|o_t(\epsilon)/\epsilon| < \delta$.) Furthermore, if it is assumed that $h = h_{\nu_1, \nu_2, \ldots, \nu_s}$ and $o(\epsilon) = o_{\nu_1, \nu_2, \ldots, \nu_s}(\epsilon)$ in (2.16) depend continuously on parameters $\nu_1, \nu_2, \ldots, \nu_s$, which may vary on intervals $a_i \leqslant \nu_i \leqslant b_i, i = 1, \ldots, s$, and moreover that $o_{\nu_1, \nu_2, \ldots, \nu_s}(\epsilon)$ is an infinitesimal of order higher than ϵ uniformly in $\nu_1, \ldots, \nu_s$ [that is, $\frac{o_{\nu_1, \ldots, \nu_s}(\epsilon)}{\epsilon}$ tends to zero uniformly with respect to $\nu_1, \ldots, \nu_s$ as $\epsilon \to 0$] then (2.17) remains valid, and moreover, $\delta x(t) = \delta x_{\nu_1, \ldots, \nu_s}(t)$ depends now on the parameters $\nu_1, \ldots, \nu_s$, and $o(\epsilon) = o_{\nu_1, \ldots, \nu_s}(\epsilon)$ in (2.17), which depends now not only on ϵ but also on t, $\nu_1, \ldots, \nu_s$, is an infinitesimal of order higher than ϵ *uniformly* in t, $\nu_1, \ldots, \nu_s$.

[The above conclusions are easily deduced from the fact that the solution $y(t, y_0)$ has continuous derivatives with respect to the variables $y_0^1, \ldots, y_0^n$, see Theorem 2.5.]

Let us apply Corollary 2.6 to the controlled object (1.2). Suppose that we are given an initial state x_0 and an admissible control $u(t)$ defined on the interval $t_0 \leqslant t \leqslant t_1$. Let us denote by $\theta_1, \ldots, \theta_k$ all the points of discontinuity of the control $u(t)$. Finally, let $x(t)$ denote the phase trajectory of the object (1.2) corresponding to the control $u(t)$ which emanates (at time t_0) from the point x_0, and let $y(t)$ denote the phase trajectory of this object corresponding to the same control $u(t)$ but emanating (at time t_0) from the point (2.16). The trajectory $x(t)$ is assumed to be defined on the entire interval $t_0 \leqslant t \leqslant t_1$. Then $x(t)$ and $y(t)$ are solutions of one and the same system of equations (1.4) [recall that the same control $u(t)$ is used for both trajectories]. The right-hand sides of this system are continuously differentiable with respect to $x^1, \ldots, x^n$ but depend discontinuously on t (they and their derivatives have discontinuities at the points $t = \theta_1, \ldots,$

$t = \theta_k$). However, Corollary 2.6 may be applied to system (1.4) on the entire interval $t_0 \leqslant t \leqslant t_1$; it suffices to apply this corollary to (1.4) first on the interval $t_0 \leqslant t \leqslant \theta_1$, then on the interval $\theta_1 \leqslant t \leqslant \theta_2$, and so on. As a result we obtain the following statement:

Corollary 2.7.   Let $x(t)$ be the phase trajectory of the object (1.2) defined on the interval $t_0 \leqslant t \leqslant t_1$ corresponding to the initial condition $x(t_0) = x_0$ and to the control $u(t)$. Furthermore, let $y(t)$ be the phase trajectory of (1.2) corresponding to the same control $u(t)$ but to the initial condition

$$y(t_0) = y_0 = x_0 + \epsilon h + o(\epsilon). \tag{2.19}$$

Then, for sufficiently small ϵ, the trajectory $y(t)$ is defined on the entire interval $t_0 \leqslant t \leqslant t_1$, and has the form

$$y(t) = x(t) + \epsilon \delta x(t) + o(\epsilon), \tag{2.20}$$

where $\delta x(t) = (\delta x^1(t), \ldots, \delta x^n(t))$ is the solution of the linear system of equations

$$\frac{d(\delta x^i)}{dt} = \sum_{\alpha=1}^{n} \frac{\partial f^i(x^1(t), \ldots, x^n(t), u^1(t), \ldots, u^r(t))}{\partial x^\alpha} \delta x^\alpha, \tag{2.21}$$

$$i = 1, \ldots, n,$$

with initial condition $\delta x(t_0) = h$.

As before, $o(\epsilon)$ in (2.20) depends on t and is an infinitesimal of order higher than ϵ uniformly in t. Similarly, if the quantities h and $o(\epsilon)$ in (2.19) depend continuously on the parameters $\nu_1, \ldots, \nu_s$ varying on some intervals, then $o(\epsilon)$ in (2.20) depends also on the variables $t, \nu_1, \ldots, \nu_s$, and is an infinitesimal of order higher than ϵ uniformly in all these variables.

16. Adjoint Linear Systems.   Let us consider the linear homogeneous system

$$\begin{aligned}
\dot{x}^1 &= a_1^1(t)x^1 + a_2^1(t)x^2 + \cdots + a_n^1(t)x^n, \\
\dot{x}^2 &= a_1^2(t)x^1 + a_2^2(t)x^2 + \cdots + a_n^2(t)x^n, \\
&\qquad\qquad \cdots \\
\dot{x}^n &= a_1^n(t)x^1 + a_2^n(t)x^2 + \cdots + a_n^n(t)x^n,
\end{aligned} \tag{2.22}$$

whose coefficients $a_j^i(t)$ are defined and continuous on an interval $t_0 \leqslant t \leqslant t_1$. The linear homogeneous system, whose matrix is obtained from the matrix $(a_j^i(t))$ of (2.22) by transposing and changing sign, is said to be adjoint to (2.22). In other words, the adjoint system has the form (for the sake of distinction, the unknowns in the adjoint system will be denoted by $\psi_1, \psi_2, \ldots, \psi_n$)

$$\dot{\psi}_1 = -a_1^1(t)\psi_1 - a_1^2(t)\psi_2 - \cdots - a_1^n(t)\psi_n,$$
$$\dot{\psi}_2 = -a_2^1(t)\psi_1 - a_2^2(t)\psi_2 - \cdots - a_2^n(t)\psi_n,$$
$$\cdots \cdots \cdots \tag{2.23}$$
$$\dot{\psi}_n = -a_n^1(t)\psi_1 - a_n^2(t)\psi_2 - \cdots - a_n^n(t)\psi_n.$$

Using the summation sign, systems (2.22) and (2.23) can be written in the form

$$\dot{x}^i = \sum_{j=1}^n a_j^i(t)x^j, \qquad i = 1, \ldots, n;$$

$$\dot{\psi}_i = -\sum_{j=1}^n a_i^j(t)\psi_j, \qquad i = 1, \ldots, n.$$

Theorem 2.8. *Let* $x(t) = (x^1(t), \ldots, x^n(t))$ *be an arbitrary solution of the system* (2.22) *and let* $\psi(t) = (\psi_1(t), \ldots, \psi_n(t))$ *be an arbitrary solution of the adjoint system* (2.23) (*both solutions being defined on the entire interval* $t_0 \leqslant t \leqslant t_1$). *Then the scalar product*

$$x(t)\psi(t) = x^1(t)\psi_1(t) + x^2(t)\psi_2(t) + \cdots + x^n(t)\psi_n(t)$$

is constant (*that is, it does not depend on* t).

The proof of this theorem reduces to a very simple calculation. In fact, since the functions $x(t)$ and $\psi(t)$ are continuous, the scalar product $x(t)\psi(t)$ is also continuous and has a continuous derivative for $t_0 < t < t_1$. Therefore, it need only be verified that the derivative of this scalar product is equal to zero for $t_0 < t < t_1$. In fact,

$$\frac{d}{dt}(x(t)\psi(t)) = \frac{d}{dt}\sum_{\alpha=1}^n x^\alpha(t)\psi_\alpha(t) = \sum_{\alpha=1}^n \dot{x}^\alpha(t)\psi_\alpha(t) + \sum_{\alpha=1}^n x^\alpha(t)\dot{\psi}_\alpha(t)$$

$$= \sum_{\alpha,j=1}^n a_j^\alpha(t)x^j(t)\psi_\alpha(t) + \sum_{\alpha,j=1}^n x^\alpha(t)(-a_\alpha^j(t)\psi_j(t)) = 0.$$

Theorem 2.8 (and its proof) remains valid for the case in which the coefficients $a_j^i(t)$ are not continuous but piecewise continuous functions of t. In fact, the scalar product $x(t)\psi(t)$ turns out to be a continuous function in this case also since each of the solutions $x(t)$ and $\psi(t)$ is continuous, and its derivative $d(x(t)\psi(t))/dt$ vanishes everywhere except possibly for a finite number of values of t at which the coefficients $a_j^i(t)$ undergo discontinuities.

This remark makes it possible to apply Theorem 2.8 to the system of variational equations (2.21). In fact, this system is linear and its coefficients, that is, the functions

$$\frac{\partial f^i(x^1(t), \ldots, x^n(t), u^1(t), \ldots, u^r(t))}{\partial x^\alpha}$$

are piecewise continuous on the interval $t_0 \leqslant t \leqslant t_1$. The system of equations adjoint to (2.21) has the following form:

$$\dot{\psi}_i = -\sum_{\alpha=1}^{n} \frac{\partial f^\alpha(x^1(t), \ldots, x^n(t), u^1(t), \ldots, u^r(t))}{\partial x^i} \psi_\alpha, \qquad (2.24)$$

$$i = 1, \ldots, n.$$

Thus, we obtain the following proposition:

Corollary 2.9. Let $\delta x(t) = (\delta x^1(t), \ldots, \delta x^n(t))$ be an arbitrary solution of the system of variational equations (2.21) and let $\psi(t) = (\psi_1(t), \ldots, \psi_n(t))$ be an arbitrary solution of the adjoint system (2.24). Then the scalar product

$$\psi(t)\delta x(t) = \psi_1(t)\delta x^1(t) + \psi_2(t)\delta x^2(t) + \cdots + \psi_n(t)\delta x^n(t)$$

is constant [on the entire interval on which both solutions $\psi(t)$, $\delta x(t)$ are defined].

In the following, system (2.24) will play a very important role in the study of optimal processes for the object (1.2).

§ 5. THE MAXIMUM PRINCIPLE (TIME-OPTIMAL CASE)

17. Variations of Controls. Let us suppose that, under the action of a specified admissible control $u(t)$, the phase point, moving according to the law (1.2), is transferred from a given initial state x_0 to a prescribed terminal state x_1. Furthermore, suppose we suspect that $u(t)$ is actually a time-optimal control, that is, that the transition from x_0 to x_1 is accomplished in the shortest time. How does one verify that this control is in fact optimal? Of course, the best way to do this would be to convince oneself that any other control is worse, that is, that under the action of any other control, the phase point, even if it reaches the prescribed terminal state, requires more time to do so. However, it is very difficult to compare the chosen control $u(t)$ with an arbitrary control. Therefore, we shall derive only a necessary condition for optimality by comparing $u(t)$ not with all other controls but only with those that are "close" to $u(t)$, that is, we shall find conditions which must be satisfied by $u(t)$ in order that it be better than any other control arbitrarily "close" to $u(t)$. It is clear that the *optimal control* (which is better than any control, not merely those that are "close") must satisfy the condition which will be thus obtained.

Various meanings can be given to the concept of a "close" control. We shall define this concept in the following manner. Let us take a point v in the control region U and an interior point τ of the interval $t_0 \leqslant t \leqslant t_1$ [on which the control $u(t)$ is to be considered], which is not a point of discontinuity of the control $u(t)$. Furthermore, let us choose a positive number l and consider a positive arbitrarily small parameter ϵ. Finally, let us denote by I an interval of length $l\epsilon$ with right-hand endpoint at τ, namely, the interval $\tau - l\epsilon \leqslant t < \tau$. Now, we replace the control $u(t)$ on the interval

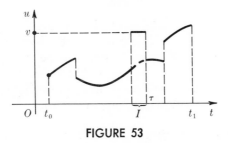

FIGURE 53

I by the constant control v without changing the control $u(t)$ outside the interval I (Figure 53). In other words, we consider a new control

$$u^*(t) = \begin{cases} u(t) & \text{outside the interval } I; \\ v & \text{on the interval } I. \end{cases}$$

Such a modification of the control $u(t)$ [that is, a transition from the control $u(t)$ to the control $u^*(t)$] will be called a *variation* of this control, more specifically, a *variation near the point τ*. One may also construct a variation of the control $u(t)$ not only in the vicinity of one point τ, but also near two or more points. For example, one can select three points τ_1, τ_2, τ_3 interior to the interval $t_0 \leqslant t \leqslant t_1$ [which are not points of discontinuity of the control $u(t)$], three points v_1, v_2, v_3 in the control region U, and three positive numbers l_1, l_2, l_3. Now let I_i denote an interval of length l_i with right-hand endpoint at τ_i, namely, the interval $\tau_i - l_i\epsilon \leqslant t < \tau_i$, where $i = 1, 2, 3$, and replace the control $u(t)$ on the interval I_i by the constant value v_i without changing the control $u(t)$ outside these intervals (Figure 54). The control $u^*(t)$ is obtained by perturbing $u(t)$ near the three points τ_1, τ_2, τ_3. The variation of $u(t)$ near any (finite) number of distinct points interior to the interval $t_0 \leqslant t \leqslant t_1$ is defined in an analogous manner.

The above construction provides, in fact, a precise definition of the term "close" controls. Namely, a control $u^*(t)$ obtained by perturbing the control $u(t)$ near several points of the interval $t_0 \leqslant t \leqslant t_1$ will be regarded as "close" to $u(t)$. We note that the control $u^*(t)$, obtained by perturbing $u(t)$, depends on the parameter ϵ. It should also be noted that varying the

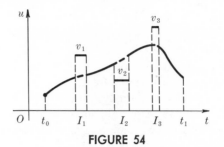

FIGURE 54

control $u(t)$ (see Figure 54) can be accomplished only if the intervals I_1, I_2, I_3 do not intersect one another and lie entirely in the basic interval $t_0 \leqslant t \leqslant t_1$. These requirements can always be assumed to be satisfied, since τ_1, τ_2, τ_3 are distinct points lying in the interior of the interval $t_0 \leqslant t \leqslant t_1$, and the lengths of I_1, I_2, I_3 are infinitesimals (that is, in the following the value of ϵ will be considered to be sufficiently close to zero).

18. Perturbation of Trajectories. Let $x(t) = (x^1(t), \ldots, x^n(t))$ denote the solution of (1.2) corresponding to the control $u(t)$ and emanating (at time t_0) from the initial point x_0. We shall assume that $x(t)$ is defined on the entire interval $t_0 \leqslant t \leqslant t_1$ on which $u(t)$ is specified. If $u(t)$ is replaced by a varied control $u^*(t)$, then the phase trajectory $x(t)$ will be replaced by a trajectory $x^*(t)$ "close" to $x(t)$, emanating from the same point x_0 but terminating at $x^*(t_1)$ which may not coincide with $x(t_1)$. We consider the *set of all points of the phase space X which can be reached from the initial point x_0 in a time equal to or somewhat less than $t_1 - t_0$ utilizing all possible varied controls $u^*(t)$.* More precisely, we consider points of the form $x^*(t_1 - \epsilon\delta t)$, where δt is an arbitrary non-negative number.

We assume that the perturbed control $u^*(t)$ is obtained by varying $u(t)$ near s points $\tau_1, \ldots, \tau_s$ [which are points of continuity of the control $u(t)$]; moreover, corresponding to these points, we select points $v_1, \ldots, v_s$ of the control region U and non-negative numbers $l_1, \ldots, l_s$. The intervals on which $u(t)$ is perturbed are denoted by $I_1, I_2, \ldots, I_s$ [that is, the interval I_i is defined by the inequalities $\tau_i - \epsilon l_i \leqslant t < \tau_i$ and on this interval the control $u(t)$ is replaced by the constant value v_i]. A displacement of the trajectory's endpoint is caused, in the first place, by perturbing the control $u(t)$ on each of the intervals $I_1, \ldots, I_s$ and, secondly, by changing the terminal time ($t_1 - \epsilon\delta t$ instead of t_1). Let us examine the effect of each of these factors on the displacement of the trajectory's endpoint. In this connection, we will make use of the vector form (1.3) of system (1.2), namely,

$$\dot{x} = f(x, u), \tag{2.25}$$

but first we present an intuitive argument.

If the control $u(t)$ is not perturbed at all [that is, assuming that $u^*(t) \equiv u(t)$ and consequently $x^*(t) \equiv x(t)$], then the point $x^*(t_1 - \epsilon \delta t)$ will coincide with the point $x(t_1 - \epsilon \delta t)$. It is easily seen that in this case

$$x^*(t_1 - \epsilon \delta t) = x(t_1 - \epsilon \delta t) = x(t_1) - \epsilon \delta t f(x(t_1), u(t_1)) + o(\epsilon). \quad (2.26)$$

In fact, (2.25) shows that the phase point, moving along the trajectory $x(t)$, has, at time t, a velocity equal to $f(x(t), u(t))$ (this is the velocity of motion of a point in the phase space X, that is, the so-called *phase velocity*). In the infinitely small time interval from $t_1 - \epsilon \delta t$ to t_1, the velocity of motion differs little from $f(x(t_1), u(t_1))$ [since the control $u(t)$ is continuous at the point t_1]; therefore, the displacement of the point during this time interval, that is, $x(t_1) - x(t_1 - \epsilon \delta t)$, is approximately equal to the product of the duration of this interval and the velocity:

$$x(t_1) - x(t_1 - \epsilon \delta t) \approx \epsilon \delta t f(x(t_1), u(t_1)).$$

Actually, the velocity changes somewhat during the time interval under consideration, and therefore, in order to write the last relation accurately, it is necessary to add the quantity $o(\epsilon)$ to the right-hand side. This, in fact, yields (2.26) which shows the effect of changing the terminal time on the displacement of the endpoint of the trajectory.

Now suppose that we vary the control on only one of the intervals I_i, leaving the terminal time t_1 unchanged. Then for $t_0 \leqslant t < \tau_i - \epsilon l_i$ we have $u^*(t) \equiv u(t)$, and therefore $x^*(t) \equiv x(t)$. We set

$$x^*(\tau_i - \epsilon l_i) = x(\tau_i - \epsilon l_i). \quad (2.27)$$

Furthermore, since the trajectory $x(t)$ on the interval I_i is the solution of (2.25) with the control $u = u(t)$, and the trajectory $x^*(t)$ is the solution of the same equation (2.25) with control $u \equiv v_i$, we have

$$x(\tau_i) - x(\tau_i - \epsilon l_i) = \epsilon l_i f(x(\tau_i), u(\tau_i)) + o(\epsilon), \quad (2.28)$$
$$x^*(\tau_i) - x^*(\tau_i - \epsilon l_i) = \epsilon l_i f(x(\tau_i), v_i) + o(\epsilon). \quad (2.29)$$

In fact, the time interval I_i has an infinitesimal length ϵl_i, and, on this interval, the point $x(t)$ moves with the velocity $f(x(t), u(t)) \approx f(x(\tau_i), u(\tau_i))$ [recall that τ_i is a point of continuity of the control $u(t)$], and the point $x^*(t)$ moves with the velocity $f(x^*(t), v_i) \approx f(x(t), v_i) \approx f(x(\tau_i), v_i)$. Comparing (2.27), (2.28), and (2.29), we obtain

$$x^*(\tau_i) = x(\tau_i) + \epsilon l_i [f(x(\tau_i), v_i) - f(x(\tau_i), u(\tau_i))] + o(\epsilon),$$

that is,

$$x^*(\tau_i) = x(\tau_i) + \epsilon h_i + o(\epsilon), \quad (2.30)$$

where

$$h_i = l_i [f(x(\tau_i), v_i) - f(x(\tau_i), u(\tau_i))]. \quad (2.31)$$

We now observe that on the interval $\tau_i \leqslant t \leqslant t_1$ we have $u^*(t) \equiv u(t)$, so that by virtue of Corollary 2.7,

$$x^*(t) = x(t) + \epsilon\delta x_i(t) + o(\epsilon), \qquad \tau_i \leqslant t \leqslant t_1,$$

where $\delta x_i(t) = (\delta x_i^1(t), \ldots, \delta x_i^n(t))$ is the solution of (2.21) with initial condition

$$\delta x_i(\tau_i) = h_i. \tag{2.32}$$

In particular, for $t = t_1$ we obtain

$$x^*(t_1) = x(t_1) + \epsilon\delta x_i(t_1) + o(\epsilon). \tag{2.33}$$

This yields the displacement of the trajectory's endpoint produced solely by a variation on the interval I_i.

Thus, a change in the end of the time interval by the amount $-\epsilon\delta t$ results in the addition of the quantity $-\epsilon\delta t f(x(t_1), u(t_1)) + o(\epsilon)$ to $x(t_1)$ [see (2.26)], and a variation on the interval I_i results in the addition of $\epsilon\delta x_i(t_1) + o(\epsilon)$, $i = 1, 2, \ldots, s$. But the individual displacements produced by several factors, each of which causes a displacement which is of order ϵ, may be added to obtain the displacement caused by the simultaneous action [to within $o(\epsilon)$] of these several factors. Consequently, *perturbing the control on all the intervals* $I_1, \ldots, I_s$ *and, in addition, changing the terminal time, we obtain*

$$x^*(t_1 - \epsilon\delta t) = x(t_1) - \epsilon\delta t f(x(t_1), u(t_1)) + \epsilon \sum_{i=1}^{s} \delta x_i(t_1) + o(\epsilon). \tag{2.34}$$

Of course, the derivation of (2.34) presented here is not rigorous. We shall give below an accurate derivation of this formula, but first, let us cast it in a somewhat different form. Namely, let τ be an arbitrary point of the interval $t_0 \leqslant t \leqslant t_1$ and let h be an arbitrary vector. Let $\delta x(t)$ denote the solution of (2.21) with initial condition $\delta x(\tau) = h$, and set

$$\delta x(t_1) = \Delta(\tau, h). \tag{2.35}$$

The vector $\Delta(\tau, h)$ is defined for any τ, $t_0 \leqslant \tau \leqslant t_1$, and for any h. In particular, recalling the definition of the solution $\delta x_i(t)$ [see (2.32), (2.33)], we can write

$$\delta x_i(t_1) = \Delta(\tau_i, h_i), \qquad i = 1, \ldots, s.$$

Thus, (2.34) takes the following form:

$$x^*(t_1 - \epsilon\delta t) = x(t_1) - \epsilon\delta t f(x(t_1), u(t_1)) + \epsilon \sum_{i=1}^{s} \Delta(\tau_i, h_i) + o(\epsilon). \tag{2.36}$$

Recall that this relation is written for the trajectory $x^*(t)$ corresponding to the control $u^*(t)$ which is obtained by variation of the control $u(t)$ on the intervals $I_1, \ldots, I_s$ (using the points $\tau_1, \ldots, \tau_s$, the points $v_1, \ldots, v_s$, and the numbers $l_1, \ldots, l_s$).

We now present a precise derivation of (2.36) based on an induction on the number s of intervals $I_1, \ldots, I_s$. For $s = 0$ we have $u^*(t) \equiv u(t)$, $x^*(t) \equiv x(t)$, and therefore (2.36) reduces to (2.26); we prove this. Integrating the ith relation of (1.4) over the interval $t_1 - \epsilon\,\delta t \leqslant t \leqslant t_1$, we obtain

$$x^i(t_1) - x^i(t_1 - \epsilon\delta t) = \int_{t_1 - \epsilon\delta t}^{t_1} f^i(x^1(t), \ldots, x^n(t), u^1(t), \ldots, u^r(t))\, dt$$
$$= \epsilon\delta t f^i(x^1(\theta), \ldots, x^n(\theta), u^1(\theta), \ldots, u^r(\theta)),$$

where θ is a point in the interval $t_1 - \epsilon\delta t \leqslant t \leqslant t_1$ [the mean value theorem for integrals; recall that for a sufficiently small ϵ, the function $u(t)$ is continuous on the interval $t_1 - \epsilon\delta t \leqslant t \leqslant t_1$]. This relation can be rewritten as follows:

$$x^i(t_1) - x^i(t_1 - \epsilon\delta t) = \epsilon\delta t f^i(x^1(t_1), \ldots, x^n(t_1), u^1(t_1), \ldots, u^r(t_1))$$
$$+ \epsilon\delta t\{f^i(x^1(\theta), \ldots, x^n(\theta), u^1(\theta), \ldots, u^r(\theta)) \qquad (2.37)$$
$$- f^i(x^1(t_1), \ldots, x^n(t_1), u^1(t_1), \ldots, u^r(t_1))\}.$$

The last term here is $o(\epsilon)$ because the expression in braces tends to zero as $\epsilon \to 0$ (since for $\epsilon \to 0$ we have $\theta \to t_1$). Thus,

$$x^i(t_1) - x^i(t_1 - \epsilon\delta t) = \epsilon\delta t f^i(x^1(t_1), \ldots, x^n(t_1), u^1(t_1), \ldots, u^r(t_1)) + o(\epsilon),$$

and this, in fact, is the component form of (2.26). Thus, formula (2.36) holds for $s = 0$.

Let us assume that formula (2.36) [or, equivalently, formula (2.34)] has already been proved when the number of intervals $I_1, I_2, \ldots$ is less than s, and let us consider a variation of the control on the s intervals $I_1, I_2, \ldots, I_s$. Let us first consider the control $u^*(t)$ and the trajectory $x^*(t)$ on the interval $t_0 \leqslant t < \tau_s - \epsilon l_s$. Since there are only $s - 1$ variation intervals $I_1, I_2, \ldots, I_{s-1}$ on this interval, we can write, by the inductive assumption,

$$x^*(\tau_s - \epsilon l_s) = x(\tau_s) - \epsilon l_s f(x(\tau_s), u(\tau_s)) + \epsilon \sum_{i=1}^{s-1} \delta x_i(\tau_s) + o(\epsilon).$$

Furthermore, we have

$$x^*(\tau_s) - x^*(\tau_s - \epsilon l_s) = \epsilon l_s f(x(\tau_s), v_s) + o(\epsilon)$$

[the proof of this relation is completely analogous to the proof of (2.26), see

(2.37)]. Adding the last two relations, we obtain

$$x^*(\tau_s) = x(\tau_s) + \epsilon l_s[f(x(\tau_s), v_s) - f(x(\tau_s), u(\tau_s))] + \epsilon \sum_{i=1}^{s-1} \delta x_i(\tau_s) + o(\epsilon)$$

$$= x(\tau_s) + \epsilon h_s + \epsilon \sum_{i=1}^{s-1} \delta x_i(\tau_s) + o(\epsilon)$$

$$= x(\tau_s) + \epsilon \delta x_s(\tau_s) + \epsilon \sum_{i=1}^{s-1} \delta x_i(\tau_s) + o(\epsilon)$$

$$= x(\tau_s) + \epsilon \sum_{i=1}^{s} \delta x_i(\tau_s) + o(\epsilon)$$

[see (2.31), (2.32)].

Since $u^*(t) \equiv u(t)$ on the interval $\tau_s \leqslant t \leqslant t_1$, we have, by virtue of Corollary 2.7,

$$x^*(t) = x(t) + \epsilon \delta x(t) + o(\epsilon), \qquad \tau_s \leqslant t \leqslant t_1, \tag{2.38}$$

where $\delta x(t)$ is the solution of (2.21) with initial condition

$$\delta x(\tau_s) = \sum_{i=1}^{s} \delta x_i(\tau_s). \tag{2.39}$$

But the system (2.21) is linear and homogeneous, and therefore, the function

$$\delta x(t) = \sum_{i=1}^{s} \delta x_i(t), \qquad \tau_s \leqslant t \leqslant t_1, \tag{2.40}$$

is a solution of this system [since each term $\delta x_i(t)$ is a solution]; moreover, this solution obviously satisfies the initial condition (2.39). Thus, the function $\delta x(t)$ in formula (2.38) has the form (2.40), that is,

$$x^*(t) = x(t) + \epsilon \sum_{i=1}^{s} \delta x_i(t) + o(\epsilon), \qquad \tau_s \leqslant t \leqslant t_1.$$

In particular, for $t = t_1$, we obtain

$$x^*(t_1) = x(t_1) + \epsilon \sum_{i=1}^{s} \delta x_i(t_1) + o(\epsilon) = x(t_1) + \epsilon \sum_{i=1}^{s} \Delta(\tau_i, h_i) + o(\epsilon). \tag{2.41}$$

Finally, noting that

$$x^*(t_1) - x^*(t_1 - \epsilon \delta t) = \epsilon \delta t f(x(t_1), u(t_1)) + o(\epsilon)$$

[this formula is proved in the same way as (2.26); compare (2.37)], and subtracting this relation from (2.41), we obtain (2.36). Thus, by induction, (2.36) is valid for any s.

It follows readily from the remarks made on p. 54 (following Corollary 2.7), that the *quantity $o(\epsilon)$ in* (2.36) *is an infinitesimal of order higher than ϵ (uniformly in δt, l_1, l_2, . . . , l_s) if δt, l_1, . . . , l_s vary on intervals*

$$0 \leqslant \delta t \leqslant \gamma, \qquad 0 \leqslant l_i \leqslant \beta_i.$$

As a matter of fact, the quantity $o(\epsilon)$ in formula (2.26) is an infinitesimal of order higher than ϵ uniformly in δt [this follows readily from (2.37)]. Furthermore, tracing the induction carried out on pp. 61, 62, we can easily convince ourselves (with the aid of remarks on p. 54) that the quantity $o(\epsilon)$ in (2.36) is an infinitesimal of order higher than ϵ uniformly in δt, l_1, . . . , l_s.

19. The Fundamental Lemma.   The coefficient ϵ in the right-hand side of (2.36) multiplies a vector of the form

$$-f(x(t_1),\, u(t_1))\delta t + \sum_{i=1}^{s} \Delta(\tau_i,\, h_i), \tag{2.42}$$

where h_i is given by (2.31). We note that the points $\tau_1, \tau_2, \ldots, \tau_s$ in the interval $t_0 < t < t_1$ are, by construction, points of continuity of the control $u(t)$. We now consider vectors of the form (2.42) (assuming, as before, that the process $(u(t), x(t))$ is given on the interval $t_0 \leqslant t \leqslant t_1$ without requiring that the points $\tau_1, \tau_2, \ldots, \tau_s$ be pairwise distinct [but considering them, as before, to be points of continuity of the control $u(t)$].

Every vector of this form will be called a *displacement vector*. Thus, the term displacement vector will refer to an arbitrary vector of the form (2.42), where δt is a non-negative number, s is an arbitrary natural number, τ_1, . . . , τ_s are arbitrary (not necessarily pairwise distinct) points of continuity of the control $u(t)$ in the interval $t_0 < t < t_1$, and the vectors h_i are defined by (2.31) (moreover, l_1, . . . , l_s are arbitrary non-negative numbers, and v_1, . . . , v_s are arbitrary points of the control region U).

Let Q denote the right-hand endpoint of the trajectory $x(t)$, that is, the point $x(t_1)$, and let us measure every displacement vector from this point. The set formed by the endpoints of all possible displacement vectors measured from the point Q will be denoted by K. In other words, the point A belongs to the set K if and only if $\overrightarrow{QA}$ is a displacement vector [that is, there exist quantities δt, τ_1, . . . , τ_s, l_1, . . . , l_s, v_1, . . . , v_s, such that the vector (2.42) coincides with $\overrightarrow{QA}$].

We shall show that the set K is a *convex cone with vertex Q*. It should be noted, first of all, that if we multiply each of the quantities δt, l_1, . . . , l_s by the same non-negative number k (leaving τ_1, . . . , τ_s and v_1, . . . , v_s unchanged) then the displacement vector (2.42) will also be multiplied by this number k. In fact, by multiplying the quantities δt, l_1, . . . , l_s by k, the vector $-f(x(t_1),\ u(t_1))\ \delta t$ and each of the vectors h_i [see (2.31)] will be multiplied by k. Therefore, by virtue of the linearity and homogeneity of (2.21), every solution $\delta x_i\ (t)$ will be multiplied by k [see (2.32)] which means that every vector $\Delta(\tau_i,\ h_i)$ will also be multiplied by k. (The number k must be non-negative, since δt, l_1, . . . , l_s must be non-negative and therefore, they can be multiplied only by non-negative numbers.) Thus we have proved that a multiple of a displacement vector (2.42) by an arbitrary non-negative number k is again a displacement vector. In other words, if $\overrightarrow{QA}$ is a displacement vector, then, for any $k \geqslant 0$, the vector $\overrightarrow{QB} = k\overrightarrow{QA}$ is also a displacement vector. This means that, together with the point A different from Q, the entire ray emanating from Q and passing through A belongs to the set K, that is, the *set K is a cone with vertex Q.*

Now, let $\overrightarrow{QA}$ and $\overrightarrow{QB}$ be two displacement vectors, let the quantities $\delta t'$, τ_1, . . . , τ_s, l_1, . . . , l_s, v_1, . . . , v_s correspond to the vector $\overrightarrow{QA}$, and let the quantities $\delta t''$, τ_{s+1}, . . . , τ_r, l_{s+1}, . . . , l_r, v_{s+1}, . . . , v_r correspond to the vector $\overrightarrow{QB}$, that is,

$$\overrightarrow{QA} = -f(x(t_1),\ u(t_1))\ \delta t' + \sum_{i=1}^{s} \Delta(\tau_i, h_i),$$

$$\overrightarrow{QB} = -f(x(t_1),\ u(t_1))\ \delta t'' + \sum_{i=s+1}^{r} \Delta(\tau_i, h_i).$$

We now set $\delta t = \delta t' + \delta t''$ and combine the quantities τ_i, l_i, v_i, corresponding to the subscripts $i = 1,$ . . . , s and $i = s + 1,$ . . . , r, that is, we consider the displacement vector $\overrightarrow{QD}$ which corresponds to the quantities $\delta t = \delta t' + \delta t''$, τ_1, . . . , τ_r, l_1, . . . , l_r, v_1, . . . , v_r:

$$\overrightarrow{QD} = -f(x(t_1),\ u(t_1))(\delta t' + \delta t'') + \sum_{i=1}^{r} \Delta(\tau_i, h_i).$$

From the above expressions for the vectors $\overrightarrow{QA}$, $\overrightarrow{QB}$, $\overrightarrow{QD}$ it is immediately seen that

$$\overrightarrow{QD} = \overrightarrow{QA} + \overrightarrow{QB},$$

that is, the *sum of any two displacement vectors is also a displacement vector.* [Note that this conclusion would be incorrect if we did not assume that there could be points which coincide among the points τ_i defining the vectors (2.42); in fact the points τ_1, . . . , τ_s corresponding to the vector $\overrightarrow{QA}$ and the points $\tau_{s+1}, \ldots, \tau_r$, corresponding to the vector $\overrightarrow{QB}$ are chosen independently of one another, and among these points there may be some which quite possibly coincide.]

Finally, let A, $B \in K$, and let C be an arbitrary point of the segment AB. Then

$$\overrightarrow{QC} = (1 - \lambda)\overrightarrow{QA} + \lambda\overrightarrow{QB},$$

where λ is a number satisfying the inequalities $0 \leqslant \lambda \leqslant 1$. Since both vectors $\overrightarrow{QA}$, $\overrightarrow{QB}$ are displacement vectors $(A, B \in K)$ and since the numbers $1 - \lambda$ and λ are non-negative, the vectors

$$(1 - \lambda)\overrightarrow{QA} \quad \text{and} \quad \lambda\overrightarrow{QB}$$

are also displacement vectors by virtue of what has been proved above; therefore, their sum, that is, the vector $\overrightarrow{QC}$, is also a displacement vector. Consequently, $C \in K$. Thus, if A, $B \in K$, then any point C of the segment AB also belongs to K, that is, K is convex.

Thus K is a convex cone with vertex Q.

We can now formulate and prove the following lemma which provides the basis for all subsequent constructions.

Fundamental Lemma.   Let $u(t)$ be an admissible control which transfers an object from a given initial state x_0 to a prescribed terminal state x_1 during the time interval $t_0 \leqslant t \leqslant t_1$. The corresponding phase trajectory is denoted by $x(t)$. If the cone K, constructed above, coincides with the entire phase space X, then the process $(u(t), x(t))$ is not optimal.

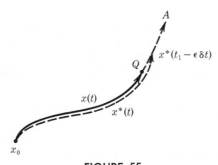

FIGURE 55

The validity of this lemma can be made plausible with the aid of a very descriptive although not quite rigorous construction. Let $\overrightarrow{QA}$ denote the vector $f(x(t_1), u(t_1))$. This vector is tangent to the trajectory $x(t)$ at the point Q (Figure 55), since $f(x(t_1), u(t_1))$ is the velocity of motion of the point $x(t)$ in the phase space X at Q. Since the cone K fills out the entire phase space X, the vector $\overrightarrow{QA}$ is a displacement vector, and therefore, according to (2.36), there exists a variation of the control $u(t)$ such that the vector (2.42) coincides with $\overrightarrow{QA}$, and consequently the corresponding variational trajectory $x^*(t)$ satisfies the condition

$$x^*(t_1 - \epsilon \delta t) = x(t_1) + \epsilon \overrightarrow{QA} + o(\epsilon)$$

[see (2.36), (2.42)]. In other words, in time $t_1 - \epsilon \delta t$ which does not exceed t_1, it is possible (to within terms of order higher than ϵ) to reach a point of the segment QA, that is, to go somewhat farther than the point $Q = x_1$ in the direction of the trajectory $x(t)$. But then the trajectory $x^*(t)$ must have passed even sooner through the point Q, that is, the point Q can be reached in a time less than $t_1 - t_0$, and consequently the process $(u(t), x(t))$ is not optimal. It is true that these arguments are carried out only to within $o(\epsilon)$, so that in reality the trajectory $x^*(t)$ may not pass through the point Q but reach only an $o(\epsilon)$-proximity of this point (see Figure 55). This, apparently, can be rectified by a rotation of the vector $\overrightarrow{QA}$: that is, by a suitable rotation, it is possible to cause the trajectory $x^*(t)$ to pass exactly through the point Q.

This geometrical argument is confirmed by the following simple calculation. Let the quantities $\delta t, \tau_1, \ldots, \tau_s, l_1, \ldots, l_s, v_1, \ldots, v_s$ be chosen so that the vector (2.42) coincides with $\overrightarrow{QA}$, that is,

$$-f(x(t_1), u(t_1)) \, \delta t + \sum_{i=1}^{s} \Delta(\tau_i, h_i) = f(x(t_1), u(t_1)) \tag{2.43}$$

(such quantities exist since, by assumption, K coincides with X, that is, any vector is a displacement vector). We use these quantities $\tau_1, \ldots, \tau_s$, $l_1, \ldots, l_s, v_1, \ldots, v_s$ to vary the control $u(t)$ and the trajectory $x(t)$. Then, according to (2.36), we find that for the perturbed trajectory $x^*(t)$,

$$x^*(t_1 - (1 + \delta t)\epsilon) = x(t_1) - \epsilon(1 + \delta t)f(x(t_1), u(t_1)) + \epsilon \sum_{i=1}^{s} \Delta(\tau_i, h_i) + o(\epsilon).$$

$$\tag{2.44}$$

By virtue of (2.43), this relation takes the form

$$x^*(t_1 - (1 + \delta t)\epsilon) = x(t_1) + o(\epsilon),$$

that is, the trajectory $x^*(t)$ passes "to within $o(\epsilon)$" through the point x_1 at time $t_1 - (1 + \delta t)\epsilon < t_1$.

These arguments, of course, can not serve as a rigorous proof. In the first place, the concluding assertion that "by a suitable rotation of the vector

$\overrightarrow{QA}$ it is possible to cause the trajectory $x^*(t)$ to pass exactly through the point Q'' is by no means justified. Secondly, the use of (2.36) in obtaining (2.44) is, strictly speaking, not valid, since some of the points $\tau_1, \tau_2, \ldots, \tau_s$, used in obtaining (2.43) may coincide, in which case (2.36) is inapplicable. We now present a rigorous proof of the fundamental lemma. Since it is not simple, we recommend that this proof be skipped by those readers who found the above intuitive arguments sufficiently convincing.

20. Proof of the Fundamental Lemma. We first carry out the proof for $n = 2$ (that is, in a phase plane); we then indicate the changes which must be made in the case of arbitrary n.

We denote the endpoint x_1 of the trajectory $x(t)$, as before, by Q, and consider an arbitrary triangle $A^{(0)}A^{(1)}A^{(2)}$ containing the point Q in its interior (Figure 56). Since the cone K coincides with the entire phase plane

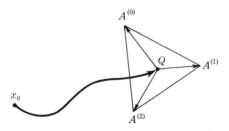

FIGURE 56

X, $\overrightarrow{QA}^{(0)}$, $\overrightarrow{QA}^{(1)}$ and $\overrightarrow{QA}^{(2)}$ are displacement vectors, that is, there exist quantities

$$
\begin{array}{ll}
\delta t^{(0)}, \tau_i^{(0)}, l_i^{(0)}, v_i^{(0)}, & i = 1, \ldots, s^{(0)}, \\
\delta t^{(1)}, \tau_i^{(1)}, l_i^{(1)}, v_i^{(1)}, & i = 1, \ldots, s^{(1)}, \\
\delta t^{(2)}, \tau_i^{(2)}, l_i^{(2)}, v_i^{(2)}, & i = 1, \ldots, s^{(2)},
\end{array}
\tag{2.45}
$$

such that the corresponding displacement vectors coincide with $\overrightarrow{QA}^{(0)}$, $\overrightarrow{QA}^{(1)}$, $\overrightarrow{QA}^{(2)}$; consequently,

$$
\overrightarrow{QA}^{(0)} = -f(x(t_1), u(t_1))\, \delta t^{(0)} + \sum_{i=1}^{s^{(0)}} \Delta(\tau_i^{(0)}, h_i^{(0)}),
\tag{2.46}
$$

$$
\overrightarrow{QA}^{(1)} = -f(x(t_1), u(t_1))\, \delta t^{(1)} + \sum_{i=1}^{s^{(1)}} \Delta(\tau_i^{(1)}, h_i^{(1)}),
\tag{2.47}
$$

$$
\overrightarrow{QA}^{(2)} = -f(x(t_1), u(t_1))\, \delta t^{(2)} + \sum_{i=1}^{s^{(2)}} \Delta(\tau_i^{(2)}, h_i^{(2)}),
\tag{2.48}
$$

where the vectors $h_i^{(0)}$, $h_i^{(1)}$, $h_i^{(2)}$ are defined by the formulas [compare (2.31)]

$$h_i^{(\alpha)} = l_i^{(\alpha)}[f(x(\tau_i^{(\alpha)}), v_i^{(\alpha)}) - f(x(\tau_i^{(\alpha)}), u(\tau_i^{(\alpha)}))]. \tag{2.49}$$

We now note that the vector $\Delta(\tau, h)$ [see (2.35)] depends continuously on the quantities τ and h (this follows directly from the theorem on the continuous dependence of the solutions of differential equations on the initial values). The vector $h_i^{(\alpha)}$ also depends continuously on $\tau_i^{(\alpha)}$ if (for fixed $l_i^{(\alpha)}$, $v_i^{(\alpha)}$) we allow $\tau_i^{(\alpha)}$ to vary on an interval which does not contain points of discontinuity of the control $u(t)$ [see (2.49)]. Since (by definition of displacement vectors) the $\tau_i^{(\alpha)}$ are points of continuity of the control $u(t)$, the vectors $h_i^{(\alpha)}$ also depend continuously on $\tau_i^{(\alpha)}$ [see (2.49)]; therefore, the vectors $\Delta(\tau_i^{(\alpha)}, h_i^{(\alpha)})$ and consequently, the vectors $\overrightarrow{QA}^{(0)}$, $\overrightarrow{QA}^{(1)}$, $\overrightarrow{QA}^{(2)}$ also depend continuously on $\tau_i^{(\alpha)}$ [see (2.46), (2.47), (2.48)]. Thus, the vectors $\overrightarrow{QA}^{(0)}$, $\overrightarrow{QA}^{(1)}$, $\overrightarrow{QA}^{(2)}$ change little with any sufficiently small displacement of the points $\tau_i^{(\alpha)}$. Let us shift the points $\tau_i^{(\alpha)}$ (without changing $\delta t^{(\alpha)}$, $l_i^{(\alpha)}$, $v_i^{(\alpha)}$) so that all the points $\tau_i^{(0)}$, $\tau_i^{(1)}$, $\tau_i^{(2)}$ become mutually distinct and, moreover, so that the vectors $\overrightarrow{QA}^{(0)}$, $\overrightarrow{QA}^{(1)}$, $\overrightarrow{QA}^{(2)}$ change so little that the point Q remains inside the triangle $A^{(0)}A^{(1)}A^{(2)}$. We assume (without changing the values of $\tau_i^{(\alpha)}$, $A^{(0)}$, $A^{(1)}$, $A^{(2)}$) that such a shift has already been performed so that all the points $\tau_i^{(0)}$, $\tau_i^{(1)}$, $\tau_i^{(2)}$ are mutually distinct.

Now let $k^{(0)}$, $k^{(1)}$, $k^{(2)}$ be arbitrary non-negative numbers. We set

$$\delta t = k^{(0)} \delta t^{(0)} + k^{(1)} \delta t^{(1)} + k^{(2)} \delta t^{(2)} \quad \text{and} \quad \bar{l}_i^{(\alpha)} = k^{(\alpha)} l_i^{(\alpha)}. \tag{2.50}$$

Then the displacement vector $\overrightarrow{QB}$ corresponding to the quantities

$$\delta t, \tau_i^{(\alpha)}, \bar{l}_i^{(\alpha)}, v_i^{(\alpha)} \qquad (\alpha = 0, 1, 2; i = 1, 2, \ldots, s^{(\alpha)}), \tag{2.51}$$

is easily seen to be equal to

$$\overrightarrow{QB} = k^{(0)}\overrightarrow{QA}^{(0)} + k^{(1)}\overrightarrow{QA}^{(1)} + k^{(2)}\overrightarrow{QA}^{(2)}. \tag{2.52}$$

In fact, multiplying the quantities $l_i^{(\alpha)}$ by $k^{(\alpha)}$ the vector $h_i^{(\alpha)}$ will also be multiplied by $k^{(\alpha)}$ [see (2.31)] and therefore, by virtue of the linearity and homogeneity of (2.21), the vector $\Delta(\tau_i^{(\alpha)}, h_i^{(\alpha)})$ will also be multiplied by $k^{(\alpha)}$ [see (2.35)]. Thus, the displacement vector $\overrightarrow{QB}$ corresponding to the quantities (2.51) has the form

$$\overrightarrow{QB} = -f(x(t_1), u(t_1))(k^{(0)} \delta t^{(0)} + k^{(1)} \delta t^{(1)} + k^{(2)} \delta t^{(2)})$$
$$+ \sum_{\alpha=0}^{2} \sum_{i=1}^{s^{(\alpha)}} k^{(\alpha)}\Delta(\tau_i^{(\alpha)}, h_i^{(\alpha)}),$$

and this, in fact, yields (2.52) [by virtue of (2.46), (2.47), (2.48)].

Since all the points $\tau_i^{(\alpha)}$ are mutually distinct, we can consider the variation of the control $u(t)$ and of the trajectory $x(t)$ which corresponds to the quantities (2.51); moreover, for the corresponding perturbed trajectory $x^*(t)$ we have, by virtue of (2.36),

$$x^*(t_1 - \epsilon \delta t) = x(t_1) + \epsilon \overrightarrow{QB} + o(\epsilon), \tag{2.53}$$

where $\overrightarrow{QB}$ is the displacement vector [recall that this vector satisfies (2.52)]. Thus, if the quantities (2.51) are defined by (2.50), then the corresponding perturbed trajectories $x^*(t)$ will satisfy (2.53), where $\overrightarrow{QB}$ is defined by (2.52). We note that if the quantities $k^{(0)}$, $k^{(1)}$, $k^{(2)}$ vary continuously (remaining non-negative), then the quantities δt and $\bar{l}_i^{(\alpha)}$ also vary continuously [see (2.50)], that is, the quantity δt and the lengths $\epsilon \bar{l}_i^{(\alpha)}$ of the intervals on which the control $u(t)$ is perturbed vary continuously. Consequently, the trajectory $x^*(t)$ and, in particular, the point $x^*(t_1 + \epsilon \delta t)$, also vary continuously. Thus, the trajectory $x^*(t)$ and the point $x^*(t_1 - \epsilon \delta t)$ depend continuously on the quantities $k^{(0)}$, $k^{(1)}$, $k^{(2)}$.

Now, let C be an arbitrary point of the phase plane X. If the point C lies inside or on the sides of the angle $A^{(1)}QA^{(2)}$, then we set $k^{(0)}(C) = 0$, and let $k^{(1)}(C)$ and $k^{(2)}(C)$ denote numbers such that

$$\overrightarrow{QC} = k^{(1)}(C) \cdot \overrightarrow{QA}^{(1)} + k^{(2)}(C) \cdot \overrightarrow{QA}^{(2)};$$

in other words, $k^{(1)}(C)$ and $k^{(2)}(C)$ are the coordinates of the vector $\overrightarrow{QC}$ with respect to the basis $\overrightarrow{QA}^{(1)}$, $\overrightarrow{QA}^{(2)}$ (Figure 57). The quantities $k^{(1)}(C)$

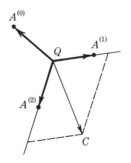

FIGURE 57

and $k^{(2)}(C)$ are non-negative since the point C lies inside or on the sides of the angle $A^{(1)}QA^{(2)}$. Similarly, if C lies inside or on the sides of the angle $A^{(0)}QA^{(1)}$, then let $k^{(0)}(C)$ and $k^{(1)}(C)$ denote numbers which satisfy

$$\overrightarrow{QC} = k^{(0)}(C) \cdot \overrightarrow{QA}^{(0)} + k^{(1)}(C) \cdot \overrightarrow{QA}^{(1)},$$

and set $k^{(2)}(C) = 0$. Finally, if C lies inside or on the sides of the angle $A^{(0)}QA^{(2)}$, then let $k^{(0)}(C)$ and $k^{(2)}(C)$ denote numbers which satisfy

$$\overrightarrow{QC} = k^{(0)}(C) \cdot \overrightarrow{QA^{(0)}} + k^{(2)}(C) \cdot \overrightarrow{QA^{(2)}},$$

and set $k^{(1)}(C) = 0$.

Thus, for any position of the point C on the plane, we have

$$\overrightarrow{QC} = k^{(0)}(C) \cdot \overrightarrow{QA^{(0)}} + k^{(1)}(C) \cdot \overrightarrow{QA^{(1)}} + k^{(2)}(C) \cdot \overrightarrow{QA^{(2)}}, \qquad (2.54)$$

where $k^{(0)}(C)$, $k^{(1)}(C)$, $k^{(2)}(C)$ are non-negative numbers. We note that $k^{(0)}(C)$, $k^{(1)}(C)$, $k^{(2)}(C)$ depend continuously on C (that is, these quantities change slightly with a small displacement of C). In fact, inside and on the sides of each of the triangles $A^{(0)}QA^{(1)}$, $A^{(1)}QA^{(2)}$, $A^{(2)}QA^{(0)}$ these quantities obviously depend continuously on C (two of them are the coordinates of the vector $\overrightarrow{QC}$, and the third is equal to zero). In the transition from one angle to another, these quantities also change continuously. For example, $k^{(0)}(C)$ approaches zero as the point C approaches the ray $QA^{(1)}$ from within the angle $A^{(0)}QA^{(1)}$ and equals zero inside the angle $A^{(1)}QA^{(2)}$. Thus, the quantity $k^{(0)}(C)$ changes continuously in crossing the ray $QA^{(1)}$. The continuity of $k^{(0)}(C)$ [and also of $k^{(1)}(C)$, $k^{(2)}(C)$] when crossing the rays $QA^{(0)}$ and $QA^{(2)}$ is treated in a similar manner.

Thus, $k^{(0)}(C)$, $k^{(1)}(C)$, $k^{(2)}(C)$ are continuous non-negative functions of C which satisfy (2.54) for any position of the point C. Let us now determine, with the aid of

$$k^{(0)} = k^{(0)}(C), \qquad k^{(1)} = k^{(1)}(C), \qquad k^{(2)} = k^{(2)}(C), \qquad (2.55)$$

and (2.50), the quantities (2.51). The corresponding perturbed trajectory $x^*(t)$ and the quantity δt will be denoted by $x_C^*(t)$ and δt_C, respectively, in order to emphasize their dependence on C. We note that $x_C^*(t)$ depends continuously on the pair of variables t, C, and that δt_C depends continuously on C since the functions (2.55) are continuous. The relation

$$x_C^*(t_1 - \epsilon \delta t_C) = x(t_1) + \epsilon \overrightarrow{QB} + o(\epsilon) \qquad (2.56)$$

holds for the perturbed trajectory $x_C^*(t)$ [see (2.53)], where the vector $\overrightarrow{QB}$ is defined by (2.52). It follows from a comparison of (2.54) and (2.52) that $\overrightarrow{QB} = \overrightarrow{QC}$, so that (2.56) takes the form

$$x_C^*(t_1 - \epsilon \delta t_C) = x(t_1) + \epsilon \overrightarrow{QC} + o(\epsilon).$$

In other words, letting C_ϵ denote a point such that $\overrightarrow{QC_\epsilon} = \epsilon\overrightarrow{QC}$ (Figure 58), we can write

$$x_C^*(t_1 - \epsilon\delta t_C) = C_\epsilon + o(\epsilon) \tag{2.57}$$

[since the point $x(t_1)$ coincides with Q].

Let us now consider a circumference S of radius 1 with center at Q. As the point C traverses S, the point C_ϵ traverses a circumference of smaller radius, S_ϵ, which is obtained from S by a similitude with center Q and coefficient ϵ (Figure 59); the point $x_C^*(t_1 - \epsilon\delta t_C)$ [see (2.57)] traverses a closed

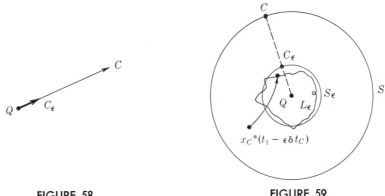

FIGURE 58   **FIGURE 59**

curve L_ϵ which is close to S_ϵ [recall that the point $x_C^*(t_1 - \epsilon\delta t_C)$ depends continuously on C]. Furthermore, since $k^{(0)}$, $k^{(1)}$, $k^{(2)}$ depend continuously on C [see (2.55)], the quantities (2.50) also depend continuously on C. Consequently, as the point C traverses S, the quantities (2.50) remain bounded (since every function which is continuous on a circumference is bounded), that is, there exist numbers γ, $\beta_i^{(\alpha)}$ such that

$$0 \leqslant \delta t \leqslant \gamma, \qquad 0 \leqslant \bar{l}_i^{(\alpha)} \leqslant \beta_i^{(\alpha)} \tag{2.58}$$

for any point $C \in S$. It follows from the remark on page 63 that the *quantity $o(\epsilon)$ in (2.57) is an infinitesimal of order higher than ϵ uniformly in $C \in S$.*

Thus far, ϵ has been considered to be a variable (an infinitesimal). We now choose a value for ϵ which will not be changed. Namely, the quantity ϵ will be chosen small enough to ensure that, in the first place, the intervals near the points $\tau_i^{(\alpha)}$ (on which the variation of the control takes place) are mutually disjoint [this is possible since all the points $\tau_i^{(\alpha)}$ are pairwise distinct and the quantities $\bar{l}_i^{(\alpha)}$ are bounded; see (2.58)], and secondly, that the quantity $o(\epsilon)$ in (2.57) is less than $\frac{1}{10}\epsilon$ (for all points $C \in S$). This means that the distance between $x_C^*(t_1 - \epsilon\delta t_C)$ and C_ϵ is less than $\frac{1}{10}\epsilon$. However, the distance between Q and C_ϵ is equal to ϵ (since the radius QC is unity and therefore the length of the vector $\overrightarrow{QC_\epsilon} = \epsilon\overrightarrow{QC}$ equals ϵ). Thus, as C traverses S, the point C_ϵ describes the circumference S_ϵ of radius ϵ with center at Q; the point $x_C^*(t_1 - \epsilon\delta t_C)$ describes the closed curve L_ϵ, being always at a distance $< \frac{1}{10}\epsilon$ from the point C_ϵ. Hence it follows that Q is inside the closed curve L_ϵ (Figure 59).

Now, let us introduce the parameter σ which will take on values $0 \leqslant \sigma \leqslant 1$, and let us consider the point

$$x_C^*((1 - \sigma)t_0 + \sigma(t_1 - \epsilon\delta t_C)). \qquad (2.59)$$

This point depends continuously on the pair of variables C, σ since $x_C^*(t)$ depends continuously on the pair t, C. If we fix a value of $\sigma(0 \leqslant \sigma \leqslant 1)$ and compel the point C to describe S, then the point (2.59) will describe a continuous closed curve (possibly having self-intersections or other singularities, which is not essential in the following) which will be denoted by $L^{(\sigma)}$. For $\sigma = 1$, the point (2.59) coincides with the point (2.57), that is, the curve $L^{(1)}$ (obtained for $\sigma = 1$) coincides with L_ϵ and therefore, Q lies inside $L^{(1)}$. If the parameter σ varies continuously, then $L^{(\sigma)}$ also varies (deforms) continuously in the plane X (several positions of the curve $L^{(\sigma)}$ are depicted in Figure 60). Finally, for $\sigma = 0$, the point (2.59) coincides with $x_C^*(t_0)$,

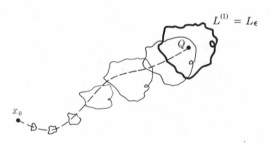

FIGURE 60

that is, with x_0, and therefore the curve $L^{(0)}$ (obtained for $\sigma = 0$) degenerates into the single point x_0. Consequently, for values of σ close to zero, $L^{(\sigma)}$ lies near x_0, and therefore $Q = x_1$ is not inside the curve $L^{(\sigma)}$ (since x_0 does not coincide with x_1). Thus, moving continuously in the phase plane X, $L^{(\sigma)}$ moves from the position $L^{(1)}$ in which it contains the point Q, to a position $L^{(\sigma)}$ in which Q is not contained inside the contour. Consequently, there must be an intermediate value $\sigma = \sigma_0$ for which the contour $L^{(\sigma_0)}$ passes through the point Q (Figure 61). In other words, there exists a number σ_0, $0 < \sigma_0 < 1$, and a point $C \in S$ such that the point (2.59) coincides with Q:

$$x_C^*((1 - \sigma_0)t_0 + \sigma_0(t_1 - \epsilon\delta t_C)) = x_1.$$

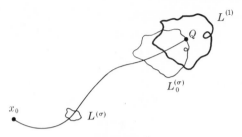

FIGURE 61

But we also have

$$(1 - \sigma_0)t_0 + \sigma_0(t_1 - \epsilon\delta t_C) = t_1 - (1 - \sigma_0)(t_1 - t_0) - \sigma_0\epsilon\delta t_C < t_1.$$

Thus, the perturbed trajectory $x_C^*(t)$ reaches the point x_1 before time t_1, and consequently, the original process $(u(t), x(t))$ was not optimal.

Accordingly, the fundamental lemma has been completely proved for $n = 2$.

REMARK. In the concluding part of the proof we used such concepts as "the point lies inside the closed curve" and "the contour deforms continuously"; moreover, we have made use of the fact that if the closed curve containing the point Q deforms continuously and moves into a position in which it does not contain the point Q, then this curve must pass through Q at some intermediate position. Precise definitions of these concepts and the proof of the above fact requires the use of some information from topology which falls outside the scope of this book. Therefore, we confine ourselves to the above statements assuming that they are "intuitively evident."

The proof for $n > 2$ is similar to the above procedure. We indicate the changes to be made in this case. Instead of the triangle $A^{(0)}A^{(1)}A^{(2)}$, containing Q, we must consider now an n-dimensional *simplex* $A^{(0)}A^{(1)} \ldots A^{(n)}$ in the space X containing the point Q. (For $n = 3$, a simplex is a tetrahedron $A^{(0)}A^{(1)}A^{(2)}A^{(3)}$.) The number of vectors $\overrightarrow{QA^{(0)}}, \ldots, \overrightarrow{QA^{(n)}}$ is now $n + 1$; formulas (2.45)–(2.49) are generalized in an obvious manner. As before, making use of the continuous dependence of the vector $\Delta(\tau, h)$ on τ and h we can achieve the situation in which all the points $\tau_i^{(\alpha)}$ are mutually distinct. The relations (2.50)–(2.53) are generalized for the n-dimensional case in an obvious manner. Let us turn now to the discussion given on pp. 69, 70. The aim of those arguments was the proof (for $n = 2$) of the following fact:

Lemma 2.10.   There exist continuous non-negative functions $k^{(\alpha)}(C)$, $\alpha = 0, 1, \ldots, n$ such that

$$\overrightarrow{QC} = k^{(0)}(C) \cdot \overrightarrow{QA^{(0)}} + k^{(1)}(C) \cdot \overrightarrow{QA^{(1)}} + \cdots + k^{(n)}(C) \cdot \overrightarrow{QA^{(n)}}$$

for any point $C \in X$.

[Moreover, the functions $k^{(0)}(C)$, $k^{(1)}(C)$, $k^{(2)}(C)$, constructed geometrically on pp. 69, 70, have the property that at least one of these functions vanishes at any point C; this property, however, was never used.] The geometrical method for constructing the required functions $k^{(\alpha)}(C)$, presented on pp. 69, 70, admits a generalization to n-dimensional space, but here we shall provide another method for constructing these functions.

PROOF OF THE LEMMA.   Let us resolve the vector $\overrightarrow{QA}^{(0)}$ with respect to the basis $\overrightarrow{QA}^{(1)}$, $\overrightarrow{QA}^{(2)}$, . . . , $\overrightarrow{QA}^{(n)}$, that is, let us find numbers λ^1, λ^2, . . . , λ^n, such that

$$\overrightarrow{QA}^{(0)} = \lambda^1 \overrightarrow{QA}^{(1)} + \lambda^2 \overrightarrow{QA}^{(2)} + \cdots + \lambda^n \overrightarrow{QA}^{(n)}. \tag{2.60}$$

It is easily seen that all the numbers λ^1, . . . , λ^n are negative (since the ray QM, being a continuation of the vector $\overrightarrow{QA}^{(0)}$, intersects the face $A^{(1)}A^{(2)}$ . . . $A^{(n)}$, that is, lies in the "positive coordinate angle" defined by the vectors $\overrightarrow{QA}^{(1)}$, $\overrightarrow{QA}^{(2)}$, . . . , $\overrightarrow{QA}^{(n)}$, see Figure 62). Now, let C be an

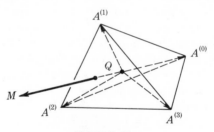

FIGURE 62

arbitrary point of X, and let x^1, . . . , x^n be the components of the vector $\overrightarrow{QC}$ with respect to the basis $\overrightarrow{QA}^{(1)}$, . . . , $\overrightarrow{QA}^{(n)}$, that is,

$$\overrightarrow{QC} = x^1 \overrightarrow{QA}^{(1)} + x^2 \overrightarrow{QA}^{(2)} + \cdots + x^n \overrightarrow{QA}^{(n)}. \tag{2.61}$$

We set

$$\varphi(C) = \left| \frac{x^1}{\lambda^1} \right| + \left| \frac{x^2}{\lambda^2} \right| + \cdots + \left| \frac{x^n}{\lambda^n} \right|.$$

It is obvious that the function $\varphi(C)$ is non-negative and continuous in X; moreover, for any $i = 1, 2, . . . , n$, we have

$$-\lambda^i \varphi(C) = |\lambda^i| \varphi(C) \geqslant |\lambda^i| \left| \frac{x^i}{\lambda^i} \right| = |x^i| \geqslant -x^i;$$

consequently (for any point $C \in X$),

$$\varphi(C) \geqslant 0, \qquad x^1 - \lambda^1\varphi(C) \geqslant 0,$$
$$x^2 - \lambda^2\varphi(C) \geqslant 0, \ldots, x^n - \lambda^n\varphi(C) \geqslant 0. \qquad (2.62)$$

Now, subtracting (2.60) multiplied by $\varphi(C)$ from (2.61), we obtain

$$\overrightarrow{QC} = \varphi(C) \cdot \overrightarrow{QA}^{(0)} + (x^1 - \lambda^1\varphi(C)) \cdot \overrightarrow{QA}^{(1)} + \cdots$$
$$+ (x^n - \lambda^n\varphi(C)) \cdot \overrightarrow{QA}^{(n)}.$$

The functions

$$k^{(0)}(C) = \varphi(C),$$
$$k^{(1)}(C) = x^1 - \lambda^1\varphi(C), \ldots, k^{(n)}(C) = x^n - \lambda^n\varphi(C)$$

are continuous, and, by virtue of (2.62), non-negative. Thus, Lemma 2.10 has been proved.

The concluding part of the proof of the fundamental lemma can now be extended without difficulty to the n-dimensional case. Instead of a circumference, a sphere S of radius 1 with center at Q must be considered in the space X. As C describes the sphere S, C_ϵ [see (2.57)] describes the sphere S_ϵ of radius ϵ with center Q, and $x_C^*(t_1 - \epsilon\delta tc)$ describes a "closed surface" L_ϵ which is close to S_ϵ. For a sufficiently small ϵ, the "surface" L_ϵ contains the point Q which makes it possible to complete the proof (see pp. 71—73).

21. The Maximum Principle. Using the above fundamental lemma, we derive in this section the necessary conditions for optimality called the maximum principle.

Let $(u(t), x(t))$ be a time-optimal process which transfers an object from the state x_0 to the state x_1 during the time $t_0 \leqslant t \leqslant t_1$. Let us construct for this process the set K considered in § 5.19. The set K is a convex cone with vertex at $Q = x_1$; since the process $(u(t), x(t))$ is optimal, it follows, by virtue of the fundamental lemma, that the cone K does not coincide with the entire phase space X. Consequently, there exists a nonzero vector n such that

$$n \cdot \overrightarrow{QP} \leqslant 0$$

for any point P of K (see the conclusion of § 4.13). But the point P belongs to K if and only if $\overrightarrow{QP}$ is a displacement vector [see (2.42)]. Thus, n has the property that its scalar product with any vector of the form (2.42) is non-positive. In particular,

$$n \cdot (-f(x(t_1), u(t_1))) \leqslant 0$$

[this is obtained by setting $s = 0$, $\delta t = 1$ in (2.42)], that is,

$$n \cdot f(x(t_1), u(t_1)) \geqslant 0. \tag{2.63}$$

Furthermore,

$$n \cdot \Delta(\tau, h) \leqslant 0, \tag{2.64}$$

where τ is any point of continuity of the control $u(t)$ and the vector h has the form

$$h = f(x(\tau), v) - f(x(\tau), u(\tau)), \qquad v \in U \tag{2.65}$$

[this is obtained by setting $\delta t = 0$, $s = 1$ in (2.42), so that the displacement vector takes the form $\Delta(\tau, h)$; moreover, we have set $l = 1$ in formula (2.31)].

Thus, we have established that there exists a vector n which satisfies the conditions (2.63) and (2.64), where τ is an arbitrary point of continuity of the control $u(t)$ and h is a vector defined by (2.65). Now, let $\delta x(t)$ denote the solution of (2.21) with initial condition

$$\delta x(\tau) = h, \tag{2.66}$$

and consider this solution on the interval $\tau \leqslant t \leqslant t_1$. Then, by definition,

$$\Delta(\tau, h) = \delta x(t_1)$$

[see (2.35)], and (2.64) takes the form

$$n \delta x(t_1) \leqslant 0. \tag{2.67}$$

Let us now recall Corollary 2.9 (p. 56). This corollary suggests that we represent the vector n in the form $\psi(t_1)$, where $\psi(t)$ is a solution of (2.24). Indeed, let us do just this. Namely, we let $\psi(t) = (\psi_1(t), \ldots, \psi_n(t))$ denote the solution of the linear system (2.24) with the initial (or, it would be better to call it "terminal") condition

$$\psi(t_1) = n. \tag{2.68}$$

By virtue of the linearity of (2.24), the solution $\psi(t)$ is defined on the entire interval $t_0 \leqslant t \leqslant t_1$ on which $u(t)$ and $x(t)$ in the right-hand sides of (2.24) are given. We can now conclude, by virtue of Corollary 2.9, that the scalar product $\psi(t) \delta x(t)$ is constant; in particular,

$$\psi(\tau) \delta x(\tau) = \psi(t_1) \delta x(t_1),$$

and therefore,

$$\psi(\tau) \delta x(\tau) \leqslant 0$$

[see (2.67), (2.68)]. This relation can be written in the form $\psi(\tau) h \leqslant 0$ [see (2.66)] and this, by (2.65), gives us

$$\psi(\tau) \{ f(x(\tau), v) - f(x(\tau), u(\tau)) \} \leqslant 0,$$

or, finally,

$$\psi(\tau)f(x(\tau), v) \leqslant \psi(\tau)f(x(\tau), u(\tau)), \qquad v \in U. \tag{2.69}$$

We note that by (2.68) the inequality (2.63) can be rewritten in the form

$$\psi(t_1)f(x(t_1), u(t_1)) \geqslant 0. \tag{2.70}$$

Thus we have arrived at the following result:

Theorem 2.11. If the process $(u(t), x(t))$, $t_0 \leqslant t \leqslant t_1$, is optimal, then there exists a solution $\psi(t)$ of (2.24) satisfying conditions (2.69) and (2.70) [where τ is an arbitrary point of continuity of the control $u(t)$ and v is an arbitrary point of the control region U]. Moreover, the solution $\psi(t)$ is nontrivial [since $n \neq 0$, see (2.68)].

This, indeed, is the necessary condition for optimality of interest to us. We now present it in another, more convenient form.

We note that (2.69) and (2.70) contain exclusively scalar products of the vector ψ and the vector $f(x, u)$. This fact suggests that it is appropriate to introduce the function

$$H(\psi, x, u) = \psi f(x, u) = \psi_1 f^1(x, u) + \psi_2 f^2(x, u) + \cdots + \psi_n f^n(x, u). \tag{2.71}$$

The function H depends on the $2n + r$ arguments $\psi_1, \ldots, \psi_n, x^1, \ldots, x^n, u^1, \ldots, u^r$. The relation (2.69) can be rewritten, with the aid of this function H, in the form

$$H(\psi(\tau), x(\tau), v) \leqslant H(\psi(\tau), x(\tau), u(\tau))$$

(for any $v \in U$) or, equivalently,

$$H(\psi(\tau), x(\tau), u(\tau)) = \max_{v \in U} H(\psi(\tau), x(\tau), v). \tag{2.72}$$

The relation (2.70) can be rewritten in the form

$$H(\psi(t_1), x(t_1), u(t_1)) \geqslant 0. \tag{2.73}$$

We note, finally, that (2.24) can be represented, with the aid of the function H, in the form

$$\dot{\psi}_i = - \frac{\partial H(\psi, x(t), u(t))}{\partial x^i}, \qquad i = 1, \ldots, n \tag{2.74}$$

[this follows directly by comparison of the right-hand sides of (2.24) with the function (2.71)].

Replacing (2.24) by (2.74) and conditions (2.69) and (2.70) by conditions (2.72) and (2.73) in the theorem we have just proved, we arrive at the final form of the theorem which gives a necessary condition for optimality.

This theorem is called the maximum principle [this is explained by (2.72) which constitutes the basic "nucleus" of the theorem].

Theorem 2.12 (The Maximum Principle). *Consider an object whose motion is described by the system of equations*

$$\dot{x}^i = f^i(x^1, \ldots, x^n, u^1, \ldots, u^r) = f^i(x, u), \qquad i = 1, \ldots, n$$

[*compare* (1.2)] *or, in vector form,*

$$\dot{x} = f(x, u). \tag{A}$$

A set U *(the control region) is given in* $u^1, \ldots, u^r$-*space; an arbitrary piecewise continuous function* $u(t) = (u^1(t), \ldots, u^r(t))$, *with values in* U *and continuous at the endpoints of the interval on which it is defined, is to be regarded as an admissible control. Furthermore, two points* x_0 *and* x_1 *(initial and terminal states) are given in the phase space* X *of the variables* $x^1, \ldots, x^n$. *Finally, a process* $(u(t), x(t))$, $t_0 \leqslant t \leqslant t_1$, *is considered which transfers the object from the state* x_0 *to the state* x_1; *this means that* $x(t)$ *is the solution of system* (A) *corresponding to the admissible control* $u = u(t)$, *which satisfies the initial and terminal conditions*

$$x(t_0) = x_0, \qquad x(t_1) = x_1.$$

Thus, the process under consideration requires the time $t_1 - t_0$ *for the transition from state* x_0 *to* x_1. *The process* $(u(t), x(t))$ *is called optimal (in the sense of rapidity of action) if there exists no process which transfers the object from* x_0 *to* x_1 *in less time.*

In order to formulate the necessary condition for optimality, let us introduce the function H *which depends on the variables* $x^1, \ldots, x^n, u^1, \ldots, u^r$ *and on the auxiliary variables* $\psi_1, \ldots, \psi_n$ [*compare* (2.71)]:

$$H(\psi, x, u) = \psi \cdot f(x, u) = \sum_{\alpha=1}^{n} \psi_\alpha f^\alpha(x, u). \tag{B}$$

Making use of this function H, *we write the following system of differential equations for the auxiliary variables:*

$$\dot{\psi}_i = - \frac{\partial H(\psi, x(t), u(t))}{\partial x^i}, \qquad i = 1, \ldots, n, \tag{C}$$

where $(u(t), x(t))$ *is the process under consideration* [*compare* (2.74)].

In order that the process $(u(t), x(t))$ **be optimal it is necessary that there exist a nontrivial solution** $\psi(t)$, $t_0 \leqslant t \leqslant t_1$, **of** (C) **satisfying the maximum condition**

$$H(\psi(\tau), x(\tau), u(\tau)) = \max_{v \in U} H(\psi(\tau), x(\tau), v) \tag{D}$$

at any point of continuity τ of the control $u(t)$ [see (2.72)], and the condition

$$H(\psi(t_1),\, x(t_1),\, u(t_1)) \geqslant 0 \tag{E}$$

at the terminal time t_1 [compare (2.73)].

This theorem is completely proved by the previous considerations.

We note, in conclusion, that if the control $u(t)$ is assumed to be continuous from the right, namely,

$$u(\tau) = u(\tau + 0)$$

at every point of discontinuity (see p. 12) then, as is easily proved, condition (D) is satisfied at all points τ of the interval $t_0 \leqslant t \leqslant t_1$, including the points of discontinuity. We shall not prove this fact, since the value of the control $u(t)$ at a discontinuity is of no importance from the point of view of either the applications or the mathematical considerations [since the solution of (1.2) does not depend on the finite number of values of the control $u(t)$ at its points of discontinuity].

We note also that if the control $u(t)$, $t_0 \leqslant t \leqslant t_1$, is not assumed to be continuous at the point t_1, then (E) must be replaced by the condition

$$\lim_{\substack{t \to t_1 \\ t < t_1}} H(\psi(t),\, x(t),\, u(t)) \geqslant 0.$$

These refinements will not be discussed here.

22. The Constancy of the Function H. In conclusion, we prove a theorem which will not be used until much later (§ 15.60).

Theorem 2.13.   If the functions

$$u(t),\, x(t),\, \psi(t),\, t_0 \leqslant t \leqslant t_1,$$

satisfy (A), (C), and (D), then the function

$$M(t) = H(\psi(t),\, x(t),\, u(t))$$

of the variable t is constant on the entire interval $t_0 \leqslant t \leqslant t_1$.

[Hence it follows, in particular, that the verification of (E) in Theorem 2.12 need not necessarily be carried out at time t_1, but at any t, $t_0 \leqslant t \leqslant t_1$.]

PROOF.   Since $u(t)$ is piecewise continuous, and $x(t)$ and $\psi(t)$ are continuous, $M(t)$ is piecewise continuous. We show, first of all, that $M(t)$ is continuous even at the points of discontinuity of the control $u(t)$, that is, that $M(t)$ is continuous on the entire interval $t_0 \leqslant t \leqslant t_1$. In fact, let θ, $t_0 < \theta < t_1$, be one of the points of discontinuity of the control $u(t)$. We must prove the validity of the equation $M(\theta - 0) = M(\theta + 0)$. Assume that this relation is not satisfied: for example, that $M(\theta - 0) > M(\theta + 0)$.

In other words,

$$H(\psi(\theta), x(\theta), u(\theta - 0)) > H(\psi(\theta), x(\theta), u(\theta + 0)).$$

Hence, for t', t'' sufficiently close to θ and satisfying $t' < \theta < t''$, it is easy to conclude that

$$H(\psi(t''), x(t''), u(t')) > H(\psi(t''), x(t''), u(t'')).$$

This, however, contradicts the maximum condition (D) and thus proves the continuity of the function $M(t)$.

It remains to prove that $M(t)$ is constant on every interval of continuity of the function $u(t)$. Let us assume the contrary: $M(\tau') \neq M(\tau)$, and the function $u(t)$ is continuous on the interval $\tau \leqslant t \leqslant \tau'$. We set

$$K = \left| \frac{M(\tau') - M(\tau)}{\tau' - \tau} \right|,$$

that is, K is the absolute value of the slope of the corresponding chord drawn on the graph of the function $M(t)$ (Figure 63). Now, dividing the

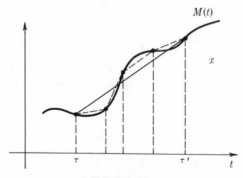

FIGURE 63

interval $[\tau, \tau']$ into several parts and considering the inscribed broken line (dotted line in Figure 63), we find that at least one segment of this broken line has a slope whose absolute value is $\geqslant K$. Hence, it is clear that, for any natural number m, there exist two distinct points τ_m, τ'_m on the interval $[\tau, \tau']$ such that

$$|\tau_m - \tau'_m| < \frac{1}{m}$$

and

$$|M(\tau'_m) - M(\tau_m)| \geqslant K|\tau'_m - \tau_m|. \tag{2.75}$$

Moreover, we can assume (going over, if necessary, to a subsequence) that the points τ_m and τ'_m have a limit as $m \to \infty$:

$$\lim_{m \to \infty} \tau_m = \lim_{m \to \infty} \tau'_m = \tau_0,$$

where τ_0 is a point in the interval $[\tau, \tau']$.

Now let θ_1 and θ_2 be two points of the interval $[\tau, \tau']$. By virtue of the maximum condition (D), we have

$$H(\psi(\theta_1),\, x(\theta_1),\, u(\theta_2)) \leqslant H(\psi(\theta_1),\, x(\theta_1),\, u(\theta_1)),$$

and therefore,

$$
\begin{aligned}
M(\theta_2) - M(\theta_1) &= H(\psi(\theta_2),\, x(\theta_2),\, u(\theta_2)) - H(\psi(\theta_1),\, x(\theta_1),\, u(\theta_1)) \\
&\leqslant H(\psi(\theta_2),\, x(\theta_2),\, u(\theta_2)) - H(\psi(\theta_1),\, x(\theta_1),\, u(\theta_2)) \\
&= (\theta_2 - \theta_1) \left. \frac{dH(\psi(t),\, x(t),\, u(\theta_2))}{dt} \right|_{t=\xi},
\end{aligned}
$$

where ξ is a point lying between θ_1 and θ_2. Furthermore, by virtue of (A) and (C),

$$
\left. \frac{dH(\psi(t),\, x(t),\, u(\theta_2))}{dt} \right|_{t=\xi}
$$

$$
= \sum_{\alpha=1}^{n} \frac{\partial H(\psi(\xi),\, x(\xi),\, u(\theta_2))}{\partial \psi_\alpha} \cdot \frac{d\psi_\alpha(\xi)}{dt}
$$

$$
+ \sum_{\alpha=1}^{n} \frac{\partial H(\psi(\xi),\, x(\xi),\, u(\theta_2))}{\partial x^\alpha} \cdot \frac{dx^\alpha(\xi)}{dt}
$$

$$
= - \sum_{\alpha=1}^{n} \frac{\partial H(\psi(\xi),\, x(\xi),\, u(\theta_2))}{\partial \psi_\alpha} \cdot \frac{\partial H(\psi(\xi),\, x(\xi),\, u(\xi))}{\partial x^\alpha}
$$

$$
+ \sum_{\alpha=1}^{n} \frac{\partial H(\psi(\xi),\, x(\xi),\, u(\theta_2))}{\partial x^\alpha} \cdot \frac{\partial H(\psi(\xi),\, x(\xi),\, u(\xi))}{\partial \psi_\alpha}.
$$

The last expression vanishes for $\theta_1 = \theta_2 = \xi = \tau_0$. Consequently, if both θ_1 and θ_2 are sufficiently close to τ_0, then the above expression will, by continuity, be arbitrarily small (in absolute value). In particular, if θ_1 and θ_2 are sufficiently close to τ_0, then

$$
\left| \left. \frac{dH(\psi(t),\, x(t),\, u(\theta_2))}{dt} \right|_{t=\xi} \right| \leqslant \frac{K}{2},
$$

and therefore,

$$M(\theta_2) - M(\theta_1) \leqslant (\theta_2 - \theta_1) \left. \frac{dH(\psi(t),\ x(t),\ u(\theta_2))}{dt} \right|_{t=\xi} \leqslant |\theta_2 - \theta_1| \frac{K}{2}.$$

Since θ_1 and θ_2 are interchangeable, both of the inequalities

$$M(\theta_2) - M(\theta_1) \leqslant |\theta_2 - \theta_1| \frac{K}{2},$$

$$M(\theta_1) - M(\theta_2) \leqslant |\theta_1 - \theta_2| \frac{K}{2},$$

hold, that is,

$$|M(\theta_2) - M(\theta_1)| \leqslant |\theta_2 - \theta_1| \frac{K}{2}.$$

This relation (valid if θ_1 and θ_2 are sufficiently close to τ_0) will be satisfied, in particular, if τ_m and τ_m' are used in place of θ_1 and θ_2 for sufficiently large m:

$$|M(\tau_m') - M(\tau_m)| \leqslant \frac{K}{2} |\tau_m' - \tau_m|.$$

But this contradicts (2.75). Thus, the assumption that $M(\tau) \neq M(\tau')$ leads to a contradiction, and therefore,

$$M(t) = \text{const.}$$

chapter 3 Linear Time-Optimal Processes

§ 6. CONVEX POLYHEDRA

23. Definition of Convex Polyhedra. In this section we formulate the definition of convex polyhedra and discuss some of their properties. This material is necessary for the development of the theory of linear controlled objects.

If A and B are two convex sets, then their intersection (Figure 64)

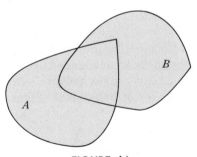

FIGURE 64

is also a convex set (provided, of course, that this intersection contains at least one point). Indeed, the intersection of any number of convex sets is convex. Since every half-space $\alpha_1 u^1 + \alpha_2 u^2 + \cdots + \alpha_r u^r + \beta \geqslant 0$ (or $\alpha_1 u^1 + \alpha_2 u^2 + \cdots + \alpha_r u^r + \beta \leqslant 0$) is convex in $u^1, u^2, \ldots, u^r$-space, the *intersection of any number of half-spaces is a convex set*. We shall be concerned with the intersection of a finite number of half-spaces.

Let us first turn to the case $r = 2$, that is, to the case in which the figures lie in the u^1, u^2-plane. In this case we shall speak of the intersection of a finite number of half-planes. The intersection of two half-planes may be an *angle* (in particular, a half-plane), a *strip* or a *straight line* (Figure 65). The intersection of three half-planes is either an *unbounded figure*

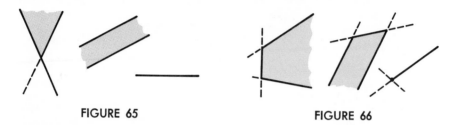

FIGURE 65   FIGURE 66

(Figure 66), a *triangle* or a *point* (Figure 67). An intersection of four half-planes may be a convex tetrahedron or a segment (in addition to the

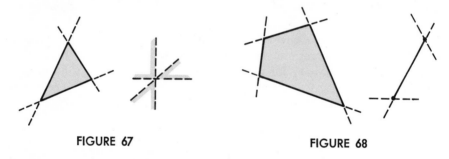

FIGURE 67   FIGURE 68

possibilities already mentioned) (Figure 68). In general, an intersection of several half-planes may be either an unbounded figure, a *point, a segment,* or a *convex polygon*. Moreover, any convex polygon may be regarded as the intersection of a finite number of half-planes (Figure 69)—it is necessary

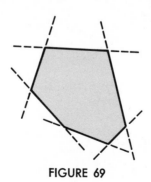

FIGURE 69

to take as many half-planes as sides in the polygon. If we agree to consider a point as a zero-dimensional polyhedron, a segment as a one-dimensional convex polyhedron, and a convex polygon as a two-dimensional polyhedron, then we can say that the *intersection of a finite number of half-planes, if it is bounded, is a convex polyhedron* (zero-, one-, or two-dimensional).

An analogous situation exists in three-dimensional space: an intersection of a finite number of half-spaces (if it is bounded) is either a zero-dimensional polyhedron (a point) or a one-dimensional convex polyhedron (a segment), or a two-dimensional convex polyhedron (that is, a convex polygon lying in a plane), or finally, a three-dimensional convex polyhedron (that is, a convex polyhedron in the usual sense).

In r-dimensional space for $r > 3$, we no longer have the graphic geometrical representations which are so helpful in considering figures in a plane and in three-dimensional space. Therefore, the term "polyhedron" does not produce in our minds a visual impression of a figure in an r-dimensional space. In this connection, the above statement on the intersection of half-spaces is taken, in the case of an r-dimensional space, as the definition of a convex polyhedron: the *intersection of a finite number of half-spaces, if bounded, is called a convex polyhedron.*

Any convex polyhedron is a convex set, that is, together with any two of its points it contains the entire segment joining these points (since a half-space is convex, and the intersection of convex sets is also convex). The converse, of course, is not true: a convex set need not be a convex polyhedron. For example, a *sphere* in r-dimensional space, that is, the set of all the points satisfying the condition

$$(u^1)^2 + (u^2)^2 + \cdots + (u^r)^2 \leqslant 1,$$

is a convex set, but is not (for $r > 1$) a convex polyhedron.

The r-dimensional parallelepiped mentioned previously is one of the simplest convex polyhedra in r-dimensional space. That it is indeed a convex polyhedron can be readily seen by writing the inequalities (1.6) in the following form:

$$u^1 \geqslant \alpha^1, u^2 \geqslant \alpha^2, \ldots, u^r \geqslant \alpha^r; \quad u^1 \leqslant \beta^1, u^2 \leqslant \beta^2, \ldots, u^r \leqslant \beta^r.$$

Each of these inequalities defines a half-space, and since the r-dimensional parallelepiped consists of the points which satisfy all these inequalities, it is the intersection of these $2r$ half-spaces. In addition, the r-dimensional parallelepiped is, obviously, a bounded set. Therefore, it is a convex polyhedron.

24. The Boundary of a Convex Polyhedron.

Let us now indicate (in most cases without proof) some of the properties of convex polyhedra. For every convex polyhedron M in an r-dimensional space there exists an integer k (one of the numbers $0, 1, 2, \ldots, r - 1, r$) such that M is contained in a *k-dimensional plane* of the *r*-dimensional space, but is not entirely contained in any $(k - 1)$-dimensional plane. Moreover, there exists only one *k*-dimensional plane containing *M*. It is called the *carrier* of the polyhedron *M*, and *k* is called the dimensionality of *M*. A zero-dimensional polyhedron is a point of *r*-dimensional space. A one-dimensional polyhedron is a segment; its carrier is the straight line on which it lies (Figure 70).

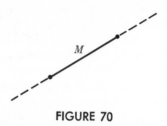

FIGURE 70

The carrier for a two-dimensional convex polyhedron is a two-dimensional plane (in *r*-dimensional space), and the polyhedron *M* is a convex polygon lying in this plane (Figure 71). We note also that the carrier of an $(r - 1)$-dimensional polyhedron is a hyperplane of *r*-dimensional space. In the case of an *r*-dimensional polyhedron, the carrier coincides with the entire space.

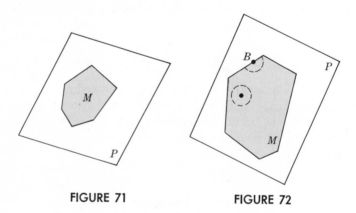

FIGURE 71   **FIGURE 72**

Let *M* be a *k*-dimensional polyhedron and let *P* be its carrier. A point *a* is called an *interior point* of *M* (with respect to its carrier) if there exists a number $\rho > 0$ such that a sphere of radius ρ with center at *a* lying in the carrier is contained entirely in *M* (Figure 72). [Recall that the distance between $a = (a^1, \ldots, a^r)$ and $u = (u^1, \ldots, u^r)$ is defined by the expression

$$\sqrt{(u^1 - a^1)^2 + (u^2 - a^2)^2 + \cdots + (u^r - a^r)^2},$$

and the sphere of radius ρ with the center at a which lies in the carrier P consists of all points u in P for which the distance between u and a does not exceed ρ.] Any point b which belongs to the polyhedron M but is not an interior point is called a boundary point of M. The totality of interior points of the polyhedron M is called its *interior;* the totality of boundary points form the *boundary* of M (Figure 73).

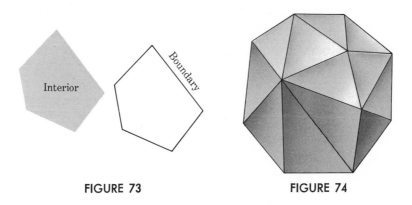

Interior

Boundary

FIGURE 73   FIGURE 74

The *boundary of any k-dimensional polyhedron* (for $k > 0$) *consists of a finite number of* $(k - 1)$-*dimensional polyhedra;* moreover, the carriers of all these $(k - 1)$-dimensional polyhedra are distinct. These $(k - 1)$-dimensional polyhedra are called the $(k - 1)$-dimensional faces of the k-dimensional polyhedron. Each of these faces [itself a $(k - 1)$-dimensional polyhedron] has, in turn, $(k - 2)$-dimensional faces. They are also considered to be faces of the original k-dimensional polyhedron [namely, its $(k - 2)$-dimensional faces]. The $(k - 3)$-dimensional faces are obtained in a similar manner, and so on. Thus, every k-dimensional polyhedron has faces of dimension $k - 1$, $k - 2$, . . . , 2, 1, 0. The zero-dimensional faces of a polyhedron are called its *vertices;* the one-dimensional faces are called *edges.*

For example, for $k = 1$, that is, in the case of a one-dimensional polyhedron (a segment, Figure 70), the zero-dimensional faces are the vertices (endpoints) of the segment. The boundary of a one-dimensional polyhedron consists of two points (the endpoints of the segment). For $k = 2$, that is, in the case of a two-dimensional convex polyhedron (convex polygon, Figure 71) there are one-dimensional faces—*edges* (sides of the polygon) and zero-dimensional faces—*vertices* of the polygon. The boundary in this case is the outline of the polygon. For $k = 3$, that is, in the case of a three-dimensional convex polyhedron (Figure 74), there are two-dimensional (that is, "usual") faces, one-dimensional faces (edges), and zero-dimensional faces (vertices). The boundary in this case is the surface of the polyhedron.

25. Convex Hull. The further properties of convex polyhedra which we shall consider are associated with the concept of the *convex hull* of a set. The convex hull of a set Q is to be understood as the smallest convex set containing Q. Such a smallest convex set certainly exists, for if we take all the convex sets containing Q, their intersection will, in fact, be the smallest convex set containing Q. In Figure 75, the set Q is shown as the cross-hatched area and its convex hull is outlined by the dotted line. The convex hull of three points not lying on a straight line is a triangle (Figure 76). In

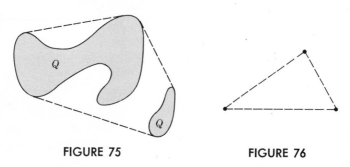

FIGURE 75   FIGURE 76

general, the *convex hull of a finite number of points* (taken in the r-dimensional $u^1, u^2, \ldots , u^r$-space) is a convex polyhedron. Each vertex of this polyhedron is one of these points, but not all of these points need be vertices: some may lie on the faces or in the interior of the polyhedron (Figure 77).

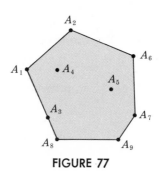

FIGURE 77

Conversely, any convex polyhedron is the convex hull of a finite set of points, namely, the set of all its vertices.

Let us now consider a *linear mapping* of the $u^1, u^2, \ldots , u^r$-space into the $v^1, \ldots , v^n$-space defined by the formulas

$$v^i = \sum_{\beta=1}^{r} b_\beta^i u^\beta, \qquad i = 1, 2, \ldots , n. \tag{3.1}$$

These formulas are to be understood in the sense that to each point $u = (u^1, u^2, \ldots, u^r)$ in r-dimensional space there corresponds a point $v = (v^1, v^2, \ldots, v^n)$ in n-dimensional space, namely, the point whose coordinates are calculated by (3.1). In other words, these formulas define a function (or, as one also says, a mapping) in which the independent variable u is a point in r-dimensional space and the values of the function are points in n-dimensional space. If the point u describes a set M in r-dimensional space, then the corresponding point v defined by (3.1) describes a set N in the n-dimensional space which is called the *image* of the set M under the mapping (3.1). It turns out that *if M is an arbitrary convex polyhedron in r-dimensional $u^1, u^2, \ldots, u^r$-space then its image under the linear mapping* (3.1) *is also a convex polyhedron N in n-dimensional $v^1, v^2, \ldots, v^n$-space.* Namely, if $u_{(1)}, u_{(2)}, \ldots, u_{(q)}$ are the vertices of the polyhedron M and $v_{(1)}, v_{(2)}, \ldots, v_{(q)}$ are the points into which the vertices are transformed under the mapping (3.1), then the polyhedron N (into which M is transformed by the mapping) is the convex hull of the points $v_{(1)}, v_{(2)}, \ldots, v_{(q)}$.

As an example, consider a three-dimensional cube defined in u^1, u^2, u^3-space by the inequalities $-1 \leqslant u^i \leqslant 1$. The vertices of this cube are the eight points with coordinates $(\pm 1, \pm 1, \pm 1)$ (where arbitrary combinations of signs are to be taken). Furthermore, consider a linear mapping of the three-dimensional u^1, u^2, u^3-space into the v^1, v^2-plane expressed by the formulas

$$v^1 = b_1^1 u^1 + b_2^1 u^2 + b_3^1 u^3, \qquad v^2 = b_1^2 u^1 + b_2^2 u^2 + b_3^2 u^3. \tag{3.2}$$

The eight vertices of the cube are transformed by this mapping into eight points in the v^1, v^2-plane, and the cube itself is transformed into the convex hull of these eight points. Figure 78 depicts the eight points into which the

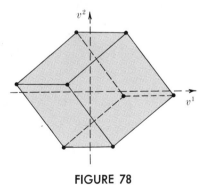

FIGURE 78

cube's vertices have been transformed by the linear mapping (3.2) as well as the convex hull of these eight points; also shown are the segments into which the edges of the cube have been transformed by this mapping.

26. Support Properties of Convex Polyhedra.   Let M be an arbitrary convex polyhedron in r-dimensional $u^1, u^2, \ldots, u^r$-space and let P be a hyperplane of support of this polyhedron (compare p. 44). It turns out that the set of all points common to the hyperplane P and the polyhedron M (that is, the intersection of M and P) either coincides with the entire polyhedron M (if the dimension of M is less than r and its carrier lies in P) or is a face of the polyhedron M. Moreover, for any face M, there exists a hyperplane of support whose intersection with M gives this particular face. If the face under consideration is $(r - 1)$-dimensional, then there exists only one hyperplane of support (namely, the carrier of this face) whose intersection with the polyhedron produces this face; if, however, the dimension of the face is less than $r - 1$, then there exists an infinite number of hyperplanes of support whose intersection with the polyhedron yields this face. For example, if $r = 2$, and the polyhedron under consideration is a convex polygon, then (Figure 79) for each one-dimensional face (side) there exists

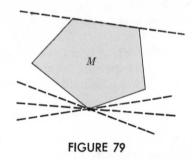

FIGURE 79

only one hyperplane of support (which in this case is a line of support) containing this face; for any zero-dimensional face (vertex) there exists an infinite number of lines of support passing through this face. Similarly, in the case of a three-dimensional polyhedron in three-dimensional space ($r = 3$) only one plane of support passes through each two-dimensional face whereas infinitely many planes of support pass through each edge and each vertex.

Closely related to support properties, and of considerable importance in the sequel, is the question of the maximum value of a linear functional on a convex polyhedron. Let $f(u)$ be a linear functional in the space of the variable $u = (u^1, u^2, \ldots, u^r)$, that is, a linear function of the r variables $u^1, u^2, \ldots, u^r$:

$$f(u) = \alpha_1 u^1 + \alpha_2 u^2 + \cdots + \alpha_r u^r. \tag{3.3}$$

In addition, let a convex polyhedron M be given in the same space. We pose the problem of finding those points at which the function $f(u)$, considered only on the polyhedron M, takes on its maximum value.

Let n denote the vector with components $\alpha_1, \alpha_2, \ldots, \alpha_r$. Then the value of the functional $f(u)$ may be written in the form of a scalar product:

$$f(u) = \alpha_1 u^1 + \alpha_2 u^2 + \cdots + \alpha_r u^r = nu.$$

Now let Γ denote the hyperplane defined by the equation

$$\alpha_1 u^1 + \alpha_2 u^2 + \cdots + \alpha_r u^r = 0,$$

and draw a hyperplane of support Γ' to the polyhedron M, which is parallel to Γ and positioned so that M and the vector n are on opposite sides of this hyperplane (Figure 80). The equation of the hyperplane Γ' has the form

$$\alpha_1 u^1 + \alpha_2 u^2 + \cdots + \alpha_r u^r + \beta = 0.$$

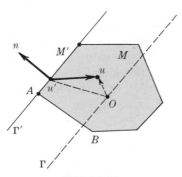

FIGURE 80

With the aid of the scalar product, this equation may be written in the form $nu + \beta = 0$. Let M' denote the convex polyhedron which is the intersection of the hyperplane of support Γ' with the polyhedron M. According to what has been said above, M' either coincides with M (if M lies entirely in the hyperplane Γ') or is a face of M. We shall prove that $f(u)$ is constant on the polyhedron M', and that M' is the set of all points at which $f(u)$, considered on M, attains its maximum value. In other words, *any linear function is either constant on the entire polyhedron M, or the set of all points at which this function (considered only on M) takes on its maximum value is a face of M.*

In fact, let u' and u'' be two points belonging to the polyhedron M'. Then, both of these points lie in the hyperplane Γ', that is, $nu' + \beta = 0$, and $nu'' + \beta = 0$. Consequently, $f(u') = nu' = -\beta$, $f(u'') = nu'' = -\beta$, that is, $f(u') = f(u'')$. Thus, the function f is constant on M'. Furthermore, let u' be a point of M', and let u be a point of M which does not belong to the face M'. Then the vector from u' to u, that is, the vector $u - u'$, and the vector n are on opposite sides of the hyperplane Γ' (that is, the vector $u - u'$ is in the negative half-space with respect to the hyperplane Γ'). Consequently [compare (2.8) on p. 41], the scalar product $n(u - u')$ is negative, that is, $n(u - u') < 0$ or $nu < nu'$. But this means that $f(u) < f(u')$. Thus, the values of $f(u)$ at the points of M which do not lie on M' are smaller than the values at points of M'.

In conclusion, we point out one more fact associated with the previous considerations. *In order that the vertex A of the polyhedron M belong to the face M' [on which the functional (3.3) takes on its maximum value], it is necessary and sufficient that $n \cdot \overrightarrow{AB} \leqslant 0$ for every edge AB of M emanating from A.*

In fact, if the vertex A belongs to M' and AB is an edge emanating from A, then the vertex B, as well as the entire polyhedron M, lies in the negative half-space with respect to the hyperplane Γ', and therefore, $n \cdot \overrightarrow{AB} \leqslant 0$. Conversely, if $n \cdot \overrightarrow{AB} \leqslant 0$ for every edge emanating from A, then all edges emanating from A lie in the negative half-space with respect to the hyperplane Γ'' passing through A parallel to Γ, and therefore, by virtue of the convexity of M, this entire polyhedron lies in the negative half-space. Consequently the hyperplane Γ'' coincides with Γ', and therefore, the point A belongs to the face M'. In particular, *in order that the linear function (3.3) attain its maximum value at only one vertex A of the polyhedron M* (Figure 81), *it is necessary and sufficient that $n \cdot \overrightarrow{AB} < 0$ for every edge AB of M emanating from A.*

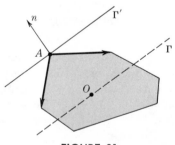

FIGURE 81

§7. THE LINEAR OPTIMAL CONTROL PROBLEM

27. Formulation of the Problem. We shall study below, in complete detail, controlled objects whose motion is described by a linear differential equation in $x^1, \ldots, x^n, u^1, \ldots, u^r$, that is, by equations of the form

$$\dot{x}^i = \sum_{\alpha=1}^{n} a_\alpha^i x^\alpha + \sum_{\beta=1}^{r} b_\beta^i u^\beta, \qquad i = 1, 2, \ldots, n, \tag{3.4}$$

where the coefficients a_α^i and b_β^i are constant.

In one of the most important cases encountered in applications, each of the quantities u^1, u^2, . . . , u^r in (3.4) represents an individual control parameter whose domain of variation does not depend on the values of the remaining control parameters and is given by

$$a^\beta \leqslant u^\beta \leqslant b^\beta, \qquad \beta = 1, \ldots, r. \tag{3.5}$$

As previously mentioned (see § 1.4), these inequalities define an r-*dimensional parallelepiped*.

However, it would be inconvenient to confine oneself exclusively to control regions which are r-dimensional parallelepipeds. In the following, when considering objects of the form (3.4), it will be assumed that the control parameter $u = (u^1, u^2, \ldots, u^r)$ ranges over a control region U which is a convex polyhedron. The reason for considering convex polyhedra rather than restricting oneself to r-dimensional parallelepipeds will not be discussed here. This question will be elucidated in complete detail in § 10.37.

In order to write (3.4) in vector form, let us introduce the matrices

$$A = \begin{pmatrix} a_1^1 & a_2^1 & \cdots & a_n^1 \\ a_1^2 & a_2^2 & \cdots & a_n^2 \\ \cdot & \cdot & & \cdot \\ a_1^n & a_2^n & \cdots & a_n^n \end{pmatrix}, \qquad B = \begin{pmatrix} b_1^1 & b_2^1 & \cdots & b_r^1 \\ b_1^2 & b_2^2 & \cdots & b_r^2 \\ \cdot & \cdot & & \cdot \\ b_1^n & b_2^n & \cdots & b_r^n \end{pmatrix}, \tag{3.6}$$

whose elements are the coefficients a_j^i, b_k^i, appearing in (3.4). As usual, the result of applying the matrix A to the vector $x = (x^1, x^2, \ldots, x^n)$ will be denoted by Ax, that is, $y = Ax$ is the n-dimensional vector whose components are defined by

$$y^i = \sum_{\alpha=1}^{n} a_\alpha^i x^\alpha, \qquad i = 1, \ldots, n. \tag{3.7}$$

Similarly, for any r-dimensional vector $u = (u^1, u^2, \ldots, u^r)$, Bu denotes the n-dimensional vector whose ith component equals $\sum_{\beta=1}^{r} b_\beta^i u^\beta$, $i = 1$, . . . , n. Thus, the matrix A defines a linear transformation of n-dimensional space into n-dimensional space, and the matrix B defines a transformation of r-dimensional space into n-dimensional space. Making use of the matrices A and B, we may now write (3.4) in vector form:

$$\dot{x} = Ax + Bu. \tag{3.8}$$

The linear problem of optimal control will denote the problem of finding a time-optimal process under the following three conditions:

(1) the equations of motion of the object are linear [see (3.4) or (3.8)];
(2) the prescribed terminal state x_1 coincides with the origin $(0, 0, \ldots, 0)$ of $x^1, x^2, \ldots, x^n$-space;

(3)   the control region U is an r-dimensional polyhedron; moreover, the origin of u^1, u^2, . . . , u^r-space belongs to, but is not a vertex of, this polyhedron [for example, the cases shown in Figure 82(a), (b) are permitted, but the case shown in Figure 82(c) is not allowed].

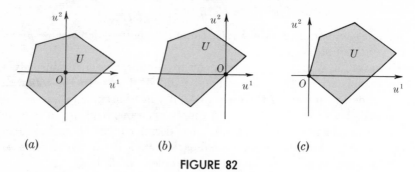

(a)   (b)   (c)

FIGURE 82

We note that the origin is an equilibrium position of the system

$$\dot{x} = \sum_{\alpha=1}^{n} a_{\alpha}^i x^{\alpha}, \qquad (3.9)$$

which is obtained from (3.4) by discarding the controls (that is, by setting $u^1 = u^2 = \cdots = u^r = 0$). Thus, condition (2) means that it is required to find a control which transfers the object from the given initial state x_0 to an equilibrium position.

We shall be concerned in this chapter with the study of the linear problem of optimal control. Moreover, *it will be assumed everywhere in the following that the linear problem satisfies an additional condition called the condition of general position*. This condition (formulated below, p. 97), is not particularly restrictive. But a number of important theorems on optimal control can be proved when this condition is satisfied.

Since the linear problem of optimal control is a special case of the more general problem considered previously, the maximum principle, proved for the general case, is certainly valid for the linear problem. However, the formulation of the maximum principle may be somewhat simplified in this case. We note, first of all, that the function H [see (B) on p. 78] takes the form

$$H = \sum_{\alpha=1}^{n} \psi_{\alpha} \left(\sum_{\gamma=1}^{n} a_{\gamma}^{\alpha} x^{\gamma} + \sum_{\beta=1}^{r} b_{\beta}^{\alpha} u^{\beta} \right) = \psi(Ax + Bu) = \psi Ax + \psi Bu. \quad (3.10)$$

(Scalar products are written on the right-hand side of this equation; for example, ψAx is the scalar product of the vectors ψ and Ax.)

Furthermore, let us consider the system of differential equations for the auxiliary variables ψ_1, ψ_2, . . . , ψ_n [see (C) on p. 78]. We have

$$\frac{\partial H}{\partial x^i} = \frac{\partial}{\partial x^i} \Big[\sum_{\alpha=1}^{n} \psi_\alpha \Big(\sum_{\gamma=1}^{n} a_\gamma^\alpha x^\gamma + \sum_{\beta=1}^{r} b_\beta^\alpha u^\beta \Big) \Big] = \sum_{\alpha=1}^{n} a_i^\alpha \psi_\alpha.$$

Consequently, the system of equations for the auxiliary variables takes the form

$$\dot{\psi}_i = - \sum_{\alpha=1}^{n} a_i^\alpha \psi_\alpha, \qquad i = 1, \ldots, n, \tag{3.11}$$

or, in vector notation,

$$\dot{\psi} = -A'\psi. \tag{3.12}$$

Here

$$A' = \begin{pmatrix} a_1^1 & a_1^2 & \cdots & a_1^n \\ a_2^1 & a_2^2 & \cdots & a_2^n \\ \cdot & \cdot & \cdots & \cdot \\ a_n^1 & a_n^2 & \cdots & a_n^n \end{pmatrix}$$

is the matrix obtained from A by *transposition* (that is, by interchanging rows and columns).

Since the first term on the right-hand side of (3.10) does not depend on u, it suffices to consider only the second term in writing relation (D) (see p. 78). Thus, in the case under consideration, (D) takes the form

$$\psi(\tau) B u(\tau) = \max_{u \in U} \psi(\tau) B u. \tag{3.13}$$

Finally, (E) (p. 79) is not needed, since in the case under consideration it is always satisfied. In fact, since $x(t_1) = (0, 0, \ldots, 0)$ [condition (2) on p. 93], the first term in $H(\psi(t_1), x(t_1), u(t_1))$ vanishes [see (3.10)]. The second term, however, is automatically non-negative [by virtue of (3.13)], since for $u^1 = \cdots = u^r = 0$ [this point belongs to the polyhedron U by virtue of the condition (3), p. 94] we have $\psi(\tau) B u = 0$, and therefore the maximum value of the expression $\psi(\tau) B u$ is non-negative. Thus, $H(\psi(t_1), x(t_1), u(t_1)) \geqslant 0$ is always satisfied for the linear optimal control problem.

The above may be summarized as follows. Let $u(t)$, $t_0 \leqslant t \leqslant t_1$, be an admissible control which transfers the object (3.8) from the given initial state x_0 to the equilibrium position $(0, 0, \ldots, 0)$. We shall say that the control $u(t)$ *satisfies the maximum principle* if there exists a nontrivial solution $\psi(t)$ of (3.12) for which the maximum condition (3.13) is satisfied (for each τ, $t_0 \leqslant \tau \leqslant t_1$). *In order for the control $u(t)$ to be optimal it is necessary that it satisfy the maximum principle.* This, in fact, is the simplified formulation of the maximum principle in the case of the linear problem of optimal control.

28. The Maximum Principle—A Necessary and Sufficient Condition for Optimality. A remarkable fact about the maximum principle is that in the case of the linear problem of optimal control, it is not only a necessary, but also a sufficient condition for optimality. However, this fact does not hold for an arbitrary linear problem; there are some unimportant exceptions. Therefore, we shall impose a certain restriction on the linear problem—the so-called *condition of general position*.

Prior to formulating the condition of general position, we recall the concept of an invariant subspace and prove a lemma which will be of importance in the sequel. As mentioned previously, the matrix A defines a linear transformation of $x^1, \ldots, x^n$-space (that is, A is a linear mapping of this space into itself). This means that it associates with each vector x a new vector $y = Ax$ defined by (3.7). The operator A can in turn be applied to this vector y and, as a result, we obtain a vector $z = Ay = A(Ax)$ which will be denoted by A^2x. If the operator A is applied to this vector, then we obtain another vector which will be denoted by A^3x, and so on. (This notation is appropriate since, as can be readily proved, the k-fold application of the matrix A to the vector x is equivalent to the application of the matrix A^k to the vector x.)

A subspace Y of the vector space X with coordinates $x^1, \ldots, x^n$ is called *invariant* if for any vector $y \in Y$ the vector Ay also belongs to the subspace Y (that is, the subspace Y is transformed into itself by the transformation A). An invariant subspace is called proper if it does not coincide with the entire space X and if it does not coincide with the trivial subspace (consisting only of the zero vector). We state the following fact (which can be readily proved): *a nonzero vector $a \in X$ belongs to a proper invariant subspace with respect to the transformation A if and only if the vectors*

$$a,\ Aa,\ A^2a,\ \ldots,\ A^{n-1}a$$

are linearly dependent. We now prove the following lemma.

Lemma 3.1. Let $\psi(t)$ be a nontrivial solution of (3.12), and let a be a nonzero vector of the space X. If $\psi(t)a = 0$ for all t on an interval $\theta_0 < t < \theta_1$, then the vector a belongs to a proper invariant subspace with respect to the transformation A.

PROOF. Let Y denote the set of all vectors $y \in X$ which satisfy $\psi(t)y = 0$ (for all t on the interval $\theta_0 < t < \theta_1$). It is clear that Y is a subspace (that is, if $y_1 \in Y$, $y_2 \in Y$, then $y_1 + y_2 \in Y$ and $ky_1 \in Y$ for any real k). Furthermore, the subspace Y contains the vector a so that it is nontrivial. In addition, the vector $\psi(t)$ is nontrivial for any t on the interval $\theta_0 < t < \theta_1$, and therefore, the subspace Y does not coincide with X. It remains to prove that Y is an invariant subspace. Let $y \in Y$. Differentiating the relation $\psi(t)y = 0$, we obtain (for all t on the interval $\theta_0 < t < \theta_1$)

$$0 = \frac{d}{dt}(\psi(t)y) = \dot{\psi}(t)y = -(A'\psi(t))y$$

$$= -\sum_{i=1}^{n}\left(\sum_{\alpha=1}^{n} a_i^\alpha \psi_\alpha(t)\right) y^i = -\sum_{\alpha=1}^{n} \psi_\alpha(t)\left(\sum_{i=1}^{n} a_i^\alpha y^i\right)$$

$$= -\psi(t)(Ay).$$

Hence, $Ay \in Y$, that is, the subspace Y is invariant. Thus the vector a belongs to the proper invariant subspace Y.

We shall now formulate the condition of general position: *if w is a vector parallel to an edge of the polyhedron U, then the vector Bw does not belong to any proper invariant subspace with respect to the transformation A.*

This condition can alternately be formulated in the following way: if w is a vector parallel to an edge of the polyhedron U, then the vectors $Bw, ABw, A^2Bw, \ldots, A^{n-1}Bw$ are linearly independent. If the condition of general position is not satisfied, then these vectors are linearly dependent for at least one edge of the polyhedron U, that is, the nth order determinant, made up of components of these vectors, vanishes. Since the polyhedron U has a finite number of edges, one need write only a finite number of such determinants. The condition of general position means that none of these determinants vanishes. Clearly, this condition is not particularly restrictive: even if some of these determinants were to vanish, then it would suffice to slightly alter the coefficients of (3.4) or the position of the polyhedron U in order to make all the determinants different from zero. Thus, the nonfulfillment of the condition of general position represents the very exceptional case in which the coefficients of (3.4) and the position of the polyhedron U are "accidentally" chosen in such a manner that one of the determinants vanishes. In other words, the condition of general position, "as a rule," must be satisfied.

We remind the reader once again that the *condition of general position is assumed to be satisfied everywhere in the following.*

Let us now return to the theorem mentioned at the beginning of this section. As a preliminary, we establish the following lemma.

Lemma 3.2. Let $u(t)$ be an arbitrary admissible control given on an interval $t_0 \leqslant t \leqslant t_1$, and let $x(t) = (x^1(t), \ldots, x^n(t))$ be the corresponding trajectory (emanating from a point x_0). Furthermore, let $\psi(t) = (\psi_1(t), \ldots, \psi_n(t))$ be an arbitrary solution of (3.12). Then,

$$\frac{d}{dt}(\psi(t)x(t)) = \psi(t)Bu(t)$$

at all points of continuity of the control $u(t)$, and therefore

$$\psi(t_1)x(t_1) - \psi(t_0)x(t_0) = \int_{t_0}^{t_1} (\psi(\tau)Bu(\tau))\, d\tau.$$

We emphasize that in this lemma the control $u(t)$ is not assumed to be optimal nor is it assumed that $u(t)$, $\psi(t)$ satisfy the maximum condition (3.13).

PROOF.   We have

$$\frac{d}{dt}(\psi(t)x(t)) = \frac{d}{dt}\left(\sum_{i=1}^{n} \psi_i(t)x^i(t)\right)$$

$$= \sum_{i=1}^{n} \psi_i(t)\dot{x}^i(t) + \sum_{i=1}^{n} \dot{\psi}_i(t)x^i(t)$$

$$= \sum_{i=1}^{n}\left(\sum_{\alpha=1}^{n} a_\alpha^i x^\alpha(t) + \sum_{\beta=1}^{r} b_\beta^i u^\beta(t)\right)\psi_i(t)$$

$$+ \sum_{i=1}^{n} x^i(t)\left(-\sum_{\alpha=1}^{n} a_i^\alpha \psi_\alpha(t)\right)$$

$$= \sum_{i=1}^{n}\left(\sum_{\beta=1}^{r} b_\beta^i u^\beta(t)\right)\psi_i(t) = \psi(t)Bu(t).$$

[This calculation is valid only at points of continuity of the control $u(t)$, since it can be asserted that $\dot{x}(t) = Ax(t) + Bu(t)$ only at such points, compare p. 49.] Furthermore, since $x(t)\psi(t)$ is a continuous function having a derivative for all but a finite number of values of t, we obtain

$$\psi(t_1)x(t_1) - \psi(t_0)x(t_0) = \int_{t_0}^{t_1}\left(\frac{d}{dt}\psi(t)x(t)\right) dt$$

$$= \int_{t_0}^{t_1} (\psi(t)Bu(t))\, dt.$$

The lemma has been proved.

Theorem 3.3.   Let $u(t)$, $t_0 \leqslant t \leqslant t_1$, be an admissible control which transfers an object from a prescribed initial state x_0 to the equilibrium position $(0, 0, \ldots , 0)$. A necessary and sufficient condition for the optimality of the control $u(t)$ is that it satisfy the maximum principle.

PROOF.   The necessity has already been established above; we prove the sufficiency. Let $x(t)$, $t_0 \leqslant t \leqslant t_1$, denote the solution of (3.8) corresponding to the control $u(t)$; thus, $x(t_0) = x_0$, $x(t_1) = (0, 0, \ldots , 0)$. Furthermore, let us select a nontrivial solution $\psi(t)$ of (3.12) which satisfies the maximum condition (3.13) for $t_0 \leqslant \tau \leqslant t_1$; such a solution exists since the control $u(t)$ satisfies the maximum principle.

Suppose that the control $u(t)$ is not optimal. Then there exists an admissible control $\tilde{u}(t) = (\tilde{u}^1(t), \tilde{u}^2(t), \ldots, \tilde{u}^r(t))$ under whose action the phase point, starting from x_0 at time t_0, reaches the origin at time $\theta < t_1$ [that is, sooner than by moving along the trajectory $x(t)$]. The phase trajectory emanating from x_0 and corresponding to the control $\tilde{u}(t)$ will be denoted by $\tilde{x}(t) = (\tilde{x}^1(t), \tilde{x}^2(t), \ldots, \tilde{x}^n(t))$. By virtue of the maximum condition (3.13), we have

$$\psi(t)Bu(t) = \max_{u \in U} \psi(t)Bu \geqslant \psi(t)B\tilde{u}(t). \qquad (3.14)$$

Since both trajectories $x(t)$ and $\tilde{x}(t)$ emanate at t_0 from one and the same point x_0,

$$\psi(t_0)x(t_0) = \psi(t_0)\tilde{x}(t_0).$$

In addition, it is obvious that

$$\psi(t_1)x(t_1) = \psi(\theta)\tilde{x}(\theta) = 0.$$

Thus, by virtue of Lemma 3.2, we obtain

$$\begin{aligned}
\psi(\theta)x(\theta) &= \psi(\theta)x(\theta) - \psi(\theta)\tilde{x}(\theta) \\
&= [\psi(\theta)x(\theta) - \psi(t_0)x(t_0)] - [\psi(\theta)\tilde{x}(\theta) - \psi(t_0)\tilde{x}(t_0)] \\
&= \int_{t_0}^{\theta} (\psi(\tau)Bu(\tau))\, d\tau - \int_{t_0}^{\theta} (\psi(\tau)B\tilde{u}(\tau))\, d\tau \geqslant 0.
\end{aligned}$$

On the other hand, $\psi(t)Bu(t) = \max_{u \in U} \psi(t)Bu \geqslant 0$ (for any t), since the point $u^1 = u^2 = \cdots = u^r = 0$ belongs to the polyhedron U. Consequently,

$$\begin{aligned}
\psi(\theta)x(\theta) &= \psi(\theta)x(\theta) - \psi(t_1)x(t_1) \\
&= -\int_{\theta}^{t_1} (\psi(\tau)Bu(\tau))\, d\tau \leqslant 0.
\end{aligned}$$

Thus $\psi(\theta)x(\theta) = 0$, and on the interval $\theta < t < t_1$, we have

$$\psi(t)Bu(t) = \max_{u \in U} \psi(t)Bu \equiv 0. \qquad (3.15)$$

Now let U_1 denote the face of the polyhedron U which contains the origin of $u^1, \ldots, u^r$-space. The polyhedron U_1 either coincides with U, or is a proper face of U, but in any case the dimension of U_1 is at least one since the origin is not a vertex of U [see condition (3) on p. 94]. Since the function $\psi(t)Bu$ takes on the value 0 at an interior point (the origin) of the polyhedron U_1 and, in addition, $\max_{u \in U} \psi(t)Bu = 0$, then $\psi(t)Bu \equiv 0$ for all the points $u \in U_1$ and for all t in the interval $\theta < t < t_1$. In particular if u' and u'' are the endpoints of an edge of the face U_1 (recall that the dimension of this face is $\geqslant 1$ so that it must have an edge), then $\psi(t)Bu' = \psi(t)Bu''$

$= 0$. Consequently, for the vector $w = u'' - u'$, which is directed along an edge of the polyhedron U, we have $\psi(t)Bw = \psi(t)Bu'' - \psi(t)Bu' = 0$ (for all t in the interval $\theta < t < t_1$). Hence it follows from Lemma 3.1 (p. 96) that the vector Bw belongs to a proper invariant subspace with respect to the transformation A, which contradicts the condition of general position.

Thus, the assumption that $\theta < t_1$ leads to a contradiction; the optimality of the control $u(t)$ has been proved.

29. Procedure for the Solution of Linear Optimal Control Problems.
The theorems proved above provide a convenient means for finding optimal controls. We note, first of all, that (3.11) [or (3.12)] is a linear system of equations with constant coefficients. It does not contain the variables x^i and u^i [compare (C) on p. 78], and therefore (3.11) may be solved independently of (3.4). As is well known (see p. 48), (3.11) may be solved uniquely if the initial value $\psi_0 = (\psi_{10}, \psi_{20}, \ldots, \psi_{n0})$ (at time t_0) of $\psi = (\psi_1, \psi_2, \ldots, \psi_n)$ is prescribed. This, in fact, is the first step in solving the linear optimal control problem; we formulate it as follows.

Problem 1. Find the solution $\psi(t)$ of (3.11) for an arbitrary initial value $\psi(t_0) = \psi_0$.

The solution of this problem is given by classical theorems on linear differential equations with constant coefficients (or by the methods of operational calculus) if the roots of the characteristic equation are known. There are also well-developed approximate methods for solving this problem, in particular, with the aid of an analog computer (compare § 8.31). Therefore, we shall assume that we are able to solve Problem 1.

Furthermore, since the maximum principle is a necessary condition for optimality, any optimal control must satisfy the maximum condition (3.13), and we arrive at the following problem.

Problem 2. Given a nontrivial solution $\psi(t)$ of (3.11), find a control $u(t)$ which satisfies the maximum condition (3.13).

This problem is an important step in finding optimal processes; the next section (§ 8.30) is devoted to its solution. We shall see there that the control $u(t)$ is uniquely defined by the maximum condition (3.13), that is, Problem 2 admits a unique solution $u(t)$ for a given function $\psi(t)$.

The next step suggests itself.

Problem 3. Given a control $u(t)$, find the corresponding trajectory $x(t)$ emanating from a prescribed initial point x_0.

This problem reduces to solving (3.4) which [if the function $u(t)$ is known] is simply an inhomogeneous system of linear ordinary differential equations with constant coefficients. Thus, Problem 3 is also a well-studied classical problem in the theory of differential equations. Thus we see that the solution of Problems 1, 2, and 3 is comparatively simple. It will be shown in § 8.31 that these three problems can be solved simultaneously with the aid of an analog computer containing relays.

Thus, let us assume that we are able to solve Problems 1, 2, 3. This means, that by choosing an arbitrary (nonzero) initial value ψ_0, we will be able to uniquely determine $\psi(t)$, then $u(t)$, and, finally, the trajectory $x(t)$ emanating from the given initial position x_0 (Figure 83). Thus, by virtue of the maximum principle, the trajectory $x(t)$ is uniquely determined by the choice of the initial value ψ_0. Generally speaking, if the initial value ψ_0 were selected at random, there would be little chance that the resulting trajectory $x(t)$ would reach the origin (Figure 83). However, for different ψ_0, we will obtain different trajectories emanating from x_0 (Figure 84). If one

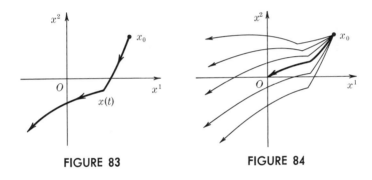

FIGURE 83   FIGURE 84

succeeds in finding an initial value ψ_0 for which the trajectory $x(t)$ *passes through the origin*, then the control $u(t)$ and the trajectory $x(t)$, obtained by the indicated procedure, will be optimal. This follows directly from the fact that if the trajectory $x(t)$ leads to the origin, then the maximum principle is a sufficient condition for optimality (p. 98).

We arrive at the problem of finding the initial value.

Problem 4.   Find the initial value ψ_0 for which the corresponding trajectory $x(t)$ passes through the origin.

A question immediately arises in connection with the statement of Problem 4: Is it always possible to find the required value ψ_0? In other words, does there exist an optimal process leading from x_0 to the origin? This question is solved by the existence (§ 8.33) and uniqueness (§ 8.32) theorems which assert (under certain reasonable limitations) that there is one and only one optimal trajectory leading from an arbitrary initial point x_0 to the origin.

The existence theorem shows that Problem 4 is meaningful, that is, that it is possible to pick out the required initial value ψ_0 for a given point x_0. But the method by which the existence theorem is proved in § 8.33 does not provide a calculation for the required initial value ψ_0. Therefore, even after the existence theorem has been proved, the full significance of Problem 4 is preserved as a computational problem.

The exact solution of Problem 4 is not known (and is hardly possible). However, there exist sufficiently reliable approximate methods for solving this problem. The idea of an approximate solution consists in the following. An arbitrary initial value ψ_0 is "improved" by a certain method in such a way that the trajectory corresponding to the "improved" initial value comes closer to the origin. The new initial value is then "improved" once again, and so on. If it turns out that the process of successive "improvements" converges reasonably rapidly to the required initial value, then we have the possibility of obtaining an approximate solution of Problem 4. "Improvement" processes which yield an approximate solution of Problem 4 will be described in § 8.35 and § 8.36.

The indicated approximate solution of Problem 4 is exclusively directed toward finding the single optimal trajectory among all trajectories emanating from the given initial point x_0 (Figure 84). Based on this method, in accordance with Problem 2, the optimal control $u = u(t)$ is found as a function of time t. For each new initial value x_0, one has to redo the entire computational process. As mentioned in the Introduction (pp. 33–36), it is substantially more convenient to solve the problem of optimal control in synthesis form, where the optimal control $u = v(x)$ is sought as a function of the point x of the phase space. The solution in synthesis form is convenient because it is suitable for any initial value x_0. We note that the construction of the synthesis function $v(x)$ is carried out by a method which is, to a certain extent, opposite to the method of "aiming at the origin" used in § 8.35 and § 8.36 for solving Problem 4. Namely, one does not consider trajectories emanating from the fixed initial position x_0 (Figure 84), but rather trajectories entering the origin (and satisfying the maximum principle). All of these trajectories are optimal, since the maximum principle is a sufficient condition for optimality for trajectories reaching the origin. Therefore, constructing a sufficiently "dense" network of such trajectories (Figure 85) by moving "backwards" along these trajectories from the origin, it is possible to record the values taken on by the optimal controls at various points of the phase space. This, in fact, yields an approximate construction of the synthesis function $v(x)$. It should be noted, however, that if the order n of (3.4) is greater than two, then the problem of finding the synthesis function $v(x)$ becomes, as a rule, extremely cumbersome from the computational point of view, and practically unrealizable. It suffices to point out that the information that must be recorded for constructing the values of the synthesis function is altogether immense. For this reason the method of "aiming" is more valuable from the practical point of view. For second order systems, however, the (exact, not merely approximate) solu-

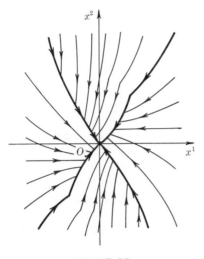

FIGURE 85

tion of the synthesis problem is comparatively simple; it is given in § 10.38 and § 10.39.

§ 8. FUNDAMENTAL THEOREMS ON LINEAR TIME-OPTIMAL PROCESSES

30. Theorems on the Number of Switchings.   In this section we shall prove the fundamental theorems on the solution of Problem 2.

Theorem 3.4.   *For each nontrivial solution* $\psi(t)$ *of* (3.12), *the relation* (3.13) *uniquely determines a control* $u(t)$ *(to within values at points of discontinuity); moreover, it turns out that the function* $u(t)$ *is piecewise constant, and that only the vertices of the polyhedron* U *may be values of this function.*

PROOF.   For each fixed t the scalar product

$$\psi(t)Bu = \sum_{\alpha=1}^{n}\sum_{\beta=1}^{r} \psi_\alpha(t)b_\beta^\alpha u^\beta \tag{3.16}$$

is obviously a linear function of the variables u^1, u^2, . . . , u^r. Therefore, according to the discussion on p. 91, the set of all points at which the scalar product (considered only on the polyhedron U) takes on its maximum value is a face of U (for convenience the polyhedron U itself is considered as its largest face). We distinguish two cases: either the scalar product (3.16) takes on its maximum value at only one vertex of U, or its maximum is attained on a face of dimensionality $\geqslant 1$. In the latter case, the function

(3.16) is constant on that face on which the maximum value is attained, and thus there exists an edge on which the function (3.16) is constant. If u', u'' are the endpoints of this edge, then $\psi(t)Bu' = \psi(t)Bu''$, and therefore, for the vector $w = u'' - u'$ directed along this edge, we have

$$\psi(t)Bw = \psi(t)(Bu'' - Bu') = \psi(t)Bu'' - \psi(t)Bu' = 0.$$

Thus, at each time t there are two possibilities: (1) the scalar product (3.16) attains its maximum at only one vertex of U; (2) an edge of U can be found such that $\psi(t)Bw = 0$ for the vector w directed along this edge. We shall show that the second case can occur only a finite number of times on any interval $t_0 \leqslant t \leqslant t_1$. As a matter of fact, let us suppose that the second case can occur for an infinite number of values of t on the interval $t_0 \leqslant t \leqslant t_1$. Since U has only a finite number of edges, we can select an infinite number of times at which $\psi(t)Bw = 0$ for one and the same edge. Let $w^1, w^2, \ldots, w^r$ be the components of the vector w directed along this edge. Then

$$\psi(t)Bw = \sum_{\alpha=1}^{n} \sum_{\beta=1}^{r} \psi_\alpha(t)b_\beta^\alpha w^\beta, \qquad (3.17)$$

that is, the scalar product is a linear combination of the functions $\psi_1(t)$, $\ldots$, $\psi_n(t)$. These functions are analytic since they constitute a solution $\psi(t)$ of (3.11) with constant coefficients. Consequently, the linear combination (3.17) is also an analytic function of t. But since the analytic function (3.17) vanishes for an infinite number of values of t on the interval $t_0 \leqslant t \leqslant t_1$, it is identically equal to zero:

$$\psi(t)Bw \equiv 0, \qquad t_0 \leqslant t \leqslant t_1.$$

Consequently, by Lemma 3.1 (p. 96), the vector w belongs to a proper invariant subspace of A, which contradicts the condition of general position. The contradiction proves our assertion.

Thus, for all but a finite number of values of t, $t_0 \leqslant t \leqslant t_1$, the function (3.16) attains its maximum (on U) at only one point which is a vertex of U. But the maximum condition (3.13) means that the function (3.16) attains its maximum at the point $u = u(t)$. Consequently, for all but a finite number of values of t, the function $u(t)$ is uniquely defined and takes on values only at the vertices of the polyhedron U.

Furthermore, let us mark on the segment $t_0 \leqslant t \leqslant t_1$ all points at which the control $u(t)$ is not uniquely defined [that is, the points at which the function (3.16) attains its maximum on a face of dimensionality $\geqslant 1$]. These points, together with the endpoints t_0 and t_1, divide the segment $t_0 \leqslant t \leqslant t_1$ into a finite number of intervals. It is not difficult to see that the function $u(t)$ is constant on each of these intervals. In fact, let J be one of these intervals. Suppose that $u(t)$ takes on different values at two points t'

and t'' of J (where $t' < t''$):

$$u(t') = e', \qquad u(t'') = e'' \neq e'.$$

In other words, the function $\psi(t')Bu$ takes on its maximum value at the vertex $u = e'$, and the function $\psi(t'')Bu$ takes on its maximum value at the vertex $u = e''$. Let us consider the graphs of the functions $\psi(t)Be$ on the interval J for all vertices e of the polyhedron U. Then, at $t = t'$ the graph of $\psi(t)Be'$ lies above all of the other graphs, while at $t = t''$ the graph of $\psi(t)Be''$ is highest. Consequently, the graphs of the functions $\psi(t)Be'$ and $\psi(t)Be''$ must intersect somewhere between t' and t''. Let us consider all the graphs which intersect the graph of the function $\psi(t)Be'$ between t' and t'' [for example, one such graph is that of the function $\psi(t)Be''$], and let e''' be a vertex such that the graph of the function $\psi(t)Be'''$ intersects $\psi(t)Be'$ between t' and t'' at a point t^* closest to t' (Figure 86). Then for t

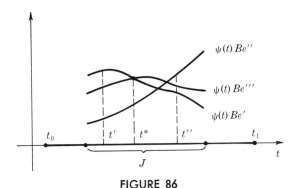

FIGURE 86

between t' and t^*, the graph of $\psi(t)Be'$ lies above all the other graphs, that is, the maximum of the function $\psi(t)Bu$ is attained at the vertex $u = e'$. At the point $t = t^*$ however, the maximum is attained at at least two vertices: $u = e'$ and $u = e'''$. But since t^* is an interior point of the interval J, then, for $t = t^*$, the maximum must be attained at only one vertex. This contradiction shows that $u(t)$ must be constant on the interval J.

Thus, $u(t)$ is piecewise constant, takes on its values at the vertices of the polyhedron U, and is uniquely defined (to within values at points of discontinuity). The theorem has been completely proved.

Each point of discontinuity of an optimal control will be called a *switching point*. More precisely, if τ is a point of discontinuity of the optimal control $u(t)$, and if $u(\tau - 0) = e_i$, $u(\tau + 0) = e_j$, where e_i and e_j are distinct vertices of U, then we shall say that a *switching* of the optimal control $u(t)$ from the vertex e_i to the vertex e_j occurs at $t = \tau$.

The theorem we have proved can be characterized as a theorem on the *finiteness of the number of switchings.* In the general case, the number of switchings, although finite, can be arbitrary. There exists, however, one case, of importance in applications, in which the number of switchings admits a precise bound. This case is considered in the following theorem due to A. A. Fel'dbaum.

Theorem 3.5.   Suppose that the polyhedron U is an r-dimensional parallelepiped (3.5) and that all the eigenvalues of the matrix $A = (a_j^i)$ of coefficients of (3.4) are real. Then each of the functions $u^\beta(t)$, $\beta = 1, \ldots, r$, in the optimal control $u(t) = (u^1(t), \ldots, u^r(t))$ is piecewise constant, takes on only the values a^β and b^β [see (3.5)], and has at most $n - 1$ switchings (that is, not more than n intervals on which it is constant), where n is the order of system (3.4).

PROOF.   In order that the function

$$\psi(t) Bu = \sum_{\beta=1}^{r} \left(\sum_{\alpha=1}^{n} \psi_\alpha(t) b_\beta^\alpha u^\beta \right)$$

take on a maximum value, it is necessary that each of the functions

$$\sum_{\alpha=1}^{n} \psi_\alpha(t) b_\beta^\alpha u^\beta, \qquad \beta = 1, 2, \ldots, r,$$

take on a maximum value (since the range of each of the quantities u^1, $\ldots$, u^r does not depend on the values of the other quantities). Consequently, $u^\beta(t)$ must take on the value a^β if the function

$$\xi_\beta = \sum_{\alpha=1}^{n} \psi_\alpha(t) b_\beta^\alpha \tag{3.18}$$

is negative, and the value b^β if this function is positive. In other words, $u^\beta(t)$ takes on only the values a^β and b^β and the number of switchings is equal to the number of times the function (3.18) changes sign. Thus, it remains to establish that the function (3.18) does not change sign more than $n - 1$ times. [Note that the function (3.18) may vanish at only a finite number of points since, according to Theorem 3.4, there exists only a finite number of instants at which $u^\beta(t)$ is not uniquely defined by the maximum condition.]

Suppose that the function (3.18) changes sign at least n times on the interval $t_0 \leqslant t \leqslant t_1$. If one modifies the coefficients a_j^i, preserving the initial value $\psi(t_0)$, the solution $\psi(t)$ of (3.11) also changes; however, the change of this solution on the interval $t_0 \leqslant t \leqslant t_1$ can be made as small as desired by a sufficiently small change of the coefficients a_j^i. Consequently, the function (3.18) will also change little on the interval $t_0 \leqslant t \leqslant t_1$. But for a sufficiently small change of the function (3.18), the number of times this function changes sign on the interval $t_0 \leqslant t \leqslant t_1$ cannot decrease. Thus, for a sufficiently small change in the coefficients a_j^i, the function (3.18) will change sign, as before, not less than n times. We now note that by a slight change in the coefficients a_j^i, we can always achieve a situation in which all the eigenvalues of the matrix $A = (a_j^i)$ remain real but become pairwise distinct. Therefore, we can consider the eigenvalues of the matrix A to be pairwise distinct (by slightly changing the coefficients a_j^i, if necessary).

Thus, let us suppose that the function (3.18) changes sign at least n times, and that all the eigenvalues of the matrix A are real and distinct (which likewise holds for the matrix $-A'$). Let $\lambda_1, \lambda_2, \ldots, \lambda_n$ denote the eigenvalues of the matrix $-A'$. Each of the functions $\psi_1(t), \ldots, \psi_n(t)$, which constitute the solution $\psi(t)$ of (3.11), has the form

$$C_1 e^{\lambda_1 t} + \cdots + C_n e^{\lambda_n t}, \tag{3.19}$$

where $C_1, \ldots, C_n$ are real constants. Consequently, the linear combination (3.18) also has this form. Thus, our assumption leads to the fact that the function (3.19) (which is not identically zero) changes sign at least n times. But, as shown in the following lemma, this is not possible, thus completing the proof of Theorem 3.5.

Lemma 3.6. If $\lambda_1, \lambda_2, \ldots, \lambda_n$ are distinct real numbers, then the function (3.19) with real coefficients $C_1, \ldots, C_n$ (not all zero) can have at most $n - 1$ real roots.

PROOF. The lemma is obviously valid for $n = 1$ (the function $e^{\lambda_1 t}$ has no real roots). We assume that the lemma has already been proved for the case in which (3.19) has less than n terms, and shall prove it for n terms. Let us suppose that the lemma does not hold so that the function (3.19) has at least n real roots. Multiplying (3.19) by $e^{-\lambda_n t}$ (which does not change its roots), we obtain the function

$$C_1 e^{(\lambda_1 - \lambda_n)t} + \cdots + C_{n-1} e^{(\lambda_{n-1} - \lambda_n)t} + C_n, \tag{3.20}$$

which also has at least n real roots. Since between each pair of real roots of a function there is at least one root of its derivative, the derivative of (3.20) has at least $n - 1$ real roots. But this derivative is

$$C_1(\lambda_1 - \lambda_n)e^{(\lambda_1 - \lambda_n)t} + \cdots + C_{n-1}(\lambda_{n-1} - \lambda_n)e^{(\lambda_{n-1} - \lambda_n)t}; \tag{3.21}$$

moreover, the numbers $\lambda_1 - \lambda_n, \ldots, \lambda_{n-1} - \lambda_n$ are obviously distinct. By the inductive hypothesis, (3.21) has at most $n - 2$ real roots, contrary to what was said before. This contradiction completes the induction.

The example considered on pp. 24–30 provides a good illustration of Theorem 3.5. In this example, $n = 2$, $r = 1$, and the eigenvalues of the matrix (a_j^i) are real ($\lambda_1 = \lambda_2 = 0$). Consequently, by Theorem 3.5, the optimal control $u(t)$ has at most one switching. This, in fact, was seen in solving that example. The examples considered in § 10.39 (where the control region is a segment or a parallelepiped) provide other illustrations of Theorem 3.5.

31. Simulation of Optimal Processes Utilizing Relays.

In this section we indicate a method of constructing an analog computer which simultaneously solves Problems 1, 2, and 3 formulated in § 7.29, that is, given the initial value ψ_0, find the corresponding trajectory $x(t)$ which satisfies the maximum principle. This analog computer consists of two linear units for (3.4) and (3.11), and of a number of relays. The precise number of relays, as well as the manner in which they are interconnected, is determined by the polyhedron U and the matrix B.

Let us proceed to the mathematical description of the analog computer. Consider the linear unit whose phase states are described by the variables $\psi_1, \ldots, \psi_n$ which vary according to (3.11). We show this unit schematically in Figure 87. Given initial values for $\psi_1, \ldots, \psi_n$ (that is, given ψ_0), the subsequent evolution of $\psi_1, \psi_2, \ldots, \psi_n$ (with time) is uniquely determined. We depict the original object [described by (3.8)] as shown in Figure 88. In order that the variation (with time) of the output (that is, the phase

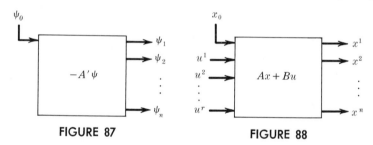

FIGURE 87   FIGURE 88

coordinates) $x^1, \ldots, x^n$ be uniquely determined, it is necessary to give the initial state x_0 of the object as well as the variation of the input $u^1, \ldots, u^r$ (that is, the control parameters) with time. The required analog computer has the form shown in Figure 89. The left-hand "box" represents the linear unit in Figure 87 and is intended for solving Problem 1 formulated in § 7.29. The right-hand "box" depicted separately in Figure 88 is intended for solving Problem 3. Finally, the middle "box" in Figure 89 is intended for solving Problem 2, that is, corresponding to the given $\psi_1(t), \ldots, \psi_n(t)$, it must compute the controls $u^1(t), \ldots, u^r(t)$ which satisfy the maximum condition (3.13). The remainder of this section is devoted to the description of this middle "box."

First of all, let us single out special cases in which the arrangement of the middle "box" is particularly simple. We consider first the case in which (3.8) contains only one control parameter u which is allowed to vary within

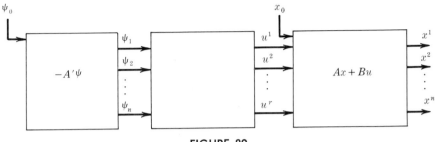

FIGURE 89

the limits $-1 \leqslant u \leqslant 1$ (that is, the case in which the polyhedron U is the closed interval $[-1, 1]$). In this case, the matrix (b_j^i) becomes the column $(b^1, b^2, \ldots, b^n)$, and the function (3.16) has the form

$$\sum_{\alpha=1}^{n} \psi_\alpha(t) b^\alpha u.$$

Therefore, from the maximum condition, we obtain

$$u = \mathrm{sign}\left(\sum_{\alpha=1}^{n} b^\alpha \psi_\alpha(t) \right). \tag{3.22}$$

In other words, if we introduce the auxiliary variable

$$\xi = \sum_{\alpha=1}^{n} b^\alpha \psi_\alpha, \tag{3.23}$$

then the control $u(t)$ satisfying the maximum condition is defined by

$$u = \mathrm{sign}\ \xi. \tag{3.24}$$

The conversion of the variables $\psi_1, \ldots, \psi_n$ into the variable ξ as defined by (3.23) is accomplished by a *summing device* shown schematically in Figure 90. A schematic representation of a *relay*, that is, a unit whose input and output are related by (3.24), is shown in Figure 91. If we now interconnect the units depicted in Figures 90 and 91, we obtain the middle "box"

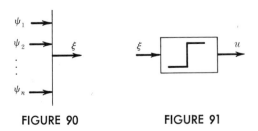

FIGURE 90   **FIGURE 91**

(Figure 92) intended for solving Problem 2, that is, the unit which transforms $\psi_1, \ldots, \psi_n$ into u according to (3.22).

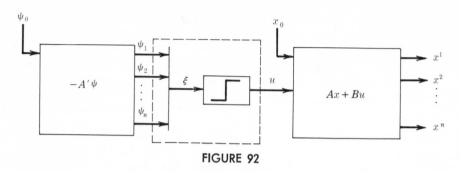

FIGURE 92

The above arguments can be readily extended to the case in which the control region U is an r-dimensional parallelepiped (3.5). In this case (see the proof of Theorem 3.5 on p. 106) the quantity u^β must take on the value a^β if the function (3.18) is negative, and the value b^β if this function is positive. In other words,

$$u^\beta = \frac{a^\beta + b^\beta}{2} + \frac{b^\beta - a^\beta}{2} \operatorname{sign} \xi_\beta, \qquad \beta = 1, \ldots, r, \qquad (3.25)$$

where ξ_β is defined by (3.18). The variables $\xi_1, \ldots, \xi_r$ are linear forms (possibly linearly dependent) in $\psi_1, \ldots, \psi_n$. The conversion of the variables ψ_i into the variables ξ_j is accomplished by a summing device shown schematically in Figure 93. The values of ξ_j are determined by the values of ψ_i, but the ξ_j have no effect on the ψ_i (this is indicated in Figure 93 by the

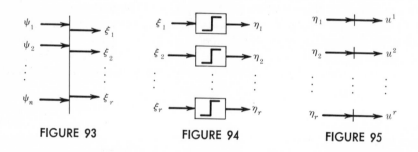

FIGURE 93   FIGURE 94   FIGURE 95

direction of the arrows). We feed each of the ξ_j into its own relay, and denote the outputs of these relays by $\eta_1, \eta_2, \ldots, \eta_r$ (Figure 94):

$$\eta_\beta = \operatorname{sign} \xi_\beta \qquad \beta = 1, \ldots, r.$$

Now, it remains to transform the η_β into the u^β according to the formulas [compare (3.25)]

$$u^\beta = \frac{a^\beta + b^\beta}{2} + \frac{b^\beta - a^\beta}{2} \eta_\beta$$

(Figure 95), that is, to multiply each of the η_β by a coefficient and to apply an additive term. As a result we obtain the desired analog computer (Figure 96).

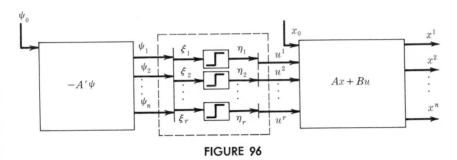

FIGURE 96

Finally, we now consider the general case in which U is an arbitrary polyhedron. Let

$$w_1, w_2, \ldots, w_\gamma \tag{3.26}$$

be vectors, no two of which are collinear, having the directions of the edges of the polyhedron U [that is, each of the vectors (3.26) is parallel to at least one edge of U, and for each edge, there is a vector in (3.26) parallel to this edge]. The components of w_j will be denoted by $w_j^i, \ldots, w_j^r$. Set

$$\xi_j = \psi B w_j = \sum_{\alpha=1}^{n} \sum_{\beta=1}^{r} \psi_\alpha b_\beta^\alpha w_j^\beta, \qquad j = 1, \ldots, \gamma \tag{3.27}$$

(Figure 97). We feed each of the quantities $\xi_1, \xi_2, \ldots, \xi_\gamma$ into its own relay (Figure 98):

$$\eta_j = \operatorname{sign} \xi_j, \qquad j = 1, \ldots, \gamma. \tag{3.28}$$

Now, let $e_1, \ldots, e_q$ be the vertices of the polyhedron U. Consider any vertex e_i and let j be one of the numbers $1, 2, \ldots, \gamma$. If the vector w_j emanating from e_i passes along one of the edges of the polyhedron U which abuts on this vertex, then we set $\epsilon_{ij} = +1$. If the vector $-w_j$ emanating from e_i passes along one of the edges of U which abuts on this vertex, then we set $\epsilon_{ij} = -1$. If neither of these two cases occurs, then the symbol ϵ_{ij} is not defined.

Fix the index $i (= 1, 2, \ldots, q)$, and consider only those j for which ϵ_{ij} is defined. Then the vectors $\epsilon_{ij} w_j$ (considered for the j we have indicated) are

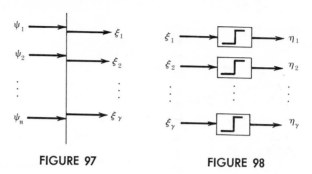

FIGURE 97   FIGURE 98

directed along the edges of U which emanate from e_i. Now, let $\psi = (\psi_1, \psi_2, \ldots, \psi_n)$ be an arbitrary nonzero vector. Let us consider the linear function

$$\psi Bu = \sum_{\alpha=1}^{n} \sum_{\beta=1}^{r} \psi_\alpha b_\beta^\alpha u^\beta = \sum_{\beta=1}^{r} \left(\sum_{\alpha=1}^{n} \psi_\alpha b_\beta^\alpha \right) u^\beta \qquad (3.29)$$

of the variables $u^1, \ldots, u^r$. Equating this linear function to zero, we obtain a hyperplane Γ in $u^1, \ldots, u^r$-space. Let n denote the vector (in $u^1, \ldots, u^r$-space) having the i-th component $\sum_{\alpha=1}^{n} b_i^\alpha \psi_\alpha$, $i = 1, \ldots, r$. This vector is orthogonal to the hyperplane Γ (compare p. 41). According to the discussion on p. 92, in order that the function (3.29), considered on the polyhedron U, take on its maximum value at the vertex e_i, it is necessary and sufficient that $nw \leqslant 0$ for every vector w emanating from e_i and passing along an edge of U, or, in other words, that $n\epsilon_{ij}w_j \leqslant 0$, that is,

$$\epsilon_{ij} \sum_{\alpha=1}^{n} \sum_{\beta=1}^{r} \psi_\alpha b_\beta^\alpha w_j^\beta \leqslant 0$$

for all indices j for which the symbol ϵ_{ij} is defined. By virtue of (3.27), the last inequality takes the form

$$\epsilon_{ij}\xi_j \leqslant 0.$$

We note that $\epsilon_{ij}\xi_j = 0$ or, equivalently, $\psi Bw_j = 0$ [see (3.27)] can be satisfied, by virtue of Theorem 3.4, for only a finite number of values of t, which will be disregarded. Thus, for the function (3.29) to attain its maximum at the vertex e_i, it is necessary and sufficient that the inequality $\epsilon_{ij}\xi_j < 0$ or, equivalently, the equality

$$\epsilon_{ij}\eta_j = -1 \qquad (3.30)$$

be satisfied for all j (for which the symbol ϵ_{ij} is defined) [see (3.28)].

We now set

$$\zeta_i = l_i - 1 + \sum_j \epsilon_{ij}\eta_j, \qquad i = 1, 2, \ldots, q, \qquad (3.31)$$

where l_i is the number of edges of U which abut on the vertex e_i and the summation is extended over all values of j for which the symbol ϵ_{ij} is defined (so that this sum contains l_i terms). The conversion of η_j into ζ_i by (3.31) is shown schematically in Figure 99. The variable ζ_i takes on the value -1 if (3.30) is satisfied for all j, and a positive value if $\epsilon_{ij}\eta_j = +1$ for at least one j. Thus (3.29) attains its maximum at the vertex $u = e_i$ if and only if $\zeta_i < 0$. If we feed $\zeta_1, \ldots, \zeta_q$ into relays and denote the outputs by $\chi_1, \ldots, \chi_q$ (Figure 100), then we find that the maximum of (3.29) is attained at the vertex $u = e_i$ if and only if $\chi_i = -1$. It is clear from the above that at any time t (except for a finite number of instants at which at least one of the quantities ξ_j vanishes) one of the quantities χ_i takes on the value -1, and the remaining quantities take on the value $+1$.

Now let $e_i^1, \ldots, e_i^r$ be the coordinates of the vertex e_i of the polyhedron U. We set

$$u^\beta = \frac{1}{2} \sum_{\alpha=1}^{q} (1 - \chi_\alpha)e_\alpha^\beta, \qquad \beta = 1, \ldots, r \qquad (3.32)$$

(Figure 101). It is clear from (3.32) that the point $(u^1, \ldots, u^r)$ coincides with the vertex e_i if $\chi_i = -1$ and the remaining χ_α equal $+1$.

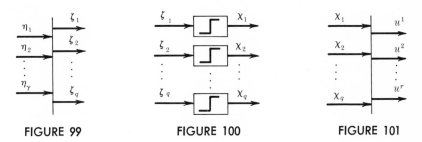

FIGURE 99   FIGURE 100   FIGURE 101

In other words, if the maximum condition (3.13) defines a unique point $u = (u^1, u^2, \ldots, u^r)$, then this point is given by (3.32).

Let us now interconnect the units depicted in Figures 87, 97–101, and 88. We obtain the schematic diagram shown in Figure 102.

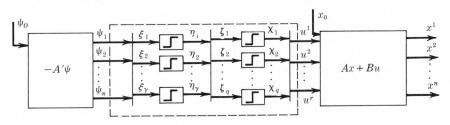

FIGURE 102

It is clear from the above that this schematic diagram represents precisely the desired analog computer, that is, the middle "box" generates the functions $u^1(t), \ldots, u^r(t)$ satisfying the maximum condition (3.13).

32. The Uniqueness Theorem

Lemma 3.7. Let $u(t)$ be an admissible control defined on the interval $t_0 \leqslant t \leqslant t_1$. Let e_i $(i = 1, \ldots, n)$ denote the vector in X whose i-th component is one and whose remaining components are zero. Furthermore, let $\varphi_i(t)$ denote the solution of $\dot{x} = Ax$ with initial condition $\varphi_i(t_0) = e_i$, and let $\psi^i(t)$ denote the solution of $\dot{\psi} = -A'\psi$ with the same initial condition $\psi^i(t_0) = e_i$ $(i = 1, \ldots, n)$. Then the trajectory $x(t)$ corresponding to the control $u(t)$ and emanating at time t_0 from the point $x_0 = (x_0^1, \ldots, x_0^n)$ is given by *

$$x(t) = \sum_{i=1}^{n} \varphi_i(t) \left[x_0^i + \int_{t_0}^{t} (\psi^i(\tau) Bu(\tau)) \, d\tau \right], \tag{3.33}$$

$$t_0 \leqslant t \leqslant t_1.$$

PROOF.   We have

$$\varphi_i(t_0)\psi^j(t_0) = e_i e_j = \begin{cases} 1 & \text{for} \quad i = j, \\ 0 & \text{for} \quad i \neq j. \end{cases}$$

Consequently, according to Theorem 2.8,

$$\varphi_i(t)\psi^j(t) = \begin{cases} 1 & i = j, \\ 0 & i \neq j \end{cases} \tag{3.34}$$

for any t.

---

* We indicate the matrix formulation of Lemma 3.7. Writing the solutions $\varphi_i(t)$ and $\psi^i(t)$ in component form

$$\varphi_i(t) = (\varphi_i^1(t), \varphi_i^2(t), \ldots, \varphi_i^n(t)),$$
$$\psi^i(t) = (\psi_1^i(t), \psi_2^i(t), \ldots, \psi_n^i(t)),$$

we obtain the matrices

$$\Phi(t) = (\varphi_i^j(t)),$$
$$\Psi(t) = (\psi_j^i(t)).$$

It is clear that $\Phi(t)$ is the *matrix solution* of the equation $\dot{X} = AX$ with initial condition $\Phi(t_0) = E$, where E is the unit matrix ([8], p. 134), and $\Psi(t)$ is the matrix solution of the equation $\dot{X} = -A'X$. The relations (3.34), proved below, show that $\Psi(t) = \Phi^{-1}(t)$, that is, the matrices $\Phi(t)$ and $\Psi(t)$ are inverses of each other. The result of applying the matrix $\Psi(t) = \Phi^{-1}(t)$ to the vector $Bu(t)$ is the vector $\Psi(t)Bu(t)$ with components

$$\psi^i(t)Bu(t), \quad i = 1, \ldots, n.$$

Thus, in matrix notation, (3.33) takes the form

$$x(t) = \Phi(t) \left[x_0 + \int_{t_0}^{t} \Phi^{-1}(\tau) Bu(\tau) \, d\tau \right].$$

The matrix notation for (3.35) is obvious.

We now show that for any vector a in X

$$\sum_{i=1}^{n} \varphi_i(t)(\psi^i(t)a) = a \qquad (3.35)$$

(for any t, $t_0 \leqslant t \leqslant t_1$). In fact, let a' denote the vector on the left-hand side of (3.35). Then for any $j = 1, \ldots, n$, we have [by (3.34)]

$$\psi^j(t)a' = \psi^j(t)\left[\sum_{i=1}^{n} \varphi_i(t)(\psi^i(t)a)\right] = \sum_{i=1}^{n} (\psi^j(t)\varphi_i(t))(\psi^i(t)a) = \psi^j(t)a,$$

that is, $\psi^j(t)(a' - a) = 0$. Since the vectors $\psi^1(t), \ldots, \psi^n(t)$ are linearly independent [they constitute a fundamental system of solutions of (3.12)], it follows from $\psi^j(t)(a' - a) = 0$, $j = 1, \ldots, n$, that $a' - a = 0$, that is. $a' = a$. Thus (3.35) has been proved.

Applying Lemma 3.2, we find

$$\psi^i(t)x(t) - \psi^i(t_0)x_0 = \int_{t_0}^{t} (\psi^i(\tau)Bu(\tau))\, d\tau,$$

or equivalently,

$$\psi^i(t)x(t) = x_0^i + \int_{t_0}^{t} (\psi^i(\tau)Bu(\tau))\, d\tau$$

[since $\psi^i(t_0)x_0 = e_i x_0 = x_0^i$]. Finally, from (3.35) we obtain the desired relation

$$x(t) = \sum_{i=1}^{n} \varphi_i(t)(\psi^i(t)x(t)) = \sum_{i=1}^{n} \varphi_i(t)\left[x_0^i + \int_{t_0}^{t} (\psi^i(\tau)Bu(\tau))\, d\tau\right].$$

Theorem 3.8 (The Uniqueness Theorem). *Let $u_1(t)$ and $u_2(t)$ be two optimal controls defined on the intervals $t_0 \leqslant t \leqslant t_1$ and $t_0 \leqslant t \leqslant t_2$, respectively, which transfer the point x_0 to the origin. Then these controls coincide, that is, $t_1 = t_2$ and $u_1(t) \equiv u_2(t)$ on the interval $t_0 \leqslant t \leqslant t_1$.*

PROOF. First of all, it is clear that $t_1 = t_2$, since for example, if $t_1 < t_2$, then the control $u = u_2(t)$, requiring a longer time than $u_1(t)$ to reach the origin, would not be optimal. Thus, both trajectories emanating from the point x_0 and corresponding to the controls $u_1(t)$ and $u_2(t)$ arrive at the same point (the origin) at $t = t_1$. Making use of (3.33), this fact can be written as follows:

$$\sum_{i=1}^{n} \varphi_i(t_1)\left[x_0^i + \int_{t_0}^{t_1} (\psi^i(\tau)Bu_1(\tau))\, d\tau\right] = \sum_{i=1}^{n} \varphi_i(t_1)\left[x_0^i + \int_{t_0}^{t_1} (\psi^i(\tau)Bu_2(\tau))\, d\tau\right].$$

Hence we obtain

$$\sum_{i=1}^{n} \varphi_i(t_1)\left[\int_{t_0}^{t_1} (\psi^i(\tau)Bu_1(\tau))\, d\tau - \int_{t_0}^{t_1} (\psi^i(\tau)Bu_2(\tau))\, d\tau\right] = 0.$$

Since the vectors $\varphi_1(t_1)$, . . . , $\varphi_n(t_1)$ are linearly independent, it follows from the last equation that

$$\int_{t_0}^{t_1} (\psi^i(\tau)Bu_1(\tau))\, d\tau = \int_{t_0}^{t_1} (\psi^i(\tau)Bu_2(\tau))\, d\tau, \qquad i = 1, \ldots, n. \quad (3.36)$$

Now let $\psi(t)$ be the solution of (3.12) corresponding to the optimal control $u_1(t)$ by virtue of the maximum principle. Since $\psi^1(t)$, . . . , $\psi^n(t)$ constitute a fundamental system of solutions of (3.12), $\psi(t)$ can be expressed linearly in terms of these functions:

$$\psi(t) = d_1\psi^1(t) + \cdots + d_n\psi^n(t).$$

Multiplying (3.36) by d_i and summing over i, we obtain

$$\int_{t_0}^{t_1} (\psi(\tau)Bu_1(\tau))\, d\tau = \int_{t_0}^{t_1} (\psi(\tau)Bu_2(\tau))\, d\tau. \quad (3.37)$$

But, by the maximum condition, we have (on the interval $t_0 \leqslant t \leqslant t_1$)

$$\psi(t)Bu_1(t) = \max_{u \in U} (\psi(t)Bu) \geqslant \psi(t)Bu_2(t),$$

and it follows from (3.37) that $\psi(t)Bu_1(t) = \psi(t)Bu_2(t)$ on the entire interval $t_0 \leqslant t \leqslant t_1$. Consequently, both controls $u_1(t)$ and $u_2(t)$ satisfy the maximum condition for the same function $\psi(t)$, and therefore (by Theorem 3.4) $u_1(t) \equiv u_2(t)$.

33. The Existence Theorem.   Let G be a set of points in the phase space X. We remind the reader that G is said to be *open* if for each of its points it is possible to find a sphere (however small) with center at this point belonging entirely to the set G (Figure 103). In other words, the set G

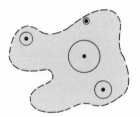

FIGURE 103

is open if it does not contain a single point of its boundary. Furthermore, a set R in X is called *closed* if its complement (consisting of all points of X not belonging to R) is open. In other words, the set R is closed if it contains all points of its boundary.

The set of all points x_0 of X from which it is possible to reach the origin with the aid of some admissible control will be called the *domain of controllability* for the process (3.8) (not to be confused with the term control region, that is, the polyhedron U!). We shall consider the origin to be in the domain of controllability. It is clear that the problem of finding optimal processes makes sense only in case the initial state x_0 belongs to the domain of controllability (in fact, it is even impossible to reach the origin from points which do not belong to the domain of controllability).

Theorem 3.9 (The Existence Theorem). *The domain of controllability is an open convex set; for any point x_0 in the domain of controllability, there exists an optimal control which transfers x_0 to the origin.*

PROOF.   Select a positive number T. Let Σ_T denote the set of all points x_0 of X for which there exists an optimal control transferring x_0 to the origin in time T.

Let $p = (p_1, \ldots, p_n)$ be a nonzero vector. Let $\psi(t, p)$ denote the solution of (3.12) with initial condition $\psi(0, p) = p$, and let $u(t, p)$ denote the control which, by virtue of the maximum condition (3.13), corresponds to the function $\psi(t, p)$; we shall consider this control on the interval $0 \leqslant t \leqslant T$. Furthermore, let $x(t)$ denote the trajectory corresponding to the control $u(t, p)$ satisfying the "terminal" condition $x(T) = 0$. The initial point $x(0)$ of this trajectory will be denoted by $\xi_T(p) = (\xi_T^1(p), \ldots, \xi_T^n(p))$. By virtue of Theorem 3.3, $x(t)$ is the optimal trajectory which transfers $x_0 = \xi_T(p)$ to the origin in time T. Therefore $\xi_T(p)$ belongs to the set Σ_T. According to (3.33), we have [considering that $x(T) = 0$]

$$\sum_{i=1}^{n} \varphi_i(T) \left[\xi_T^i(p) + \int_0^T (\psi^i(\tau) B u(\tau, p)) \, d\tau \right] = 0.$$

The expression in brackets vanishes by virtue of the linear independence of the vectors $\varphi_1(T), \ldots, \varphi_n(T)$, and therefore,

$$\xi_T^i(p) = - \int_0^T (\psi^i(\tau) B u(\tau, p)) \, d\tau, \qquad i = 1, 2, \ldots, n. \qquad (3.38)$$

Thus ξ_T is a transformation (that is, a function) which puts the point $\xi_T(p)$, defined in terms of coordinates by (3.38), into correspondence with the n-dimensional (nonzero) vector p.

The hyperplane orthogonal to the vector p passing through the point $x_0 = \xi_T(p)$ will be denoted by $\Gamma_T(p)$. The half-space defined by the hyperplane $\Gamma_T(p)$ into which the vector p is directed (Figure 104) will be considered *positive*. Finally, let V_T denote the convex hull of the set Σ_T.

We shall prove that the *origin is an interior point of the convex body V_T, and that the set Σ_T is its boundary* (Figure 105); *furthermore, all the hyperplanes $\Gamma_T(p)$, and only these hyperplanes, are hyperplanes of support for V_T and, moreover, the hyperplane of support $\Gamma_T(p)$ has only the single point $\xi_T(p)$ in common with V_T.*

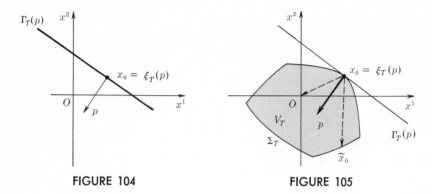

FIGURE 104   FIGURE 105

As a matter of fact, let $\tilde{x}_0$ be a point of Σ_T different from $x_0 = \xi_T(p)$. Let $x(t)$ and $\tilde{x}(t)$ denote the optimal trajectories leading from x_0 and $\tilde{x}_0$ to the origin, and let $u(t)$ and $\tilde{u}(t)$ denote the corresponding optimal controls. Thus

$$x(0) = x_0, \qquad \tilde{x}(0) = \tilde{x}_0, \qquad x(T) = \tilde{x}(T) = 0.$$

Furthermore, let $\psi(t) = \psi(t, p)$ denote the solution of (3.12) corresponding to the optimal control $u(t) = u(t, p)$. Then,

$$\psi(t)Bu(t) = \max_{u \in U} \psi(t)Bu \geqslant \psi(t)B\tilde{u}(t), \qquad 0 \leqslant t \leqslant T. \qquad (3.39)$$

Applying Lemma 3.2, we obtain

$$
\begin{aligned}
p(\tilde{x}_0 - x_0) &= \psi(0)(\tilde{x}_0 - x_0) = -\psi(0)x(0) + \psi(0)\tilde{x}(0) \\
&= [\psi(T)x(T) - \psi(0)x(0)] - [\psi(T)\tilde{x}(T) - \psi(0)\tilde{x}(0)] \\
&= \int_0^T (\psi(\tau)Bu(\tau))\,d\tau - \int_0^T (\psi(\tau)B\tilde{u}(\tau))\,d\tau \geqslant 0. \quad (3.40)
\end{aligned}
$$

This means (Figure 105) that the vector $\tilde{x}_0 - x_0$, and consequently, the point $\tilde{x}_0$, lies in the positive half-space [or on the hyperplane $\Gamma_T(p)$]. In other words, all points of Σ_T lie on one side of $\Gamma_T(p)$; in addition, the hyperplane $\Gamma_T(p)$ has a common point $\xi_T(p)$ with the set Σ_T. Thus $\Gamma_T(p)$ is a hyperplane of support of Σ_T, and consequently, of the set V_T. In particular, it follows that the point $x_0 = \xi_T(p)$ (that is, any point of the set Σ_T) is a boundary point of V_T (the hyperplane of support passes through this point).

In order to prove that Σ_T is the boundary of the convex body V_T, we must establish the converse: every boundary point of V_T belongs to Σ_T. For this purpose we shall make use of the following lemma whose proof will be provided in the next section (p. 122) so as not to interrupt the present discussion.

Lemma 3.10. *The transformation* ξ_T *defined by* (3.38) *is continuous, that is, the point* $\xi_T(p)$ *depends continuously on* p. *Furthermore, the set* Σ_T *is closed and bounded.*

Let us suppose, contrary to the fact being proved, that there exists a boundary point η of V_T which does not belong to Σ_T. We draw the hyperplane of support Γ of V_T through the point η. It is known from the theory of convex bodies that if Σ is a closed bounded set, if V is its convex hull, and if η is a boundary point of V which does not belong to Σ, then the hyperplane of support Γ of V, passing through η, contains at least two points of Σ (Figure 106). Thus, at least two points of Σ_T lie on the hyperplane Γ.

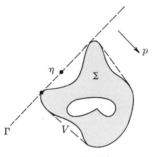

FIGURE 106

Let p denote a vector orthogonal to Γ directed into the half-space (defined by the hyperplane Γ) which contains V_T. Then, by virtue of what has been proved above, $\Gamma_T(p)$ is a hyperplane of support of V_T positioned in the same way as Γ, that is, the hyperplane Γ coincides with $\Gamma_T(p)$. Since the hyperplane Γ contains at least two points of Σ_T we can find a point $\tilde{x}_0$ on this hyperplane belonging to the set Σ_T which is distinct from $x_0 = \xi_T(p)$. We can write (3.40) for these two points; moreover, the equality sign in this relation must apply since both points x_0 and $\tilde{x}_0$ lie in the hyperplane $\Gamma = \Gamma_T(p)$, and therefore, the vector $\tilde{x}_0 - x_0$ is orthogonal to p. Consequently, (3.39) must also be a strict equality:

$$\psi(t)Bu(t) = \psi(t)B\tilde{u}(t) = \max_{u \in U} \psi(t)Bu.$$

Hence, by virtue of Theorem 3.4, we obtain $u(t) \equiv \tilde{u}(t)$, and therefore, $x(t) \equiv \tilde{x}(t)$; in particular, $x_0 = \tilde{x}_0$, which contradicts the choice of the points x_0 and $\tilde{x}_0$. The contradiction proves our assertion.

Thus, Σ_T is the boundary of the convex body V_T. At the same time, it has been established that any hyperplane of support of the set V_T has the form $\Gamma_T(p)$, and that this hyperplane of support has only one point [namely $\xi_T(p)$] in common with V_T.

It remains to prove that the origin is an interior point of the body V_T. Since the origin $(0, 0, \ldots, 0)$ of $u^1, \ldots, u^r$-space belongs to the polyhedron U but is not one of its vertices, we obtain (by Theorem 3.4) that for the optimal control $u(t)$, $\psi(t, p)Bu(t, p) > 0$ (for all but a finite number of values of t). Consequently,

$$px_0 = +\psi(0, p)x(0) = -[\psi(T, p)x(T) - \psi(0, p)x(0)]$$
$$= -\int_0^T (\psi(\tau, p)Bu(\tau, p))\, d\tau < 0,$$

that is, the vector x_0 [from the origin to the point $x_0 = \xi_T(p)$] is not orthogonal to p, and therefore (Figure 105) the origin does not lie in the hyperplane of support $\Gamma_T(p)$. But then the origin is not a boundary or an exterior point of the set V_T, since otherwise it would be possible to draw a hyperplane of support to V_T through the origin [Figures 107(a), (b)].

Let us establish further properties of the set V_T. It can be readily seen that *it is not possible to reach the origin in a time $\leqslant T$ from the points which lie outside V_T*. In fact, if this were possible, then the corresponding trajectory would intersect the boundary Σ_T of V_T at a point y_0 (Figure 108),

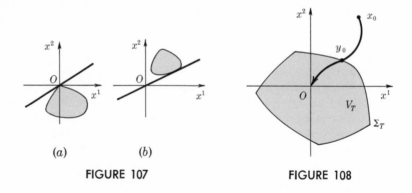

(a)   (b)

FIGURE 107   FIGURE 108

and therefore, continuing to move along this trajectory, we could reach the origin from the point y_0 in a time less than T. But this contradicts the fact that the time of optimal motion to the origin is equal to T for any point $y_0 \in \Sigma_T$.

Furthermore, *if $T' < T$, then $V_{T'}$ is contained in the interior of V_T.* In fact, if $x_0' \in \Sigma_{T'}$, then it is possible to reach the origin from x_0' in time $T' < T$, and therefore x_0' can lie neither on the boundary Σ_T of V_T nor outside V_T. This means that $\Sigma_{T'}$ and consequently $V_{T'}$ lie entirely in the interior of V_T.

Let z_0 be an arbitrary interior point of V_T. Let us draw a ray l from the origin passing through z_0, and let y_0 denote the point of the ray's intersection with the set Σ_T (Figure 109). Furthermore, let us consider the point of intersection of the ray l with $\Sigma_{T'}$ for any $T' < T$, and let $\varphi(T')$ denote the distance from this point of intersection to the origin. From the above discussion, the function φ is monotonically increasing.

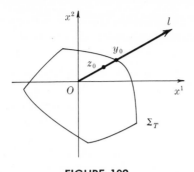

FIGURE 109

Lemma 3.11.   *The function φ is continuous.*

The proof of this lemma will be given in the next section (p. 124) so as not to interrupt the presentation. We shall now complete the proof of the existence theorem.

Note that if T' is very small, then $\varphi(T')$ is also very small [more precisely, $\varphi(T') \leqslant NT'$ if $T' < T$; see (3.45) in the proof of Lemma 3.11, p. 124]. Therefore, for small values of T', the values of the function $\varphi(T')$ are smaller than the length of the segment Oz_0. For $T' = T$ the function assumes the value $\varphi(T)$ equal to the length of Oy_0, that is, a value greater than the length of the segment Oz_0. But a continuous function takes on all intermediate values and therefore there exists a value T' (between 0 and T) such that $\varphi(T')$ equals the length of the segment Oz_0. In other words, there exists a T' such that the set $\Sigma_{T'}$ passes through the point z_0. But then there is an optimal trajectory which passes from the point z_0 to the origin (with the optimal transition time equal to T').

Thus, from any interior point of V_T there exists an optimal trajectory which leads to the origin. In addition, it is now clear that it is possible to reach the origin in time $\leqslant T$ from any (interior or boundary) point of V_T; however, as has already been proved, it is not possible to reach the origin in time $\leqslant T$ from points which do not belong to V_T. In other words, the *body* V_T *consists of only those points which can be transferred to the origin in time* $\leqslant T$.

It is now not difficult to complete the proof of the existence theorem. As T increases, V_T continuously expands. Let G denote the union of all bodies V_T (for all $T, 0 < T < \infty$). It is clear that the set G is convex. It is also open, since an arbitrary point x_0 of G belongs to some V_T, and consequently, for $T_1 > T$, x_0 is an interior point of V_{T_1} (and hence of the set G). It is possible to reach the origin from every point of G; indeed, there exists an optimal transition to the origin. This means that all points of G belong to the domain of controllability. Conversely, if x_0 is a point which belongs to the domain of controllability, that is, if it is possible to reach the origin from this point in some time T, then x_0 must belong to V_T, and consequently, also to the set G. Thus G coincides with the domain of controllability. The properties of the set G established above prove the existence theorem.

Theorem 3.12.   If the matrix A [see (3.6)] in the linear problem of optimal control is stable, that is, all of its eigenvalues have negative real parts, then the domain of controllability coincides with the entire phase space X. Consequently, for any point $x_0 \in X$ there exists an optimal control which transfers the phase point from x_0 to the origin.

PROOF.   Let us choose an arbitrary $T > 0$. Then the interior of V_T is an open set containing the origin; moreover, it is possible to reach the origin (in time $< T$) from any point of this set with the aid of some control. Let x_0 be an arbitrary point of X. At the outset, let us compel this point to move from the state x_0 under the control $u(t) \equiv 0$. Since all the eigenvalues of the matrix A have negative real parts, the moving point, after the expiration of some time, will be as close to the origin as desired. This means that after some time the moving point will reach the set V_T, after which it can be transferred to the origin. Thus, the point x_0 belongs to the domain of controllability. It follows from the arbitrariness of x_0 that the domain of controllability coincides with the entire phase space X.

34. The Proofs of the Lemmas.

In this section we shall give the previously omitted proofs of Lemmas 3.10 and 3.11 which contain certain mathematical "refinements" (on first reading, the proofs of these lemmas may be skipped).

PROOF OF LEMMA 3.10.   Let ϵ be an arbitrary positive number. Let M denote the maximum value of the expression $|\psi^i(t)Bu|$ for any t, $0 \leqslant t \leqslant T$, and for any $u \in U$. Furthermore, let $\tau_1, \ldots, \tau_q$ be all the instants of time (on the interval $0 \leqslant t \leqslant T$) at which the control $u(t, p)$ is not uniquely defined by the maximum condition (3.13). Let us enclose the points $\tau_1, \ldots, \tau_q$ by small intervals the sum of whose lengths is less than $\epsilon/2M$.

Let p^* be a vector. Then, according to (3.38),

$$\xi_T^i(p^*) + \int_0^T (\psi^i(\tau)Bu(\tau, p^*))\, d\tau = 0. \tag{3.41}$$

Eliminating the above small intervals from the interval $0 \leqslant t \leqslant T$, the change in value of each integral in (3.38), (3.41) will be less than $M \cdot \epsilon/2M = \epsilon/2$. In other words, letting Q denote the set obtained from the interval $0 \leqslant t \leqslant T$ after discarding the above-mentioned small intervals (Figure 110), we have

$$\left| \xi_T^i(p) + \int_Q (\psi^i(\tau)Bu(\tau, p))\, d\tau \right| < \frac{\epsilon}{2},$$
$$\left| \xi_T^i(p^*) + \int_Q (\psi^i(\tau)Bu(\tau, p^*))\, d\tau \right| < \frac{\epsilon}{2}. \tag{3.42}$$

Now let I be one of the intervals constituting the set Q. On the interval I, the control $u(t)$ takes on a constant value $u(t, p) = e$, where e is one of the vertices of the polyhedron U; moreover, if e' is any other vertex of U, then $\psi(t, p)Be > \psi(t, p)Be'$ (for all t on I, see Figure 111). It is clear that the

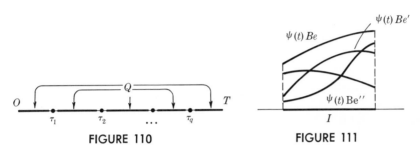

FIGURE 110   FIGURE 111

inequalities $\psi(t, p)Be > \psi(t, p)Be'$ are not violated if we replace $\psi(t, p)$ by any other function which differs slightly from $\psi(t, p)$ on the entire interval I. But if the vector p^* differs sufficiently little from p, then the functions $\psi(t, p)$ and $\psi(t, p^*)$ [that is, the solutions of (3.12) with the initial conditions $\psi(0) = p, \psi(0) = p^*$] will differ little from one another on the entire interval $0 \leqslant t \leqslant T$. Consequently, for the function $\psi(t, p^*)$ we obtain (on the interval I) $\psi(t, p^*)Be > \psi(t, p^*)Be'$ for any vertex e' different from e. In other words, by virtue of the maximum condition (3.13), $u(t, p^*) = e$ on I, that is, $u(t, p^*) = u(t, p)$. This argument can be carried out for each of the intervals constituting the set Q.

Thus, for $|p - p^*| < \delta$, where δ is a sufficiently small positive number, the equality $u(t, p) = u(t, p^*)$ holds on the entire set Q, and therefore, by virtue of (3.42), $|\xi_T^i(p) - \xi_T^i(p^*)| < \epsilon$. Accordingly, the continuity of the function $\xi_T^i(p)$ has been proved.

Furthermore, let S denote the set of all vectors p having unit length, that is, which satisfy the condition

$$|p| = \sqrt{(p_1)^2 + \cdots + (p_n)^2} = 1 \qquad (3.43)$$

(Figure 112). It can be easily seen that for any point $x_0 \in \Sigma_T$ it is possible to find a vector $p \in S$ which satisfies the condition $\xi_T(p) = x_0$; in other words, ξ_T maps the set S onto the entire set Σ_T. In fact, since $x_0 \in \Sigma_T$, there exists an optimal control $u(t)$, $0 \leqslant t \leqslant T$, which transfers x_0 to the origin (in time T). Let $\psi(t)$ be the solution of (3.12) which corresponds to the optimal control $u(t)$ by virtue of the maximum condition (3.13). Moreover, we can assume that $|\psi(0)| = 1$ [since multiplying $\psi(t)$ by a positive number does not change anything]. Then, letting p denote the vector $\psi(0)$, one obviously obtains $\xi_T(p) = x_0$.

One of the classical theorems of mathematical analysis states that if R and R' are two sets (in Euclidean spaces), with the set R moreover being

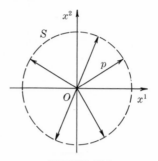

FIGURE 112

closed and bounded, and if there exists a continuous mapping of R onto the entire set R', then R' is also a closed bounded set. Since the set S [the boundary of a sphere in an n-dimensional vector space; see (3.43)] is obviously closed and bounded, and ξ_T, as we have proved, is a continuous mapping of the set S onto the entire set Σ_T it follows that Σ_T is a *closed bounded set*.

PROOF OF LEMMA 3.11.   Let $0 < T^* \leqslant T' \leqslant T'' \leqslant T$. Let us choose a number R so that the sphere of the radius R with center at the origin is contained entirely in the interior of V_{T^*} (Figure 113). The sets $\Sigma_{T'}$ and $\Sigma_{T''}$

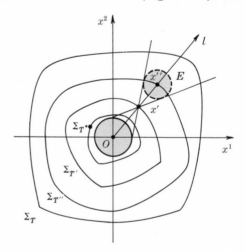

FIGURE 113

intersect the ray l at x', x''; the distance between these two points is equal to $\varphi(T'') - \varphi(T')$. Let us consider a cone with vertex x' circumscribing the sphere of radius R with center at the origin; in this cone we inscribe a sphere E with center at x''. The sphere E obviously does not contain any points of the set $\Sigma_{T'}$; the radius r of this sphere can be readily found from the proportion $r: [\varphi(T'') - \varphi(T')] = R: \varphi(T')$. Thus,

$$r = \frac{R[\varphi(T'') - \varphi(T')]}{\varphi(T')} > \frac{R[\varphi(T'') - \varphi(T')]}{\varphi(T)}. \qquad (3.44)$$

Furthermore, since V_T and U are bounded sets, the phase velocities of points in the set V_T are bounded:

$$|\dot{x}| = |Ax + Bu| \leqslant N \qquad \text{for} \qquad x \in V_T, \qquad u \in U. \qquad (3.45)$$

Let us draw the optimal trajectory $x(t)$ from x'' to the origin. The trajectory $x(t)$ intersects $\Sigma_{T'}$ at a point y'; moreover, the portion of this trajectory from y' to the origin is optimal. Therefore, the duration of the motion from y' to the origin is T', and the duration of the motion along the trajectory $x(t)$ from x'' to y' is $T'' - T'$. Consequently, by virtue of the estimate of phase velocity [see (3.45)], we have

$$|x'' - y'| < N(T'' - T').$$

Thus, there exists a point y' of $\Sigma_{T'}$ whose distance from x'' is less than $N(T'' - T')$, that is, the sphere of radius $N(T'' - T')$ with center at x'' intersects the set $\Sigma_{T'}$. Hence it follows that $r < N(T'' - T')$, and therefore, by (3.44),

$$\varphi(T'') - \varphi(T') < \frac{N\varphi(T)}{R}(T'' - T').$$

This obviously implies that the function φ is continuous on the interval from T^* to T. In view of the arbitrariness of T^* and T, we conclude that φ is continuous for all positive values of its argument.

§ 9. COMPUTATIONAL METHODS

35. Determination of the Initial Values of the Auxiliary Variables: Neustadt's Differential Equation [6].

In this and the next sections we present results of the American mathematicians Neustadt and Eaton which yield the solution of Problem 4 formulated in § 7.29. The notation introduced in the previous section will be used here.

Let p be an arbitrary nonzero vector. It is easy to see that the inequality $p\xi_{t'}(p) > p\xi_t(p)$ holds for $t' < t$. In fact, the point $\xi_{t'}(p)$ lies in the interior of V_t (Figure 114), and therefore the scalar product $p(\xi_{t'}(p) - \xi_t(p))$ is positive. Thus, the scalar product $p\xi_t(p)$ is a monotonically decreasing function of time t $(0 < t < \infty)$. We note moreover that

$$\lim_{t \to 0} p\xi_t(p) = 0 \qquad (3.46)$$

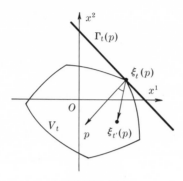

FIGURE 114

[since the point $\xi_t(p)$ belongs to the boundary of V_t which contracts to the origin O as $t \to 0$]. It follows from (3.46) that the function $p\xi_t(p)$, decreasing monotonically (with respect to t), takes on only negative values. Furthermore, it is not difficult to see that the scalar product $p\xi_t(p)$, considered as a function of the two variables t and p, is continuous. In fact, let t^* and p^* be values of the arguments close to t and p, and let ϵ be a positive number. We have

$$|\xi_t^i(p) - \xi_{t^*}^i(p^*)| \leqslant |\xi_t^i(p) - \xi_t^i(p^*)| + |\xi_t^i(p^*) - \xi_{t^*}^i(p^*)|.$$

The first term in the right-hand side can be made smaller than $\epsilon/2$ if p^* is sufficiently close to p [since for fixed t the function $\xi_t^i(p)$ depends continuously on p, see Lemma 3.10 on p. 118]. The second term, by virtue of (3.38), has the form

$$|\xi_t^i(p^*) - \xi_{t^*}^i(p^*)| = \left| \int_0^{t^*} (\psi^i(\tau)Bu^*(\tau))\, d\tau \right.$$
$$\left. - \int_0^t (\psi^i(\tau)Bu^*(\tau))\, d\tau \right| = \left| \int_t^{t^*} (\psi^i(\tau)Bu^*(\tau))\, d\tau \right|,$$

where $u^*(t)$ is the optimal control corresponding to the solution $\psi^*(t)$ of (3.12) with initial condition $\psi(0) = p^*$. Since the function $\psi^i(t)$ is continuous and the polyhedron U is bounded, the second term can also be made smaller than $\epsilon/2$ if t^* is sufficiently close to t. Thus, if t^* and p^* are sufficiently close to t and p, the inequality $|\xi_t^i(p) - \xi_{t^*}^i(p^*)| < \epsilon$ is satisfied. This means that the function $\xi_t^i(p)$ is continuous with respect to the pair of variables t and p. Consequently, the function

$$p\xi_t(p) = \sum_{i=1}^{n} p_i\xi_t^i(p)$$

is continuous.

We now fix a nonzero initial state x_* belonging to the domain of controllability G. The set of all vectors p satisfying the condition $px_* < 0$ will be denoted by D. The set D is an open half-space defined by the hyperplane orthogonal to the vector x_*.

We shall say that p is a vector *corresponding* to the point x_* if the initial condition $\psi(0) = p$ yields a solution of Problem 4 (p. 101) for the initial state x_*, that is, if $x_* = \xi_{t_*}(p)$ holds for some positive $t = t_*$. We already know that a corresponding vector p can be found for the point x_* (existence theorem). However, this does not mean that there exists only one vector p which corresponds to the point x_*. (Examples show that, as a rule, there is, to within a positive factor, a single vector p corresponding to x_*, but there may exist points having an infinite number of corresponding vectors p which are not mutually proportional.) The set of all vectors p corresponding to the point x_* is denoted by H_*. The *set H_* is entirely contained in the half-space D*. In fact, if the vector p corresponds to x_*, that is, $x_* = \xi_{t_*}(p)$, then $px_* = p\xi_{t_*}(p) < 0$, and therefore, $p \in D$.

Let us introduce the function

$$f(t, p) = p(x_* - \xi_t(p)). \tag{3.47}$$

By virtue of the above discussion, this function is continuous in t and p; for any fixed p, it is monotonically increasing in t. If $p \in D$, then by (3.46), we have

$$\lim_{t \to 0} f(t, p) = px_* < 0. \tag{3.48}$$

Now, let t_* be the duration of the optimal motion from the point x_* to the origin. If $p \in H_*$, then $x_* = \xi_{t_*}(p)$, and therefore, $f(t_*, p) = 0$. If, however, the vector p does not belong to the set H_* (that is, does not correspond to the point x_*), then $\xi_{t_*}(p) \neq x_*$, that is, $\xi_{t_*}(p)$ and x_* are two distinct points of the set Σ_{t_*}. Consequently, $p(x_* - \xi_{t_*}(p)) > 0$ [see (3.40)], that is, $f(t_*, p) > 0$. It follows from this inequality [considering (3.48) and monotonicity of the function $f(t, p)$] that there exists a unique value of t, $0 < t < t_*$, for which $f(t, p) = 0$. We denote this value of t by $F(p)$. Thus, for any $p \in D$, the function $f(t, p)$ vanishes for a unique value of time, namely for $t = F(p)$, that is,

$$f(F(p), p) = 0; \tag{3.49}$$

by virtue of the above, we have

$$\begin{aligned} F(p) &= t_* &&\text{for} &&p \in H_*, \\ 0 < F(p) &< t_* &&\text{for} &&p \notin H_*. \end{aligned} \tag{3.50}$$

It follows from the continuity of $f(t, p)$ in t and p that $F(p)$ is also continuous.

We shall indicate a geometric interpretation of the function $F(p)$. Let us draw the hyperplane Γ through the point x_* which is orthogonal to the vector p, and let us consider the convex body V_t (Figure 115). The hyperplane $\Gamma_t(p)$, which has the single point $\xi_t(p)$ in common with the body V_t, is also orthogonal to the vector p, that is, it is parallel to Γ. Since the origin lies in the positive half-space with respect to Γ, that is, in the half-space into

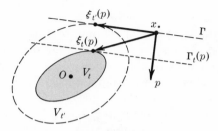

FIGURE 115

which the vector p is directed (since $px_* < 0$), it follows that for t close to zero, the entire body V_t, and in particular the point $\xi_t(p)$, lies in the positive half-space. Hence, it follows that for t close to zero, the scalar product $p(\xi_t(p) - x_*)$ is positive, that is, $f(t, p) < 0$. Moreover, the hyperplane Γ passes outside the body V_t. As t increases the body V_t expands, so that ultimately there is an instant of time t' at which Γ becomes the hyperplane of support of the body $V_{t'}$ (Figure 115). This instant of time t' is precisely $F(p)$ since the point $\xi_{t'}(p)$ lies in the hyperplane Γ, that is, the vector $\xi_{t'}(p) - x_*$ is orthogonal to p, and therefore $f(t', p) = 0$. Thus $F(p) = t'$ *is the time at which* Γ *is a hyperplane of support of the body* $V_{t'}$. If, moreover, $p \notin H_*$, then $\xi_{t'}(p) = x_*$, and therefore the point x_* lies outside $V_{t'}$ [since this body has the single point $\xi_{t'}(p)$ in common with Γ]. But this means that the duration t_* of the optimal motion from x_* to the origin is greater than t', that is, $t_* > F(p)$ [compare (3.50)]. If, however, $p \in H_*$, then $\xi_{t'}(p) = x_*$, and therefore $t' = t_*$. This, in fact, yields the geometrical interpretation of the function $F(p)$ and the relation (3.50).

Theorem 3.13. *Let us consider the differential equation*

$$\frac{dp}{d\tau} = -[x_* - \xi_{F(p)}(p)]. \tag{3.51}$$

Its right-hand side is continuous for $p \in D$, *and therefore this equation has a solution for any initial condition* $p(0) = p_0 \in D$. *It turns out that the solution* $p(\tau)$ *of (3.51) with initial condition* $p(0) = p_0$ *is defined for all positive values of* τ *and lies entirely in the half-space* D. *Furthermore, as* $\tau \to \infty$, *the solution* $p(\tau)$ *approaches the set* H_*; *more precisely, any* ω-*limiting point* of this solution belongs to the set* H_*.

---

*Let us recall the concept of an ω-limiting point. Let $\dot{x} = f(x)$ be an autonomous system of differential equations in vector form, and let $x(t)$ be a solution of this system defined for all $t > t_0$. The point y of the phase space X is called an ω-limiting point of the solution $x(t)$ if there exists a sequence $t_1, t_2, \ldots, t_k, \ldots$ such that $\lim_{k \to \infty} t_k = +\infty$ and $\lim_{k \to \infty} x(t_k) = y$. (See [8], p. 227.)

We note, as will be made clear in the proof, that for every solution of (3.51), $|p(\tau)| = $ const. For example, if $p_0 \in S$ [see (3.43)], then the entire solution $p(\tau)$ will lie on the sphere S (more precisely, on the side of the sphere S which lies in the half-space D).

We pass on to the proof of Theorem 3.13. We note, first of all, that for any p and p', we have, according to (3.40),

$$p(\xi_t(p') - \xi_t(p)) \geqslant 0, \qquad p'(\xi_t(p) - \xi_t(p')) \geqslant 0,$$

and therefore,

$$0 \leqslant p(\xi_t(p') - \xi_t(p)) \leqslant (p - p')(\xi_t(p') - \xi_t(p)).$$

Since the function $\xi_t(p)$ is continuous, we have (considering t as fixed) $\xi_t(p') - \xi_t(p) \to 0$ as $p' \to p$, and therefore

$$p(\xi_t(p') - \xi_t(p)) = |p - p'|\eta(p - p'),$$

where $\eta(p - p') \to 0$ as $p' \to p$. In particular, setting $p' = p + he^{(i)}$, where $e^{(i)}$ is the ith coordinate vector and h is a real number, we obtain

$$\lim_{h \to 0} p \frac{\xi_t(p + he^{(i)}) - \xi_t(p)}{h} = 0.$$

Taking this relation into consideration, we find

$$
\begin{aligned}
\frac{\partial}{\partial p_i}(p\xi_t(p)) &= \lim_{h \to 0} \frac{(p + he^{(i)})\xi_t(p + he^{(i)}) - p\xi_t(p)}{h} \\
&= \lim_{h \to 0} \frac{p[\xi_t(p + he^{(i)}) - \xi_t(p)]}{h} + \lim_{h \to 0} e^{(i)}\xi_t(p + he^{(i)}) \\
&= e^{(i)}\xi_t(p) = \xi_t^i(p),
\end{aligned}
$$

and finally,

$$\frac{\partial f(t, p)}{\partial p_i} = \frac{\partial}{\partial p_i}(p(x_* - \xi_t(p))) = x_*^i - \xi_t^i(p). \tag{3.52}$$

Now, let $p(\tau)$ be a solution of (3.51) with the initial condition $p(0) = p_0 \in D$. Then, by (3.52), we have

$$
\begin{aligned}
\frac{\partial f(t, p(\tau))}{\partial \tau} &= \sum_{i=1}^{n} \frac{\partial f(t, p(\tau))}{\partial p_i} \cdot \frac{dp_i(\tau)}{d\tau} \\
&= -\sum_{i=1}^{n}(x_*^i - \xi_t^i(p(\tau)))(x_*^i - \xi_{F(p(\tau))}^i(p(\tau))).
\end{aligned}
\tag{3.53}
$$

Hence it is seen that if $p(\tau) \notin H_*$ [and consequently $x_* - \xi_{F(p(\tau))}(p(\tau)) \neq 0$], then the derivative $\partial f(t, p(\tau))/\partial \tau$ is negative for $t = F(p(\tau))$, and therefore also for t close to $F(p(\tau))$. In other words, the function $f(t, p(\tau))$ decreases in τ for values of t close to $F(p(\tau))$. Therefore, for a small positive value of h, we have

$$f(F(p(\tau)), p(\tau)) = 0, \qquad f(F(p(\tau)), p(\tau + h)) < 0.$$

Comparing the relations

$$f(F(p(\tau)), p(\tau + h)) < 0, \qquad f(F(p(\tau + h)), p(\tau + h)) = 0$$

and recalling that the function $f(t, p)$ increases in t, we find

$$F(p(\tau)) < F(p(\tau + h)),$$

that is, $F(p(\tau))$ *is a monotonically increasing function of τ for $p(\tau) \notin H_*$.*

It is easy to conclude from this that the solution $p(\tau)$ of (3.51) never leaves the half-space D (if $p_0 \in D$). Otherwise, we would have $p(\tau)x_* = 0$ at a certain time τ, that is, $F(p(\tau)) = 0$, which is not possible since $F(p_0) > 0$ and the function $F(p(\tau))$ increases monotonically.

Furthermore, we have

$$\frac{d}{d\tau}(p(\tau)p(\tau)) = 2p(\tau)\frac{dp(\tau)}{d\tau}$$
$$= -2p(\tau)(x_* - \xi_{F(p(\tau))}(p(\tau))) = -2f(F(p(\tau)), p(\tau)) \equiv 0,$$

from which it follows that $|p(\tau)| = $ const, that is, the solution $p(\tau)$ lies on a sphere with center at the origin. Thus, $p(\tau)$ is defined for all $\tau > 0$ and lies in a bounded portion of space; consequently, $p(\tau)$ has ω-limiting points. Let p_* be an arbitrary ω-limiting point of $p(\tau)$. We shall prove that $p_* \in H_*$.

Let us suppose the contrary: $p_* \notin H_*$. Then $x_* - \xi_{F(p_*)}(p_*) \neq 0$, and therefore, by (3.53), there exist a positive constant M and an $r > 0$ such that $\partial f(t, p(\tau))/\partial \tau < -M$ provided $|p_* - p(\tau)| < r$ and $|t - F(p^*)| < r$. Furthermore, by the boundedness of the convex body V_{t_*} [which contains all the points $\xi_{F(p)}(p)$], there exists a positive number N such that $|x_* - \xi_{F(p)}(p)| < N$ for any p. Let us now choose a $\tau_1 > 0$ for which

$$|p_* - p(\tau_1)| < \frac{\tau}{2}, \qquad |F(p_*) - F(p(\tau_1))| < r \qquad (3.54)$$

[such a τ_1 exists, since p_* is an ω-limiting point of the trajectory $p(\tau)$]. Then the point $p(\tau)$, for which $|dp(\tau)/d\tau| < N$ [by (3.51)], will be displaced during the time interval $\tau_1 \leqslant \tau \leqslant \tau_1 + h$, where $h = r/2N$, by a distance less than $N(r/2N) = r/2$, and therefore will not leave the r-neighborhood of the point p_* [see the first relation in (3.54)]:

$$|p_* - p(\tau)| < r \qquad \text{for} \qquad \tau_1 \leqslant \tau \leqslant \tau_1 + h.$$

Considering, in addition, the second relation in (3.54), we obtain, by definition of the number r,

$$\frac{\partial f(F(p(\tau_1)),\, p(\tau))}{\partial \tau} < -M \qquad \text{for} \qquad \tau_1 \leqslant \tau \leqslant \tau_1 + h.$$

Integrating, we find

$$f(F(p(\tau_1)),\, p(\tau_1 + h)) < f(F(p(\tau_1)),\, p(\tau_1)) - Mh = -Mh$$

[since $f(F(p(\tau_1)),\, p(\tau_1)) = 0$]. Thus, if τ_1 satisfies the relations (3.54), then

$$f(F(p(\tau_1)),\, p(\tau_1 + h)) < -Mh. \tag{3.55}$$

Furthermore, by (3.38), we have

$$\frac{\partial \xi_t^i(p)}{\partial t} = -\psi^i(t) B u(t, p), \qquad i = 1, 2, \ldots, n,$$

and therefore [see (3.47)],

$$\frac{\partial f(t, p)}{\partial t} = -\sum_{i=1}^{n} p_i \frac{\partial \xi_t^i(p)}{\partial t} = \sum_{i=1}^{n} p_i(\psi^i(t) B u(t, p)). \tag{3.56}$$

It is seen from this formula that the derivative $\partial f/\partial t$ is bounded if t and p are bounded. Thus, there exists a positive constant P such that

$$\frac{\partial f(t, p)}{\partial t} < P \qquad \text{for} \qquad 0 < t < t_* \qquad \text{and any } p \in S. \tag{3.57}$$

In particular,

$$\frac{\partial f(F(p(\tau)),\, p(\tau_1 + h))}{\partial t} < P$$

from which, integrating, we obtain

$$f(F(p(\tau_1 + h)),\, p(\tau_1 + h)) - f(F(p(\tau_1)),\, p(\tau_1 + h))$$
$$< P\{F(p(\tau_1 + h)) - F(p(\tau_1))\}.$$

Thus, by (3.49) and the inequality (3.55), we have

$$P\{F(p(\tau_1 + h)) - F(p(\tau_1))\} > -f(F(p(\tau_1)),\, p(\tau_1 + h)) > Mh,$$

that is,

$$F(p(\tau_1 + h)) - F(p(\tau_1)) > \frac{M}{P} h \tag{3.58}$$

[if the conditions (3.54) are satisfied].

Since p_* is an ω-limiting point of $p(\tau)$, the relation $F(p_*) = \lim_{\tau \to \infty} F(p(\tau))$

holds by virtue of the continuity of the function $F(p)$. On the other hand, for any $\tau > 0$ one can always find a $\tau_1 > \tau$ such that the relation (3.54) will be satisfied (since p_* is an ω-limiting point), and therefore, by (3.58),

$$F(p(\tau)) < F(p(\tau_1)) < F(p(\tau_1 + h)) - \frac{Mh}{P} < F(p_*) - \frac{Mh}{P}.$$

Hence,

$$\lim_{\tau \to \infty} F(p(\tau)) \leqslant F(p_*) - \frac{Mh}{P}.$$

But this contradicts the relation $F(p_*) = \lim_{\tau \to \infty} F(p(\tau))$ which proves, in fact, that $p_* \in H_*$. The theorem has been proved.

REMARK.   It follows from the equation $f(F(p), p) = 0$ which defines the function $F(p)$, that if the derivative $(\partial f/\partial t)\big|_{t=F(p)}$ [see (3.56)] is different from zero, then the function $F(p)$ has a derivative which can be determined from the relation

$$\frac{\partial}{\partial p_i}(f(F(p), p)) = \frac{\partial f}{\partial t} \cdot \frac{\partial F(p)}{\partial p_i} + \frac{\partial f}{\partial p_i} = 0,$$

that is,

$$\frac{\partial F}{\partial p_i} = -\frac{\dfrac{\partial f}{\partial p_i}}{\dfrac{\partial f}{\partial t}} = -\frac{x_*^i - \xi_{F(p)}^i(p)}{\dfrac{\partial f(t, p)}{\partial t}}\Bigg|_{t=F(p)}$$

[see (3.52)]. Hence we obtain

$$\operatorname{grad} F(p) = \left\{ \frac{\partial F}{\partial p_1}, \frac{\partial F}{\partial p_2}, \cdots, \frac{\partial F}{\partial p_n} \right\} = -\frac{x_* - \xi_{F(p)}(p)}{\dfrac{\partial f(t, p)}{\partial t}}\Bigg|_{t=F(p)}$$

(if the denominator is different from zero). Thus (disregarding the possibility that the denominator may vanish), we see that the vector $\operatorname{grad} F(p)$ does not vanish at any point $p \notin H_*$, and that any solution of (3.51) is tangent to the vector $\operatorname{grad} F(p(\tau))$ [since the right-hand side of (3.51) is proportional to the vector $\operatorname{grad} F$]. In other words, except for the maximum at the points of H_*, the function $F(p)$ does not have any other ("false") local maxima and minima in D, and the trajectories of (3.51) are flow lines for the field of vectors $\operatorname{grad} F(p)$, that is, they correspond to the method of steepest ascent for calculating the maximum of the function $F(p)$. (The minimum of this function, which equals zero, is attained on the hyperplane bounding the half-space D.)

The meaning of the theorem proved above is that the vector $p(\tau)$ lies very close to the set H_* for a sufficiently large value of τ, and therefore the vector $p(\tau)$ may be considered as an approximation to a vector corresponding to the point x_*:

$$\xi_{t_*}(p(\tau)) \approx x_* \qquad \text{(for large values of } \tau).$$

36. Determination of the Initial Values of the Auxiliary Variables: Eaton's Iterative Process. A question naturally arises which concerns the manner in which a solution $p(\tau)$ of (3.51) may be found. A process for solving this equation can be realized by computers. Namely, the function $\xi_t(p)$ may be computed directly from (3.38). It is also possible to find $\xi_t^i(p)$ of formula (3.38) by making use of an analog computer similar to that described in § 8.31. [The use of the first two "boxes" shown in Figure 102 makes it possible to find the corresponding optimal control $u(t)$ which appears in (3.38) from the initial value $\psi_0(p)$; the vector $\psi^i(t)$ can also be computed directly by the first "box" depicted in Figure 102; finally, the scalar product and the integration can be accomplished by ordinary methods. Of course, in analog computers used for the above purpose, the course of "time" must be more rapid than for the real object (this is achieved by appropriate choice of the parameters): in fact, all the computations must be completed during a short period of time while the object has not yet deviated noticeably from the initial state x_*.]

The function $\xi_t(p)$ need be computed (for every p) only up to the value $t = F(p)$ which is the root of the equation $f(t, p) = 0$ [see (3.47)]; as a result, we obtain the expression $\xi_{F(p)}(p)$ which appears in the right-hand side of (3.51). Thus, the right-hand side of (3.51) can be calculated for every $p \in D$.

Furthermore, in solving (3.51), it is usually necessary to apply a difference approximation. Namely, instead of computing the entire continuous solution $p(\tau)$ of (3.51), a "discrete solution" is constructed, that is, a sequence $p_0, p_1, p_2, \ldots$. Moreover, each successive approximation p_{k+1} is formed from the preceding approximation p_k in the following manner. We choose a "step" $\Delta\tau_k > 0$ and determine p_{k+1} by replacing the derivative $dp/d\tau$ by the difference quotient

$$\frac{\Delta p_k}{\Delta\tau_k} = \frac{p_{k+1} - p_k}{\Delta\tau_k} = -[x_* - \xi_{F(p_k)}(p_k)]. \tag{3.59}$$

From this relation we obtain the formula for finding the next approximation:

$$p_{k+1} = p_k - \Delta\tau_k [x_* - \xi_{F(p_k)}(p_k)]. \tag{3.60}$$

It is natural to expect that if p_k is an approximate value of the solution $p(\tau)$ at time τ_k, then the vector p_{k+1}, defined by (3.60), will be an approximate value of the same solution $p(\tau)$ at time $\tau_{k+1} = \tau_k + \Delta\tau_k$. Therefore, it can be anticipated that the sequence p_0, p_1, p_2, . . . , calculated recursively from (3.60), will behave in approximately the same way as the solution $p(\tau)$ and, in particular, will approach the set H_*.

Of course, in order to justify the above anticipation, it is necessary to choose the sequence of "steps" $\Delta\tau_0$, $\Delta\tau_1$, $\Delta\tau_2$, . . . with care. In fact, if these "steps" are too large, then the difference quotient $\Delta p_k/\Delta\tau_k$ may differ substantially from the derivative $dp/d\tau$, and the "discrete" solution p_0, p_1, p_2, . . . may depart substantially from the "continuous" solution $p(\tau)$ (and even leave the half-space D). If, however, the "steps" are too small, we will move too slowly along the solution $p(\tau)$, performing unproductively a huge volume of computations (or even making no headway, like Achilles trying to catch up with a tortoise).

In addition, we note the following fact. The relation $|p(\tau)| = \text{const}$, is satisfied, as we have seen, for every solution $p(\tau)$ of (3.51), but we cannot assert that the vectors of the sequence p_0, p_1, p_2, . . . will have identical lengths since (3.59) differs from (3.51) even for a small value of the "step" $\Delta\tau_k$. This could lead to such undesirable consequences as an unbounded increase (or decrease) in the lengths $|p_0|$, $|p_1|$, $|p_2|$, . . . of the vectors of the sequence under consideration. In order to avoid these consequences, let us "improve" somewhat the recurrence relation (3.60) by replacing the vector contained in the right-hand side by the unit vector in the same direction. In other words, let us replace (3.60) by the following:

$$p_{k+1} = \frac{q_{k+1}}{|q_{k+1}|}, \text{ where } q_{k+1} = p_k - \Delta\tau_k [x_* - \xi_{F(p_k)}(p_k)], \quad (3.61)$$

where the use of (3.61) [instead of (3.60)] in constructing the sequence of approximations p_0, p_1, p_2, . . . yields a "discrete" solution lying entirely on the sphere S [as does the "continuous" solution $p(\tau)$], that is, it consists of vectors having the same length (equal to unity).

The following theorem provides an exceedingly efficient method for selecting the sequence of "steps" $\Delta\tau_0$, $\Delta\tau_1$, $\Delta\tau_2$, . . . ensuring a rapid approach of the sequence p_0, p_1, p_2, . . . to the set H_*.

Theorem 3.14. *Let p_0 be an arbitrary unit vector in the half-space D. (For example, one may take p_0 to be the vector $-x_*/|x_*|$, which obviously lies in the half-space D.) Suppose that the vectors p_0, p_1, . . . , p_k lie in D. Set*

$$q_{k+1}^{(m)} = p_k - 2^{-m}(x_* - \xi_{F(p_k)}(p_k)), \quad (3.62)$$

where m is the smallest non-negative integer for which the vector $p_{k+1} = q_{k+1}^{(m)}/|q_{k+1}^{(m)}|$ satisfies the inequality

$$p_{k+1}(x_* - \xi_{F(p_k)}(p_{k+1})) < -2^{-(m+1)}|x_* - \xi_{F(p_k)}(p_k)|^2. \quad (3.63)$$

It turns out that either $p_k \in H_*$ for some k, or the above inductive construction leads to an infinite sequence of vectors p_0, p_1, p_2, . . . which lie in the half-space D and have the following properties:

(A)   *The numbers $F(p_0)$, $F(p_1)$, $F(p_2)$, . . . form a monotonically increasing sequence converging to the number t^*.*

(B) $\lim\limits_{k \to \infty} \xi_{F(p_k)}(p_k) = x_*.$

(C)   *The sequence of vectors p_0, p_1, p_2, . . . approaches the set H_* (since any limiting vector of this sequence belongs to H_*).*

PROOF.   First of all, we shall prove that if p is a unit vector in the half-space D, and $h > 0$, then

$$f\left(F(p), \frac{p - h(x_* - \xi_{F(p)}(p))}{|p - h(x_* - \xi_{F(p)}(p))|}\right) = -h|x_* - \xi_{F(p)}(p)|^2 + o(h), \quad (3.64)$$

where $o(h)$ has the property that $o(h)/h$ tends uniformly (that is, independently of p) to zero as $h \to 0$. In fact, since the vectors p and $x_* - \xi_{F(p)}(p)$ are orthogonal [see (3.47), (3.49)],

$$|p - h(x_* - \xi_{F(p)}(p))| = 1 + o_1(h),$$

where

$$\lim_{h \to 0} \frac{o_1(h)}{h} = 0$$

(uniformly in p). Since the derivatives $\partial f(t, p)/\partial p_i$ are bounded for $0 \leqslant t \leqslant t^*$ [see (3.52)], it follows that

$$f\left(F(p), \frac{p - h(x_* - \xi_{F(p)}(p))}{|p - h(x_* - \xi_{F(p)}(p))|}\right) = f(F(p), p - h(x_* - \xi_{F(p)}(p))) + o_2(h).$$

Furthermore, making use of (3.52), we find (for some θ, $0 < \theta < 1$)

$$f(F(p), p - h(x_* - \xi_{F(p)}(p)))$$
$$= f(F(p), p - h(x_* - \xi_{F(p)}(p))) - f(F(p), p)$$
$$= -h \sum_{i=1}^{n} \frac{\partial f(F(p), p - \theta h(x_* - \xi_{F(p)}(p)))}{\partial p_i} (x_* - \xi_{F(p)}(p))^i$$
$$= -h(x_* - \xi_{F(p)}(p - \theta h(x_* - \xi_{F(p)}(p)))(x_* - \xi_{F(p)}(p))$$
$$= -h|x_* - \xi_{F(p)}(p)|^2 + o_3(h).$$

By the same token, (3.64) has been established.

It follows from (3.64) that if $x_* - \xi_{F(p)}(p) \neq 0$ (where p is fixed), then for a sufficiently small h, we have

$$f\left(F(p), \frac{p - h(x_* - \xi_{F(p)}(p))}{|p - h(x_* - \xi_{F(p)}(p))|}\right) < -\frac{1}{2}h|x_* - \xi_{F(p)}(p)|^2.$$

In particular, setting $p = p_k$, $h = 2^{-m}$, we find that if $p_k \notin H_*$ [that is, $x_* - \xi_{F(p_k)}(p_k) \neq 0$], then

$$f\left(F(p_k), \frac{q_{k+1}^{(m)}}{|q_{k+1}^{(m)}|}\right) < -2^{-m-1}|x_* - \xi_{F(p_k)}(p_k)|^2$$

for a sufficiently large m [see (3.62)]. Hence, it is clear that for $p_k \notin H_*$, (3.63) makes it possible to determine the vector p_{k+1}.

Let us show that $p_{k+1} \in D$. Since $p\xi_t(p) < 0$ (see p. 126), we have, in particular, $p_{k+1}\xi_{F(p_k)}(p_{k+1}) < 0$. But, according to (3.63), $p_{k+1}(x_* - \xi_{F(p_k)}(p_{k+1})) < 0$, and therefore $p_{k+1}x_* < p_{k+1}\xi_{F(p_k)}(p_{k+1}) < 0$, that is, $p_{k+1} \in D$.

It is clear from the above that the inductive construction described in Theorem 3.14 leads either (at some step) to a vector $p_k \in H_*$, or makes it possible to construct an infinite sequence of vectors $p_0, p_1, p_2, \ldots$ lying in the half-space D. We shall prove that in the latter case properties (A), (B), and (C) are satisfied.

We have

$$-f(F(p_k), p_{k+1}) = f(F(p_{k+1}), p_{k+1}) - f(F(p_k), p_{k+1})$$
$$= (F(p_{k+1}) - F(p_k))\frac{\partial f(\theta, p_{k+1})}{\partial t},$$

where the number θ is contained between $F(p_k)$ and $F(p_{k+1})$. Since $-f(F(p_k), p_{k+1}) > 0$ [see (3.63)] and $\partial f/\partial t \geqslant 0$ [$f(t, p)$ increases with t], the difference $F(p_{k+1}) - F(p_k)$ is positive. Therefore, according to (3.57),

$$-f(F(p_k), p_{k+1}) < P(F(p_{k+1}) - F(p_k)),$$

and, consequently,

$$F(p_{k+1}) - F(p_k) > \frac{-f(F(p_k), p_{k+1})}{P}. \tag{3.65}$$

Since $F(p_{k+1}) - F(p_k) > 0$ (for any k), the sequence $F(p_0), F(p_1), F(p_2), \ldots$ increases monotonically. Suppose that this sequence does not converge to t^*, that is, $\lim_{k\to\infty} F(p_k) = t' < t^*$. Then all the points $\xi_{F(p_k)}(p_k)$ belong to the body $V_{t'}$, which does not contain the point x_*, and therefore, for every k we have $|x_* - \xi_{F(p_k)}(p_k)| > \delta$, where δ is a positive number. Let us choose a number $h_0 > 0$ such that the inequality $o(h)/h < \delta^2/2$ is satisfied for $h < h_0$ [see (3.64)]. Then

$$f\left(F(p), \frac{p - h(x_* - \xi_{F(p)}(p))}{|p - h(x_* - \xi_{F(p)}(p))|}\right) < -h|x_* - \xi_{F(p)}(p)|^2 + h\frac{\delta^2}{2}$$

for $h < h_0$.

In particular, if μ is a positive integer such that $2^{-\mu} < h_0$, then

$$f\left(F(p_k), \frac{q_{k+1}^{(\mu)}}{|q_{k+1}^{(\mu)}|}\right) < -2^{-\mu}|x_* - \xi_{F(p_k)}(p_k)|^2 + 2^{-\mu}\frac{\delta^2}{2}$$
$$< -2^{-\mu}|x_* - \xi_{F(p_k)}(p_k)|^2 + 2^{-\mu-1}|x_* - \xi_{F(p_k)}(p_k)|^2$$
$$= -2^{-\mu-1}|x_* - \xi_{F(p_k)}(p_k)|^2.$$

Thus, according to the definition of p_{k+1} [see (3.63)], we have $p_{k+1} = q_{k+1}^{(m)}/|q_{k+1}^{(m)}|$, where $m \leqslant \mu$. Hence, we obtain on the basis of (3.63),

$$f(F(p_k), p_{k+1}) = p_{k+1}(x_* - \xi_{F(p_k)}(p_{k+1}))$$
$$< -2^{-m-1}|x_* - \xi_{F(p_k)}(p_k)|^2 < -2^{-m-1}\delta^2 < -2^{-\mu-1}\delta^2,$$

that is, $-f(F(p_k), p_{k+1}) > 2^{-\mu-1}\delta^2$. Now, it follows from (3.65) that $F(p_{k+1}) - F(p_k) > (1/P)2^{-\mu-1}\delta^2$ (for any k). But this obviously contradicts the boundedness of the sequence $F(p_0)$, $F(p_1)$, $F(p_2)$, ... The contradiction shows that $\lim_{k \to \infty} F(p_k) = t^*$. Thus, assertion (A) has been proved.

If it is assumed that statement (B) does not hold, that is, that the equation $\lim_{k \to \infty} |x_* - \xi_{F(p_k)}(p_k)| = 0$ is not valid, then there exists $\delta > 0$ such that $|x_* - \xi_{F(p_k)}(p_k)| > \delta$ for an infinite number of values of k. Hence, as before, we find that $F(p_{k+1}) - F(p_k) > (1/P)2^{-\mu-1}\delta^2$ (for an infinite number of values of k). This again contradicts the boundedness of the function $F(p)$.

Finally, let p^* be any limiting vector of the sequence p_0, p_1, p_2, . . . (that is, there exists a subsequence of this sequence which converges to p^*). Then, by virtue of the continuity of F, we have $F(p^*) = t^*$, that is, $p^* \in H_*$. Accordingly, statement (C) has been established.

Theorem 3.14 has been completely proved.

REMARK 1. Suppose that after k steps of the inductive construction described in Theorem 3.14 we have arrived at the vector p_k. Let us set $F(p_k) = t_k$ and let $u_k(t)$ denote the control which transfers the point $\xi_{t_k}(p_k)$ to the origin in time t_k [that is, the optimal control for the point $\xi_{t_k}(p_k)$], and let $x_k(t)$ denote the corresponding optimal trajectory: $x_k(0) = \xi_{t_k}(p_k)$, $x_k(t_k) = 0$. Finally, let $x_k^*(t)$ be the trajectory corresponding to the same control $u_k(t)$ and emanating from the point x_*, so that $x_k^*(0) = x_*$. Then for $0 \leqslant t \leqslant t_k$ we have

$$\frac{d}{dt}x_k(t) = Ax_k(t) + Bu_k(t), \qquad \frac{d}{dt}x_k^*(t) = Ax_k^*(t) + Bu_k(t),$$

and therefore,

$$\frac{d}{dt}(x_k^*(t) - x_k(t)) = A(x_k^*(t) - x_k(t)), \qquad 0 \leqslant t \leqslant t_k.$$

Thus, $x_k^*(t) - x_k(t)$ is a solution of the homogeneous equation $\dot{x} = Ax$. But for every solution $y(t)$ of this homogeneous equation the exponential estimate $|y(t)| \leqslant |y(0)|e^{t\|A\|}$ holds for $t > 0$, where $\|A\|$ is the norm* of the matrix A. Hence, we obtain

$$|x_k^*(t_k)| = |x_k^*(t_k) - x_k(t_k)| \leqslant |x_k^*(0) - {}_k(0)|e^{t_k\|A\|}$$

$$= |x_* - \xi_{F(p_k)}(p_k)|e^{t_k\|A\|}. \quad \textbf{(3.66)}$$

The right-hand side of (3.66) tends to zero as $k \to \infty$, since $t_k \leqslant t^*$ and $|x_* - \xi_{F(p_k)}(p_k)| \to 0$. Consequently, $x_k^*(t_k) \to 0$ as $k \to \infty$. Thus, (3.66) makes it possible to judge the accuracy of the approximation: the computational process, described in Theorem 3.14, is terminated when the trajectory $x_k^*(t)$ reaches the origin within the required accuracy. If the required accuracy is achieved after k steps, then the vector p_k may be taken as the initial value ψ_0 corresponding to the point x_*; this yields the desired optimal process.

REMARK 2. In the above theorem we first found the vectors p_0, p_1, ..., p_k, and then determined the vector p_{k+1} by (3.61) choosing $\Delta\tau_k$ as the largest of the numbers $1, \frac{1}{2}, \frac{1}{4}, \frac{1}{8}, \ldots$ which satisfy (3.63). We could, of course, use a sequence from which $\Delta\tau_k$ is chosen by some other method. For example, we could take the two-sided infinite sequence $\ldots, 8, 4, 2, 1,$ $\frac{1}{2}, \frac{1}{4}, \frac{1}{8}, \ldots$, that is, in determining $\Delta\tau_k$ one seeks the smallest integer m (not necessarily non-negative!), for which (3.63) is satisfied. In other words, if $m = 0$ satisfies the relation (3.63), then one can try to use $m = -1$, $m = -2, \ldots$. In the original work of Eaton [5], the sequence $\{\Delta\tau_k\}$ is chosen in another way: he takes $\Delta\tau_k$ (disregarding slight differences in notation) to be the largest of the numbers

$$2^{-m}\frac{|x_* - \xi_{F(p_{k-1})}(p_{k-1})|}{|x_* - \xi_{F(p_k)}(p_k)|}\Delta\tau_{k-1}, \quad m = 0, 1, 2, \ldots$$

for which the inequality

$$p_{k+1}(x_* - \xi_{F(p_k)}(p_{k+1})) < 0$$

is satisfied [in the notation of (3.61)].

However, the author of this book was not able to derive from Eaton's work a correct proof of the convergence of such a process [that is, the proof of statements of the type (A), (B), and (C)].

REMARK 3. The method presented above makes it possible to realize a control process which (for a sufficiently fast computer) approaches the optimal process. Namely, let h denote a time sufficient for a computer to

---

* The norm of a matrix A is the smallest number M such that $|Ax| \leqslant M|x|$ for any vector x. From the differential equation $\dot{x} = Ax$, it is easy to derive the inequality $d|x|/dt \leqslant \|A\| \, |x|$, which, in fact, yields the above-mentioned exponential estimate.

find the value of p_{k+1} if the vector p_k is known. We supply to the optimal regulator (described in § 8.31) the initial state x_0 of the object and the value $\psi_0 = p_0$ (since we do not have a better value for ψ_0 at this time). The optimal regulator computes the value $u(t)$ under whose action the object will move for h seconds. During this time, the computer, proceeding from the initial value x_0, finds the next approximation p_1 for the vector ψ_0. The new initial value x_0' is measured at time h, and the computed value $\psi_0 = p_1$ is supplied at the input of the optimal regulator. The motion of the object proceeds for another h seconds under the action of a newly computed control. Meanwhile, the computer, proceeding from the initial value x_0' and from the value p_1, finds the vector p_2. At time $2h$, the phase coordinate of the object, x_0'', is measured, and the subsequent motion is realized with the initial values x_0'' and $\psi_0 = p_2$, and so on (Figure 116).

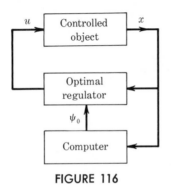

FIGURE 116

In other words, we gradually "improve" the control process without waiting for the execution of all the computations specified by Theorem 3.14.

§ 10. THE SOLUTION OF THE SYNTHESIS PROBLEM FOR A LINEAR SYSTEM OF THE SECOND ORDER

37. Simplification of the Equations of a Linear Controlled Object.
The form (3.4), in which the equation of motion of an object is generally written in the linear problem, is frequently inconvenient and it is expedient to make use of some simplifications. We shall indicate here the standard simplifications which arise from a change of coordinates.

1° First of all, let us consider the question of a change of coordinates in the phase space X of the controlled object. Suppose that the coordinates $x^1, \ldots, x^n$ in X are to be replaced by new coordinates $y^1, \ldots, y^n$

defined by the relations

$$x^i = \sum_j p^i_j y^j, \qquad y^i = \sum_j q^i_j x^j \tag{3.67}$$

[where $P = (p^i_j)$ and $Q = (q^i_j)$ are reciprocal matrices]. It is clear that under such a substitution the linear system (3.4) is transformed into the new linear system

$$\dot{y}^i = \sum_{\alpha=1}^n c^i_\alpha y^\alpha + \sum_{\beta=1}^r d^i_\beta u^\beta, \qquad i = 1, 2, \ldots, n, \tag{3.68}$$

whose coefficients can be readily calculated:

$$\dot{y}^i = \sum_{j=1}^n q^i_j \dot{x}^j = \sum_{j=1}^n q^i_j \left(\sum_{\gamma=1}^n a^j_\gamma x^\gamma + \sum_{\beta=1}^r b^j_\beta u^\beta \right)$$

$$= \sum_{j,\gamma} q^i_j a^j_\gamma x^\gamma + \sum_{j,\beta} q^i_j b^j_\beta u^\beta$$

$$= \sum_{j,\gamma,\alpha} q^i_j a^j_\gamma p^\gamma_\alpha y^\alpha + \sum_{j,\beta} q^i_j b^j_\beta u^\beta = \sum_{\alpha=1}^n c^i_\alpha y^\alpha + \sum_{\beta=1}^r d^i_\beta u^\beta.$$

Thus

$$c^i_\alpha = \sum_{j,\gamma} q^i_j a^j_\gamma p^\gamma_\alpha, \qquad d^i_\beta = \sum_j q^i_j b^j_\beta. \tag{3.69}$$

Reverting to vector notation, we can say that the above change of coordinates transforms (3.8) into the equation $\dot{y} = Cy + Du$ [see (3.68)], where the matrices C and D are expressed in terms of the matrices A, B, P, and Q by the formulas [see (3.69)]

$$C = QAP, \qquad D = QB.$$

It is obvious that under such a substitution, every process $x(t)$, $u(t)$ satisfying $\dot{x} = Ax + Bu$ turns into a process $y(t)$, $u(t)$ satisfying $\dot{y} = Cy + Du$ (and conversely). Moreover, since t is not transformed, the above substitution converts optimal processes of $\dot{x} = Ax + Bu$ into optimal processes of $\dot{y} = Cy + Du$ (and conversely). In particular, with the aid of the change of coordinates (3.67), the synthesis of optimal controls for $\dot{x} = Ax + Bu$ is converted into the synthesis of optimal controls for $\dot{y} = Cy + Du$. Thus, if $\dot{y} = Cy + Du$ turns out to be simpler and if it is possible to synthesize optimal controls for this equation, then from this synthesis, it is possible [with the aid of the affine transformation (3.67)] to obtain also the synthesis for the original equation $\dot{x} = Ax + Bu$. Indeed, the meaning of the change of coordinates (3.67) consists in this: it makes it possible to replace the matrix A by the transformed matrix $C = QAP$ while producing simply an affine distortion of the "phase portrait" of the synthesis of optimal controls.

Thus, transformation (3.67) can be used to simplify the matrix A composed of the coefficients of the phase coordinates.

2° Let us assume that the matrix A in $\dot{x} = Ax + Bu$ has already been reduced to its simplest form (by the method described above). Let us now show how to simplify the matrix B composed of the coefficients of the control parameters.

For this purpose, we set

$$v^i = \sum_{\beta=1}^{r} b_\beta^i u^\beta, \qquad i = 1, 2, \ldots, n. \tag{3.70}$$

This means that the r control parameters $u^1, \ldots, u^r$ are to be replaced by the n new parameters $v^1, \ldots, v^n$ and, as a result, (3.4) is transformed into the following:

$$\dot{x}^i = \sum_{\alpha=1}^{n} a_\alpha^i x^\alpha + v^i, \qquad i = 1, 2, \ldots, n,$$

or, in vector notation,

$$\dot{x} = Ax + v.$$

It is only necessary to clarify the limits within which the control point $v = (v^1, v^2, \ldots, v^n)$ may vary. It is convenient to assume that $v = (v^1, v^2, \ldots, v^n)$ lies in the same phase space X as the point $x = (x^1, \ldots, x^n)$.

The relations (3.70) define a linear mapping of the r-dimensional $u^1, \ldots, u^r$-space into the phase space X (compare p. 88). The image of U under the mapping (3.70) is a convex polyhedron in X which will be denoted by V.

Thus, we obtain two linear systems

$$\dot{x} = Ax + Bu, \qquad u \in U; \tag{3.71}$$
$$\dot{x} = Ax + v, \qquad v \in V. \tag{3.72}$$

We shall show that these two linear systems are equivalent, so that in searching for optimal processes one may confine oneself to considering the simpler system (3.72).

As an example, let us consider the second-order system

$$\dot{x}^1 = a_1^1 x^1 + a_2^1 x^2 + b_1^1 u^1 + b_2^1 u^2 + b_3^1 u^3,$$
$$\dot{x}^2 = a_1^2 x^1 + a_2^2 x^2 + b_1^2 u^1 + b_2^2 u^2 + b_3^2 u^3$$

with three control parameters u^1, u^2, u^3 which vary in the cube $-1 \leqslant u^i \leqslant 1$. The above simplification converts this system into

$$\dot{x}^1 = a_1^1 x^1 + a_2^1 x^2 + v^1,$$
$$\dot{x}^2 = a_1^2 x^1 + a_2^2 x^2 + v^2,$$

in which the parameters v^1, v^2 are related to u^1, u^2, u^3 by (3.2). The polyhedron V in which the control point $v = (v^1, v^2)$ may vary is the image of a cube under the linear mapping (3.2), that is, it has the form of a hexagon (Figure 78). This example provides a good illustration of the remark made previously to the effect that it is not convenient to confine ourselves only to parallelepipeds (as opposed to arbitrary convex polyhedra U).

We proceed to the proof of the equivalence of (3.71) and (3.72). This equivalence must be understood in the sense that for any process $x(t)$, $u(t)$ satisfying (3.71), a piecewise continuous function $v(t)$ with values in the polyhedron V can be found such that the process $x(t)$, $v(t)$ satisfies (3.72) (and conversely). Moreover, an optimal process $x(t)$, $u(t)$ corresponds to an optimal process $x(t)$, $v(t)$ (and conversely). Finally, if the condition of general position is satisfied for (3.71) then it is likewise satisfied for (3.72).

In this proof, the mapping defined by (3.70) will be denoted by B. In other words, if $u = (u^1, \ldots, u^r)$ is an arbitrary point, then the point $v = (v^1, \ldots, v^n)$ with coordinates defined by (3.70) will be denoted by Bu.

Let $x(t)$, $u(t)$ be an arbitrary process satisfying (3.71). We set $v(t) = Bu(t)$. Since $u(t)$ is piecewise continuous, $v(t)$ is also piecewise continuous. Furthermore, $u(t) \in U$ for all t, and therefore $v(t) \in V$. Finally, since the process $x(t)$, $u(t)$ satisfies (3.71), that is, $\dot{x}(t) = Ax(t) + Bu(t)$, it follows that $\dot{x}(t) = Ax(t) + v(t)$, that is, the process $x(t)$, $v(t)$ satisfies (3.72). Thus, the transition from (3.71) to (3.72) is accomplished directly.

It is somewhat more complicated to effect the reverse transition (because of its multivalence; for example, it is seen from Figure 78 that, generally speaking, an infinitely large number of points of the polyhedron U transforms into one point of the polyhedron V under the mapping B). In order to carry out this reverse transition, we perform the following construction. Let $v_1, v_2, \ldots, v_s$ be the vertices of the polyhedron V. For each vertex v_i ($i = 1, 2, \ldots, s$), we select a point u_i of the polyhedron U which maps into the point v_i under the mapping B (recall that B is a mapping of U onto the entire polyhedron V). We obtain the points $u_1, u_2, \ldots, u_s$ of U. Now let us divide the polyhedron V into simplices with vertices at the points $v_1, v_2, \ldots, v_s$; this is always possible, being similar to dividing a polygon into triangles (Figure 117) and a polyhedron into tetrahedrons. Let $T_1, T_2, \ldots, T_q$ denote the simplices into which the polyhedron V is divided. Let $v_{i_1}, v_{i_2}, \ldots, v_{i_k}$ be the vertices of simplex T_1; then, the simplex with vertices $u_{i_1}, u_{i_2}, \ldots, u_{i_k}$ will be denoted by S_1. Furthermore, if $v_{j_1}, v_{j_2}, \ldots, v_{j_k}$ are the vertices of T_2, then the simplex with vertices $u_{j_1}, u_{j_2}, \ldots, u_{j_k}$ will be denoted by S_2, and so on. As a result, we obtain simplices $S_1, S_2, \ldots, S_q$ lying in the polyhedron U. It is easy to see that the simplices $S_1, S_2, \ldots, S_q$, taken together, form a polyhedral surface which, under the mapping B, maps in a one-to-one fashion onto the entire polyhedron V. This polyhedral surface U will be denoted by W.

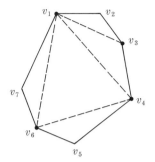

FIGURE 117

At this point it is not difficult to carry out the reverse transition from (3.72) to (3.71). Let $x(t)$, $v(t)$ be a process satisfying (3.72). Since W maps onto V in a one-to-one fashion, it follows that for any t there exists a unique point on W which goes over to the point $v(t)$ under the mapping B; this point of W will be denoted by $u(t)$. We obtain a function $u(t)$ with values in the polyhedron U which, obviously, is piecewise continuous and satisfies the relation $Bu(t) = v(t)$. Since the process $x(t)$, $v(t)$ satisfies (3.72), that is, $\dot{x}(t) = Ax(t) + v(t)$, it follows that $\dot{x}(t) = Ax(t) + Bu(t)$, and therefore, the process $x(t)$, $u(t)$ satisfies (3.71). Accordingly, the transition from (3.72) to (3.71) has been completed and the equivalence of these equations has been established.

If the process $x(t)$, $u(t)$ is optimal, then the corresponding process $x(t)$, $v(t)$ is also optimal (and conversely), since the trajectory $x(t)$ and the time t are the same for both processes.

Finally, let ρ be an arbitrary edge of the polyhedron V. Then there exists an edge π of the polyhedron U which is superimposed under the mapping B on all or part of the edge ρ. In other words, under the mapping B the vector w passing along the edge π goes over into the vector $z = Bw$ passing along the edge ρ. If the condition of general position is satisfied for (3.71), then the vectors Bw, ABw, A^2Bw, . . . , $A^{n-1}Bw$ are linearly independent. In other words, the vectors z, Az, A^2z, . . . , $A^{n-1}z$ are linearly independent, which means that the condition of general position is satisfied for (3.72) [since the unit matrix serves as the matrix B of system (3.72)]. Thus, if the condition of general position is satisfied for (3.71), it is also satisfied for (3.72).

REMARK. We have previously mentioned that the transition from (3.72) to (3.71) is not unique, that is, to a single process $x(t)$, $v(t)$, there corresponds, generally speaking, an infinitely large number of processes $x(t)$, $u(t)$ with the same trajectory $x(t)$. However, for optimal processes, the transition from (3.72) to (3.71) is unique. In fact, for an optimal process $x(t)$, $v(t)$, the control $v(t)$ takes on only values at the vertices v_1, . . . , v_s

of the polyhedron V (Theorem 3.4), and each vertex of V is the image, under B, of a single vertex of the polyhedron U (this can be readily verified from the condition of general position). (Compare Figure 78 in which it can be clearly seen that only one vertex of the cube U maps onto each of the six vertices of the hexagon V.)

38. Solution of the Synthesis Problem in the Case of Complex Eigenvalues.* We shall present here a complete solution of the problem of synthesizing optimal controls for linear objects described by second-order equations. In this case, the phase space X is a plane.

It may be assumed, according to the discussion in the previous section, that the object is described by the system of equations

$$\dot{x}^1 = a_1^1 x^1 + a_2^1 x^2 + v^1,$$
$$\dot{x}^2 = a_1^2 x^1 + a_2^2 x^2 + v^2, \tag{3.73}$$

that the matrix $A = (a_j^i)$ is reduced to its simplest form as a result of a transformation, and that the point $v = (v^1, v^2)$ ranges over a convex "polyhedron" V in the phase plane X. In other words, the "polyhedron" V can be either a segment passing through the origin [Figure 118(a)], or a convex polygon containing the origin in its interior [Figure 118(b)] or on one of its sides [Figure 118(c)]. In the sequel, V will be called a "polygon" (although

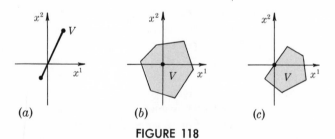

(a)   (b)   (c)

FIGURE 118

it may degenerate into a segment). As we shall see, the general picture of the synthesis depends essentially on whether the eigenvalues of the matrix A are real or complex. The case of complex eigenvalues will be considered in this section.

Let the eigenvalues of the matrix A have the form $\lambda_1 = a + bi$, $\lambda_2 = a - bi$, where $b \neq 0$. We can assume that $b > 0$. It is known [11] that in this case the matrix A can be reduced (by an appropriate change of coordi-

---

* The results in this and the next sections were obtained in the thesis of A. A. Ryvkin under the supervision of the author.

nates in the phase plane) to the form

$$\begin{pmatrix} a & -b \\ b & a \end{pmatrix}. \tag{3.74}$$

Thus, according to the previous section, we should consider the system

$$\begin{aligned} \dot{x}^1 &= ax^1 - bx^2 + v^1, \\ \dot{x}^2 &= bx^1 + ax^2 + v^2. \end{aligned} \tag{3.75}$$

First of all, we consider the corresponding homogeneous system

$$\begin{aligned} \dot{x}^1 &= ax^1 - bx^2, \\ \dot{x}^2 &= bx^1 + ax^2. \end{aligned} \tag{3.76}$$

It can be directly verified that the general solution of this system has the form

$$x^1(t) = ce^{at}\cos(bt + \alpha), \qquad x^2(t) = ce^{at}\sin(bt + \alpha), \tag{3.77}$$

where c and α are arbitrary constants. These formulas are, in fact, the parametric equations of the phase trajectory.

It is most convenient to write these formulas in polar coordinates by setting $x^1 = \rho\cos\varphi$, $x^2 = \rho\sin\varphi$. In this case, (3.77) takes the form

$$\rho = ce^{at}, \qquad \varphi = bt + \alpha. \tag{3.78}$$

Thus, the argument φ varies uniformly with time, that is, the ray from the origin through the phase point $(x^1(t), x^2(t))$ rotates uniformly in the counterclockwise direction with an angular velocity of b radians per second. Eliminating t from (3.78), we obtain the equation of the phase trajectory in polar coordinates:

$$\rho = Ke^{(a/b)\varphi}, \tag{3.79}$$

where $K = ce^{-\alpha a/b}$ is a constant. The phase trajectories have different shapes for $a < 0$, $a = 0$, $a > 0$.

For $a < 0$ we have the trajectories depicted in Figure 119. These trajectories [defined by (3.79)] are called *logarithmic spirals*. Motion along these trajectories is counterclockwise; moreover, as t increases, $\rho \to 0$ (since $a/b < 0$), that is, the phase point approaches the origin. The direction of the points' motion along the phase trajectories is indicated in Figure 119 by arrows. This phase portrait is called a *stable focus*.

For $a = 0$, (3.79) takes the form $\rho = $ const, that is, the phase trajectories are circles (on which the phase points move in the counterclockwise direction). This phase portrait is called a *center* (Figure 120).

Finally, for $a > 0$ we again have logarithmic spirals defined by (3.79). The motion along these spirals occurs, as before, in the counterclockwise

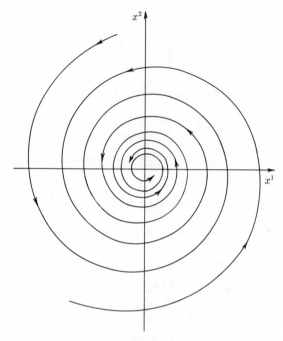

FIGURE 119

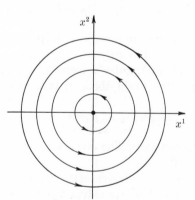

FIGURE 120

direction, but now as t increases, we have $\rho \to \infty$, that is, the phase point moves away from the origin. This phase portrait is called an *unstable focus* (Figure 121).

An important property of phase trajectories (valid in each of the cases depicted in Figures 119, 120, and 121) follows from (3.79). Namely, *any two phase trajectories of system* (3.79) *are obtained from one another by a dilation* (*homothety*) *with center at the origin*. In fact, one obtains two different phase

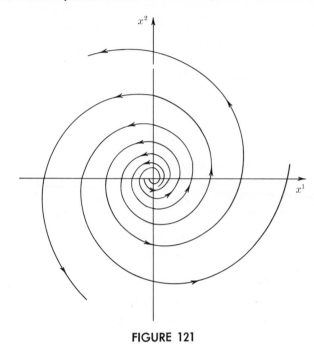

FIGURE 121

trajectories if two different values of K are considered in (3.79):

$$\rho = K_1 e^{(a/b)\varphi}, \qquad \rho = K_2 e^{(a/b)\varphi}.$$

Hence it is seen that subjecting the first trajectory to a dilation with coefficient K_2/K_1 (that is, increasing all its radius vectors ρ by the factor K_2/K_1), the first trajectory will go over into the second trajectory. It is just as simple to prove that *for any dilation with center at the origin and also for any rotation about the origin, each phase trajectory of system* (3.76) *transforms into another trajectory of the same system.*

Let us now consider system (3.75) which differs from (3.76) by the presence of the free terms v^1, v^2. Let e_1, e_2, . . . , e_s denote an enumeration of the vertices of the polygon V in traversing its boundary in the counterclockwise direction (if the "polygon" V is a segment, then the endpoints of this segment will be considered as its vertices e_1, e_2). The coordinates of the vertex e_i will be denoted by e_i^1, e_i^2 ($i = 1, 2, \ldots, s$).

Furthermore, for an arbitrary point $v = (v^1, v^2)$ of X, let $w = g(v)$ denote the point whose coordinates w^1, w^2 satisfy

$$\begin{aligned}
aw^1 - bw^2 + v^1 &= 0, \\
bw^1 + aw^2 + v^2 &= 0.
\end{aligned} \tag{3.80}$$

The point w is uniquely defined by v since the determinant of the matrix (3.74) is $a^2 + b^2 \neq 0$. The mapping g transferring the point v to the point

$w = g(v)$ is an affine mapping since it is described by the linear equations (3.80). The affine transformation g maps the convex polygon V into a convex polygon W with vertices $h_1, h_2, \ldots, h_s$ (h_i denotes the vertex into which e_i is mapped under the transformation g). Thus, the coordinates h_i^1, h_i^2 of the vertex h_i are defined in terms of the coordinates of the vertex e_i ($i = 1, 2, \ldots, s$) by

$$ah_i^1 - bh_i^2 + e_i^1 = 0,$$
$$bh_i^1 + ah_i^2 + e_i^2 = 0. \tag{3.81}$$

We know that the optimal control $v = (v^1, v^2)$ takes on only values which are vertices of the polygon V. Let us assume that the optimal control takes on the value $v = e_i$ on some time interval. Then, the equation of motion of the object on this time interval takes the form [see (3.75)]

$$\dot{x}^1 = ax^1 - bx^2 + e_i^1,$$
$$\dot{x}^2 = bx^1 + ax^2 + e_i^2.$$

By (3.81), these equations can be written as follows:

$$(x^1 - h_i^1)^{\cdot} = a(x^1 - h_i^1) - b(x^2 - h_i^2),$$
$$(x^2 - h_i^2)^{\cdot} = b(x^1 - h_i^1) + a(x^2 - h_i^2). \tag{3.82}_i$$

Thus, if $v = e_i$, then the corresponding portion of the trajectory is a piece of the phase trajectory of system $(3.82)_i$, that is, of the system which differs from the homogeneous system (3.76) only by a shift of the equilibrium position to the point h_i.

Let us now draw s rays from the origin having the directions of outer normals to the sides of the polyhedron V (Figures 122, 123). The angle

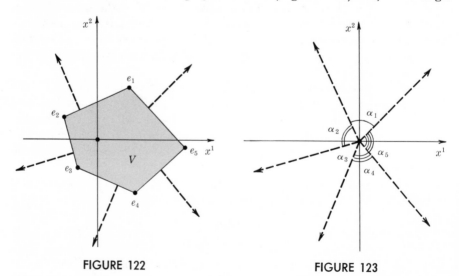

FIGURE 122   FIGURE 123

between the rays which are perpendicular to the sides concurrent at the vertex e_i will be denoted by α_i (Figure 123). Thus, the complete angle is divided into s angles $\alpha_1, \alpha_2, \ldots, \alpha_s$; moreover, it is clear that $\alpha_i = \pi - \gamma_i$, where γ_i is the interior angle of the polygon V at the vertex e_i.

Let ψ be an arbitrary nonzero vector lying in the interior of α_i. Then the vector ψ makes obtuse angles with both of the sides emanating from the vertex e_i (Figure 124). Hence it follows (compare p. 92) that as the vector v traverses the polygon V the scalar product ψv attains its maximum at $v = e_i$.

In case V is a segment with the endpoints e_1 and e_2, both angles α_1 and α_2 become equal to π (Figure 125); everything that has been said above remains valid also for this case.

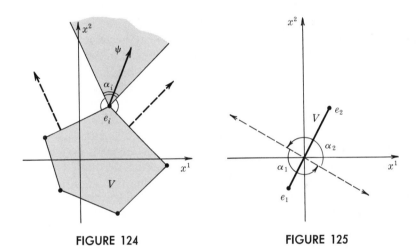

FIGURE 124   FIGURE 125

We shall now write the function H for the system (3.75) and apply the maximum principle. We have

$$H = \psi_1(ax^1 - bx^2 + v^1) + \psi_2(bx^1 + ax^2 + v^2)$$
$$= \cdots + \psi_1 v^1 + \psi_2 v^2 = \cdots + \psi v,$$

where dots denote the terms which do not depend on v. Thus, the function H attains its maximum (in v) simultaneously with the scalar product ψv, that is, it attains its maximum at the vertex e_i if the vector ψ lies in the interior of the angle α_i.

The system of equations for the auxiliary variables ψ_1, ψ_2 has the following form:

$$\dot{\psi}_1 = -a\psi_1 - b\psi_2$$
$$\dot{\psi}_2 = b\psi_1 - a\psi_2.$$

Its general real solution is given by

$$\psi_1 = c'e^{-at} \cos(bt + \alpha'),$$
$$\psi_2 = c'e^{-at} \sin(bt + \alpha'),$$

where c' and α' are arbitrary constants. Hence it is seen that the vector $\psi = (\psi_1, \psi_2)$ rotates uniformly in the counterclockwise direction with angular velocity b radians per second (it is immaterial that its length is changing). Therefore, the vector ψ varies in the following manner: for a time interval of duration α_i/b it is inside the angle α_i, then for a time interval of duration α_{i+1}/b it is inside the angle α_{i+1}, then for a time interval of duration α_{i+2}/b it is inside the angle α_{i+2}, and so on (moreover, α_{s+1} must be understood, of course, as the angle α_1). Hence it is clear that the optimal control v must be as follows: for a time interval of duration α_i/b the vector v takes on the value e_i, then for a time interval of duration α_{i+1}/b it takes on the value e_{i+1}, then for a time interval of duration α_{i+2}/b it takes on the value e_{i+2}, and so on. Finally, it is not difficult to see the structure of the optimal trajectory (more precisely, the trajectory satisfying the maximum principle): for a time interval of duration α_i/b the point moves along a trajectory of $(3.82)_i$, then for a time interval of duration α_{i+1}/b it moves along a trajectory of $(3.82)_{i+1}$, then for a time interval of duration α_{i+2}/b it moves along a trajectory of $(3.82)_{i+2}$, and so on. The only possible exceptions are the first and the last portions of the trajectory, that is, the first and the last time intervals may be shorter than the corresponding quantities α_i/b (the motion may not begin at an instant of switching and may terminate by reaching the origin prior to the instant of the next switching).

We note that an arc of the trajectory of system (3.76) traced by the phase point during the time α/b subtends an angle α at the equilibrium position (Figure 126); this follows readily from (3.78). This likewise pertains to $(3.82)_i$ which is obtained from (3.78) by a shift of the equilibrium position. Therefore, the structure of the optimal trajectory found above may be described as follows. The phase point moves along a trajectory of $(3.82)_i$ describing an arc which subtends the angle α_i at the equilibrium position, then travels along an arc of a trajectory of $(3.82)_{i+1}$ subtending the angle α_{i+1}, then moves along a trajectory of $(3.82)_{i+2}$, and so on. (The first and the last portions of the trajectory may subtend smaller angles at the corresponding equilibrium positions.) According to what has been said in § 7.28, any optimal trajectory has this form and, conversely, any trajectory of this form terminating at the origin is optimal.

Now, it is not difficult to construct the "switching curves" in the phase plane X which define the synthesis of optimal controls. Let A_iO ($i = 1, 2, \ldots, s$) denote the arc of the trajectory of $(3.82)_i$ which terminates at the origin O and subtends at the equilibrium position the angle α_i (Figures 127, 128). It is clear that the concluding stage of the optimal motion

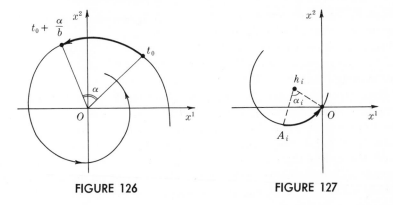

FIGURE 126   FIGURE 127

of the phase point takes place along one of the arcs A_iO; moreover, the point may not cover the entire arc, but only a portion of this arc, X_iO (since the last portion of the optimal trajectory may subtend an angle less than α_i). Furthermore, since the "switching" took place at the point X_i, and after the "switching" the phase point began moving according to the system $(3.82)_i$, it follows that prior to the instant of switching, the phase point was moving according to the law $(3.82)_{i-1}$. Thus, the preceding segment of the optimal trajectory Y_iX_i is an arc of the trajectory of system $(3.82)_{i-1}$ terminating at the point X_i and subtending the angle α_{i-1} at the equilibrium position. As the point X_i traverses the entire arc A_iO, the arcs Y_iX_i of the above form fill out a "curvilinear quadrangle" (Figure 129) one "side" of which coincides

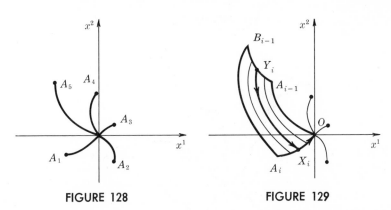

FIGURE 128   FIGURE 129

with the arc $A_{i-1}O$ (since for $X_i = O$, the arc Y_iX_i coincides with $A_{i-1}O$). Thus, three vertices of the curvilinear quadrangle under consideration are at the points A_i, O, A_{i-1}; the fourth vertex will be denoted by B_{i-1}. Then the arc $B_{i-1}A_{i-1}$ is the set of all the points Y_i, that is, of all the points at which the switching [from system $(3.82)_{i-2}$ to system $(3.82)_{i-1}$] takes place.

Let P_i denote the dilation with center at h_i and coefficient $e^{-a\alpha_i/b}$, combined with a clockwise rotation through the angle α_i. It can be readily derived from (3.78) that the arc $B_{i-1}A_{i-1}$ is obtained from the arc A_iO with the aid of the transformation P_{i-1} (that is, the point Y_i is obtained from the corresponding point X_i by the above transformation).

Prior to the switching at the point Y_i, the phase point was moving according to the law $(3.82)_{i-2}$, having described, along the corresponding phase trajectory, the arc Z_iY_i subtending the angle α_{i-2} at the equilibrium position (Figure 130). The point Z_i is obtained from the corresponding point Y_i with the aid of the transformation P_{i-2}. As the point Y_i traverses the entire arc $A_{i-1}B_{i-1}$, the arcs Z_iY_i (of the above form) fill out a "curvilinear quadrangle" two sides of which are the arcs $A_{i-1}B_{i-1}$ and $A_{i-1}B_{i-2}$. The

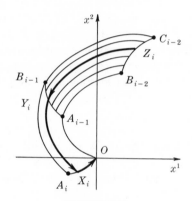

FIGURE 130

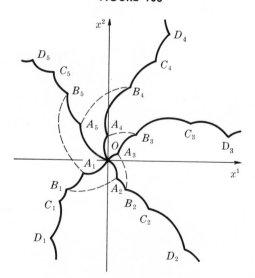

FIGURE 131

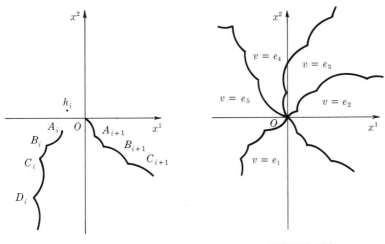

FIGURE 132   FIGURE 133

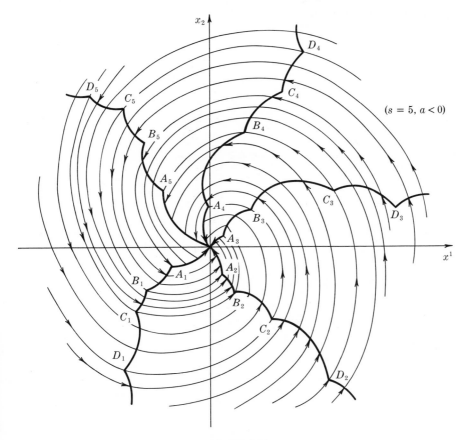

FIGURE 134

fourth vertex of this quadrangle will be denoted by C_{i-2}. Thus, the arc $B_{i-2}C_{i-2}$ (the set of all switching points Z_i) is obtained from the arc $A_{i-1}B_{i-1}$ with the aid of the transformation P_{i-2}.

Proceeding in this manner, we trace s curves $OA_iB_iC_iD_i \ldots$ ($i = 1, 2, \ldots, s$) emanating from the origin which together represent the set of all switching points (Figure 131). The transformation P_i transforms the curve $OA_{i+1}B_{i+1}C_{i+1} \ldots$ into the curve $A_iB_iC_iD_i \ldots$ (Figure 132). This makes it possible to successively trace portions of the curves $OA_iB_iC_iD_i \ldots$ knowing the first pieces of these curves $OA_1, OA_2, \ldots, OA_s$ (the determination of these pieces was given above). It remains to note that the control parameter v takes on the value e_i in the interior of the "angle" between the curves $OA_{i+1}B_{i+1}C_{i+1} \ldots$ and $OA_iB_iC_iD_i \ldots$ and on the arc A_iO. This, in fact, yields the synthesis of optimal controls (Figure 133). The form of the optimal trajectories is shown in Figure 134.

Naturally, the synthesis "portrait" changes depending on the values of the coefficients a, b, and on the shape of the polygon V. Several different cases are depicted in Figures 134–137. For $a < 0$ (that is, in the case in

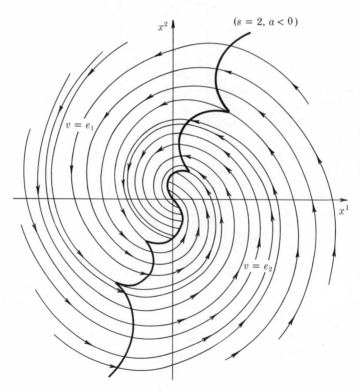

FIGURE 135

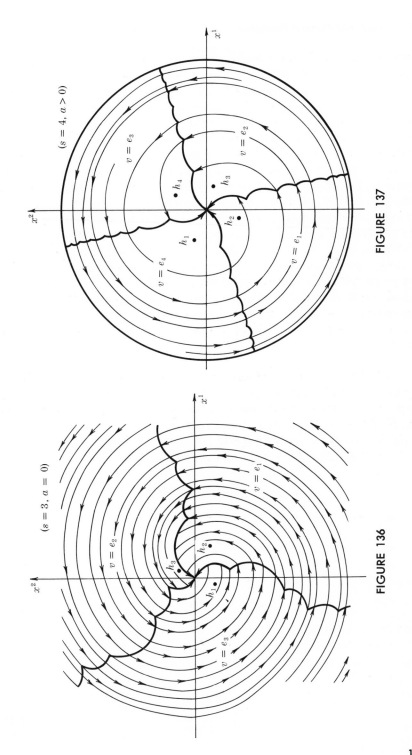

$(s = 4, a > 0)$

$v = e_3$

$v = e_2$

h_4

h_3

h_1

h_2

x^2

x^1

$v = e_4$

$v = e_1$

FIGURE 137

$(s = 3, a = 0)$

$v = e_2$

$v = e_1$

h_3

h_2

h_1

x^2

x^1

$v = e_3$

FIGURE 136

155

which the eigenvalues of the matrix A have negative real parts), the lengths of the arcs OA_i, A_iB_i, B_iC_i, $\ldots$ increase since in this case the transformations P_i have positive dilation coefficients. The synthesis of optimal controls is realized on the entire plane X (Figures 134, 135), and the separate portions of optimal trajectories are logarithmic spirals. For $a = 0$, the lengths of the arcs do not change, that is, $OA_i = B_{i-1}A_{i-1} = C_{i-2}B_{i-2} = \cdots$ (since the transformations P_i consist only of rotations); the synthesis of optimal controls, as before, is realized on the entire plane X (Figure 136), and the optimal trajectories consist of arcs of circles. Finally, for $a > 0$, the lengths of the arcs OA_i, $B_{i-1}A_{i-1}$, $C_{i-2}B_{i-2}$, $\ldots$ decrease in geometric progression, and the synthesis of optimal controls is realized only on a bounded portion of the plane X (Figure 137).

We note also that the condition of general position is always satisfied for the system under consideration (since the transformation A does not have proper invariant subspaces). It should also be mentioned that we have studied only system (3.75) which was obtained as a result of the simplifications presented in § 10.37; for an arbitrary system of the second order [not reduced to the form (3.75)] the synthesis portrait undergoes an affine distortion.

39. Solution of the Synthesis Problem in the Case of Real Eigenvalues.

We shall assume at the outset that the eigenvalues of the matrix A of system (3.73) (see p. 144) are distinct and different from zero. In this case, the matrix A can be reduced to diagonal form (by an appropriate change of coordinates in the phase plane). Thus, according to the discussion in § 10.37, we consider the system

$$\begin{aligned} \dot{x}^1 &= \lambda_1 x^1 + v^1, \\ \dot{x}^2 &= \lambda_2 x^2 + v^2, \end{aligned} \qquad (\lambda_1, \lambda_2 \neq 0). \qquad (3.83)$$

Let us first consider the corresponding homogeneous system

$$\begin{aligned} \dot{x}^1 &= \lambda_1 x^1, \\ \dot{x}^2 &= \lambda_2 x^2. \end{aligned} \qquad (3.84)$$

The general solution of this system has the form

$$x^1(t) = c_1 e^{\lambda_1 t}, \qquad x^2(t) = c_2 e^{\lambda_2 t}, \qquad (3.85)$$

where c_1, c_2 are arbitrary constants. These formulas are parametric equations of the phase trajectories. Each of the four coordinate semi-axes represents a phase trajectory. The equations of the remaining phase trajectories (in an implicit form) are obtained by eliminating the parameter t from relations (3.85):

$$\frac{(x_1)^{\lambda_2}}{(x_2)^{\lambda_1}} = \text{const.} \qquad (3.86)$$

The general "portrait" of phase trajectories of system (3.84) depends essentially on the magnitude and sign of the numbers λ_1, λ_2 (Figures 138–140).

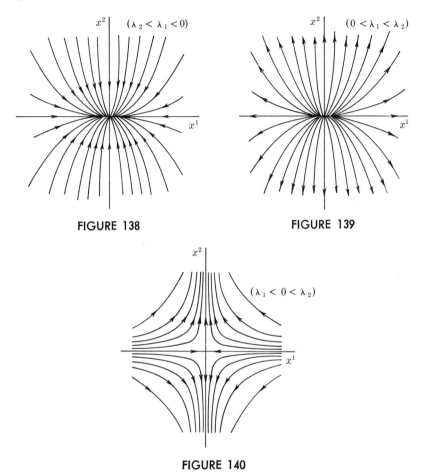

FIGURE 138   FIGURE 139

FIGURE 140

Let us now return to the consideration of system (3.83) which differs from system (3.84) by the presence of the free terms v^1, v^2. The enumeration of the vertices of V will remain the same as on p. 147.

For an arbitrary point $v = (v^1, v^2)$ of the phase plane X, let $w = g(v)$ denote the point with coordinates

$$w^1 = -\frac{1}{\lambda_1} v^1, \qquad w^2 = -\frac{1}{\lambda_2} v^2. \tag{3.87}$$

The mapping g transferring the point v to the point $w = g(v)$ is affine. Under this affine mapping, V is transformed into a convex polygon W with vertices h_1, h_2, . . . , h_s [where $h_i = g(e_i)$]. Thus, the coordinates h_i^1, h_i^2 of h_i are related to the coordinates of e_i ($i = 1, 2, . . . , s$) by the relation

$$\lambda_1 h_i^1 + e_i^1 = 0, \qquad \lambda_2 h_i^2 + e_i^2 = 0. \tag{3.88}$$

Suppose that the optimal control takes on the value $v = e_i$ on some time interval. Then the equation of motion of the object on this interval takes the form [see (3.83)]

$$\dot{x}^1 = \lambda_1 x^1 + e_i^1, \qquad \dot{x}^2 = \lambda_2 x^2 + e_i^2.$$

By virtue of (3.88), these equations can be written as follows:

$$(x^1 - h_i^1)^{\cdot} = \lambda_1(x^1 - h_i^1),$$
$$(x^2 - h_i^2)^{\cdot} = \lambda_2(x^2 - h_i^2). \tag{3.89}_i$$

Thus, if $v = e_i$, then the corresponding portion of the trajectory is a piece of a phase trajectory of system $(3.89)_i$, that is, of the system which differs from the homogeneous system (3.84) only in that the equilibrium position is shifted to the point h_i.

The discussion on p. 149 pertaining to the angles α_1, α_2, . . . , α_s applies here without change; the function H attains its maximum (in v) simultaneously with the scalar product ψv, that is, it attains its maximum at the vertex e_i if the vector ψ lies in the interior of the angle α_i.

We now note that the coordinate axes are proper invariant subspaces of the transformation A with matrix

$$\begin{pmatrix} \lambda_1 & 0 \\ 0 & \lambda_2 \end{pmatrix}$$

(where $\lambda_1 \neq \lambda_2$). Therefore, by virtue of the condition of general position (which we assume to be satisfied), *none of the sides of the polygon V is parallel to either one of the coordinate axes.* Consequently, none of the rays which are the sides of the angles α_1, α_2, . . . , α_s (Figure 123) is parallel to either one of the coordinate axes.

The system of equations for the auxiliary variables ψ_1, ψ_2 has the following form:

$$\dot{\psi}_1 = -\lambda_1 \psi_1, \qquad \dot{\psi}_2 = -\lambda_2 \psi_2.$$

It can be readily seen from this system that if $\psi_1 = 0$ (or $\psi_2 = 0$) at the initial time, then $\psi_1 \equiv 0$ (or $\psi_2 \equiv 0$). If however, at the initial time, we have $\psi_1 \neq 0$, $\psi_2 \neq 0$, then ψ_1 and ψ_2 do not vanish during the entire motion. In other words, the vector ψ either maintains a constant direction parallel to one of the coordinate axes throughout the entire motion, or remains at all times in the same quadrant (that is, its components ψ_1, ψ_2 maintain a constant sign). Moreover, we have

$$\frac{d}{dt}\left(\frac{\psi_2}{\psi_1}\right) = \frac{\psi_1\dot{\psi}_2 - \dot{\psi}_1\psi_2}{(\psi_1)^2} = \frac{-\lambda_2\psi_1\psi_2 + \lambda_1\psi_1\psi_2}{(\psi_1)^2} = (\lambda_1 - \lambda_2)\frac{\psi_2}{\psi_1}. \quad (3.90)$$

Hence it is seen that $\frac{d}{dt}\left(\frac{\psi_2}{\psi_1}\right)$ maintains a constant sign in each of the quadrants, that is, the slope of the vector ψ varies monotonically, and therefore, in each of the quadrants, the vector ψ rotates continuously in one direction (Figure 141). The rule by which the quantity ψ_2/ψ_1 changes with time can be found from (3.90) by direct integration:

$$\ln\left|\frac{\psi_2}{\psi_1}\right| = (\lambda_1 - \lambda_2)t + \text{const.} \quad (3.91)$$

We must consider four cases corresponding to the motion of the vector ψ in each of the four quadrants. The investigation is carried out similarly in all four cases. We shall consider in complete detail the case in which the vector ψ moves in the first quadrant; moreover, for the sake of definiteness, we shall assume that $\lambda_1 - \lambda_2 > 0$ [Figure 141(a)]. Let $\alpha_p, \alpha_{p+1}, \ldots, \alpha_q$

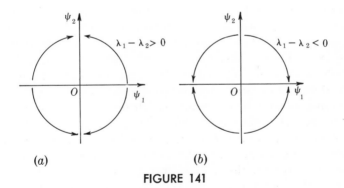

(a)   (b)

FIGURE 141

denote the angles contained entirely in the first quadrant (so that only part of the angles α_{p-1} and α_{q+1} lie in this quadrant), and let $l_p, l_{p+1}, \ldots, l_q, l_{q+1}$ denote the sides of these angles (Figure 142). The tangents of the angles formed by these straight lines with the x^1-axis will be denoted by $k_p, k_{p+1}, \ldots, k_{q+1}$. It follows directly from formula (3.91) that the time during which the vector ψ is inside the angle α_i (that is, the difference between the times at which the relations $\psi_2/\psi_1 = k_{i+1}$ and $\psi_2/\psi_1 = k_i$ are satisfied) is

$$T_i = \frac{1}{\lambda_1 - \lambda_2}\ln\frac{k_{i+1}}{k_i}.$$

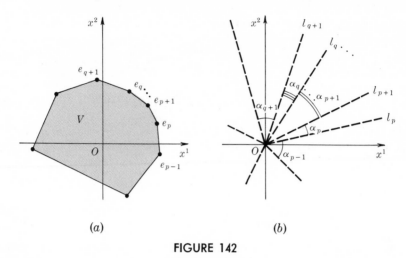

(a)   (b)

FIGURE 142

Thus, the change of ψ with time is described as follows. The vector ψ may remain inside the angle a_{p-1} for an indefinite duration of time, then rotates inside the angle a_p for a time T_p, then rotates inside the angle a_{p+1} for a time $T_{p+1}, \ldots$, then rotates inside the angle a_q for a time T_q, and finally, may remain inside the angle a_{q+1} for an indefinite duration of time. Hence we obtain the form of the optimal control. The control parameter v may take on the value e_{p-1} for an indefinite duration of time, then takes on the value e_p for a time T_p, then takes on the value e_{p+1} for a time $T_{p+1}, \ldots$, has the value e_q for a time T_q, and finally, v may take on the value e_{q+1} for an indefinite duration of time. Altogether we obtain $q - p + 2$ switchings (for ψ varying in the first quadrant) which take place when the vector ψ passes through each of the straight lines $l_p, l_{p+1}, \ldots , l_q, l_{q+1}$. Of course, the number of switchings may turn out to be less than $q - p + 2$. For example, the motion could begin at a time at which v has assumed the value e_p. Then we would find that v has the value e_p during a time interval whose duration does not exceed T_p, after which v has the value e_{p+1} for a time T_{p+1}, and so on. In exactly the same way, it could happen, for example, that the motion will stop (the origin having been reached) before the switching from the vertex e_q to the vertex e_{q+1}.

It is now easy to construct the "switching curves" on the X plane which determine the synthesis of optimal controls. First, let us plot the trajectories for which the final stage of the motion corresponds to the value of the parameter $v = e_{q+1}$, that is, the motion takes place along the arc $A_{q+1}O$ of the trajectory of system $(3.89)_{q+1}$ which terminates at the origin (Figure 143). Before reaching $A_{q+1}O$ the motion proceeded under system $(3.89)_q$. Thus, $A_{q+1}O$ is the switching curve from e_q to e_{q+1}. Let X be a point of $A_{q+1}O$. Then, the section YX of the optimal trajectory which precedes X is an arc of a trajectory of system $(3.89)_q$ corresponding to a time interval of duration T_q. Since the solutions of system (3.84) have the form $x^1 = c_1 e^{\lambda_1 t}$ and $x^2 = c_2 e^{\lambda_2 t}$, a point, after moving along the trajectory of this system for a time T_q, has its abscissa multiplied by $e^{\lambda_1 T_q}$ and its ordinate multiplied by $e^{\lambda_2 T_q}$. The system $(3.89)_q$ differs from system (3.84) only in that the equilibrium position has been translated. Thus, the points X are obtained from the corresponding points Y (Figure 143) by means of an affine transformation L_q. In the coordinate system with origin at h_q and axes parallel to the x^1- and x^2-axes, L_q consists in the abscissas being multiplied by $e^{\lambda_1 T_q}$, and the ordinates by $e^{\lambda_2 T_q}$. Consequently, the set of the points Y is the curve $A_q B_q$ which goes into the curve $A_{q+1} O$ under the affine transformation L_q (Figure 144). If

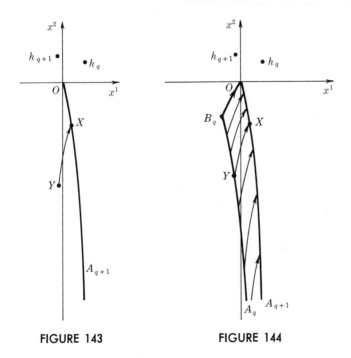

FIGURE 143   FIGURE 144

we also plot the curve $B_q O$, which is an arc of the trajectory of system $(3.89)_q$ terminating at the origin and corresponding to a time interval of duration T_q, we find that the entire "band" $A_{q+1} O B_q A_q$ is filled out by pieces of trajectories of system $(3.89)_q$ which originate on $A_q B_q$, terminate on $A_{q+1} O$, and correspond to a time interval of duration T_q.

Thus, the curve $A_q B_q$ is the set of points Y at which the switching from the vertex e_{q-1} to the vertex e_q occurs. To this curve, we must add also the arc $B_q O$, since the optimal trajectories which terminate upon reaching the origin prior to the instant of switching from e_q to e_{q+1} must also be considered [for such trajectories the final stage of the motion is described by system $(3.89)_q$ and occurs during a time interval of duration $\leqslant T_q$, that is, this final stage is represented by a section YO of the arc B_qO]. Up to the point Y on A_qB_qO, the motion proceeded along the arc ZY of the trajectory of system $(3.89)_{q-1}$ during the time interval T_{q-1}. Hence, we obtain in a similar manner that the locus of the point Z is the curve $A_{q-1}B_{q-1}C_{q-1}$ which goes into the curve A_qB_qO under the affine transformation L_{q-1} (which in the coordinate system with origin at h_{q-1} consists in the multiplication of the abscissas by $e^{\lambda_1 T_{q-1}}$, and the ordinates by $e^{\lambda_2 T_{q-1}}$). By plotting also the curve $C_{q-1}O$, which is the arc of the trajectory of system $(3.89)_{q-1}$ terminating at O and corresponding to a time interval of duration T_{q-1}, we find that the entire "band" $A_qB_qOC_{q-1}B_{q-1}A_{q-1}$ is filled out by pieces of the trajectories of system $(3.89)_{q-1}$ which originate on the curve $A_{q-1}B_{q-1}C_{q-1}$, terminate on A_qB_qO, and correspond to a time interval of duration T_{q-1} (Figure 145).

Continuing in this fashion we can construct the "bands" filled out by pieces of the trajectories of systems $(3.89)_q$, $(3.89)_{q-1}$, . . . , $(3.89)_p$ (Figure 146). The "extreme" curve $A_pB_pC_p$ . . . K_pO of the last of these "bands"

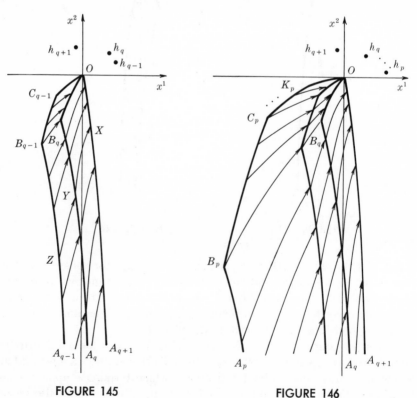

FIGURE 145   FIGURE 146

is the set of all the points at which switching from the vertex e_{p-1} to the vertex e_p occurs. Prior to reaching this curve, the phase point was moving along a trajectory of system $(3.89)_{p-1}$ (Figure 147). As a result, we obtain

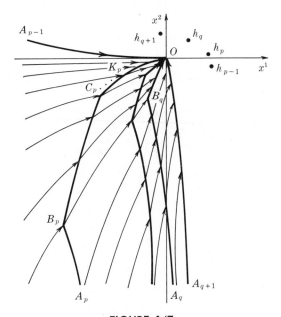

FIGURE 147

all the optimal trajectories for which the vector ψ varies in the first quadrant. They fill out an "angle" $A_{q+1}OA_{p-1}$ (in the plane), where $A_{p-1}O$ is the trajectory of system $(3.89)_{p-1}$ which terminates at O.

Optimal trajectories corresponding to variation of the vector ψ in the three other quadrants are plotted in a similar manner. This yields three more "angles" which are analogous to the "angle" $A_{q+1}OA_{p-1}$. Together, these four angles fill out the entire phase plane X, which yields the desired synthesis of optimal controls (Figure 148). If both eigenvalues are positive, then the synthesis is realized in a bounded convex set (Figures 149, 150), but if the eigenvalues have different signs, then the synthesis is realized in some band (Figure 151).

We have considered in detail the case in which the eigenvalues λ_1, λ_2 are real, different from zero, and distinct. Let us dwell briefly upon the changes which take place in the construction for the remaining cases.

First, let us assume that only one eigenvalue vanishes, that is, we have the system

$$\begin{aligned} \dot{x}^1 &= \lambda x^1 + v^1, \\ \dot{x}^2 &= v^2, \end{aligned} \qquad (\lambda \neq 0).$$

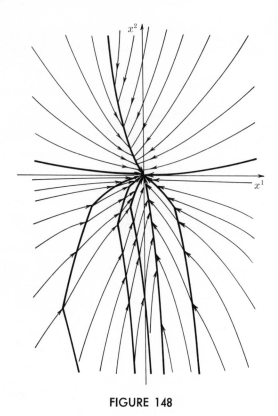

FIGURE 148

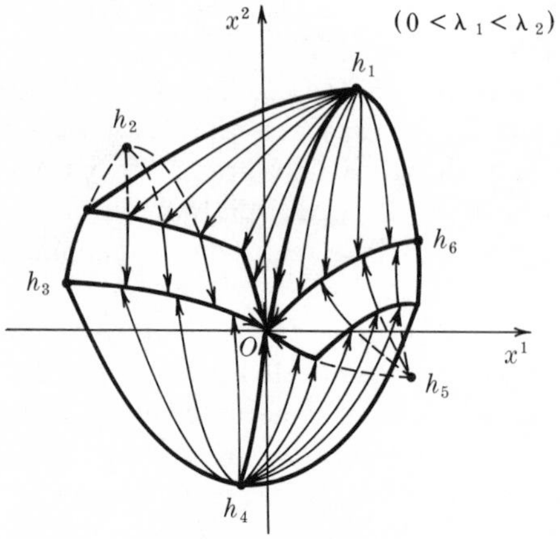

$(0 < \lambda_1 < \lambda_2)$

FIGURE 149

$(0 < \lambda_1 < \lambda_2)$

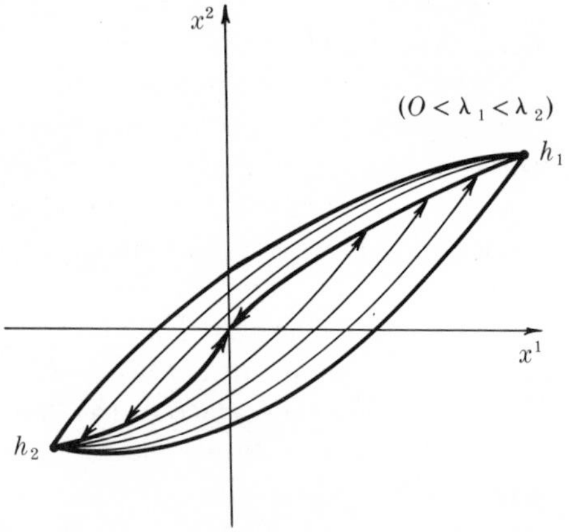

FIGURE 150

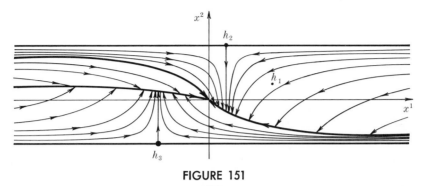

FIGURE 151

The axes of coordinates are proper invariant subspaces, so that the condition of general position requires that the polygon V does not have sides parallel to the coordinate axes. The arguments on the character of changes affecting the vector ψ [compare (3.90), (3.91)] are also fully preserved. However, instead of (3.89) we now have the system

$$\begin{aligned}
\dot{x}^1 &= \lambda x^1 + e_i^1, \\
\dot{x}^2 &= e_i^2,
\end{aligned} \qquad (3.92)_i$$

whose phase trajectories are the curves

$$x^1 = -\frac{e_i^1}{\lambda} + c_1 e^{\lambda t}, \qquad x^2 = e_i^2 t + c_2$$

(these are exponential curves for $e_i^2 \neq 0$, and rays parallel to the axis of abscissas for $e_i^2 = 0$). The synthesis of optimal controls is realized the same way as for system (3.83) but substituting the phase trajectories of system $(3.89)_i$ by those of system $(3.92)_i$. An example of this case is shown in Figure 152.

It remains to consider the case in which the eigenvalues λ_1, λ_2 coincide. In this case, the matrix A can be reduced to one of the following two forms:

$$\begin{pmatrix} \lambda & 0 \\ 0 & \lambda \end{pmatrix}, \qquad \begin{pmatrix} \lambda & 1 \\ 0 & \lambda \end{pmatrix}.$$

The first of these matrices is excluded from consideration since any line passing through the origin is an invariant subspace for this matrix and the condition of general position is not satisfied. The second matrix leads us to the system of equations

$$\begin{aligned}
\dot{x}^1 &= \lambda x^1 + x^2 + v^1, \\
\dot{x}^2 &= \lambda x^2 + v^2.
\end{aligned}$$

The x^1-axis is the only proper invariant subspace, and therefore, the condition of general position is satisfied if the polygon V does not have a side parallel to the x^1-axis.

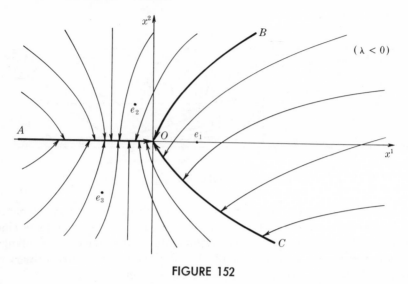

FIGURE 152

The system of equations for the auxiliary variables has, in this case, the form

$$\dot{\psi}_1 = -\lambda\psi_1,$$
$$\dot{\psi}_2 = -\psi_1 - \lambda\psi_2.$$

It is seen from this system that if $\psi_1 = 0$ at the initial time, then $\psi_1 \equiv 0$; if, however, $\psi_1 \neq 0$ at the initial time, then ψ_1 does not vanish throughout the entire motion. In other words, the vector ψ either maintains a fixed direction parallel to the axis of ordinates during the entire motion, or it lies in the right or left half-plane all the time. Moreover,

$$\frac{d}{dt}\left(\frac{\psi_2}{\psi_1}\right) = \frac{\dot{\psi}_1\psi_2 - \psi_1\dot{\psi}_2}{(\psi_1)^2} = -1; \tag{3.93}$$

hence it is seen that the vector ψ rotates (in the left as well as in the right half-plane) in the clockwise direction (Figure 153). The time interval T_i during which the vector ψ remains in the interior of the angle α_i can be readily found from (3.93). Instead of considering four quadrants, it is sufficient now to consider two cases corresponding to the motion of the vector ψ in the right or in the left half-plane. Letting $\alpha_p, \alpha_{p+1}, \ldots, \alpha_q$ denote the angles contained entirely in the right half-plane and numbering these angles in the clockwise direction (Figure 154), we construct, in exactly the same manner as before, the "angle" $A_{q+1}OA_{p-1}$ filled out by the optimal trajectories and corresponding to the variation of the vector ψ in the right half-plane. An analogous "angle" is formed also in the left half-plane. Together both of these "angles" define the synthesis of optimal controls in the entire plane X. In other respects the construction of the synthesis does not differ from that considered above.

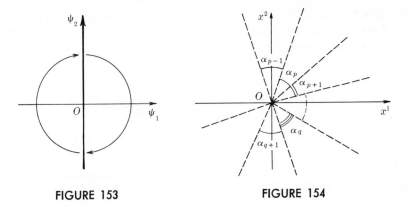

FIGURE 153   FIGURE 154

40. Synthesis of Optimal Controls for an Equation of the Second Order.   Let us consider the second-order equation

$$\ddot{x} + 2\,\delta\dot{x} + \omega^2 x = u$$

with non-negative coefficients δ and ω. The control parameter u (scalar) will be assumed to vary within the limits $-1 \leqslant u \leqslant 1$ (compare p. 7). The slightly more general case in which the control parameter u varies within $\alpha \leqslant u \leqslant \beta$ (where $\alpha < 0$, $\beta > 0$) is left to the reader. Setting $x = x^1$, $\dot{x} = x^2$, we rewrite the above equation in the form of the following normal system:

$$\begin{aligned}
\dot{x}^1 &= x^2, \\
\dot{x}^2 &= -\omega^2 x^1 - 2\delta x^2 + u.
\end{aligned} \qquad (3.94)$$

In the x^1, x^2-plane, the "polyhedron" U will be represented by the segment $[-1, 1]$ lying on the x^2-axis. It is easy to see that the x^2-axis is not a proper invariant subspace of the matrix

$$A = \begin{pmatrix} 0 & 1 \\ -\omega^2 & -2\delta \end{pmatrix},$$

and therefore the condition of general position is always satisfied.

First, let the eigenvalues of the matrix A, that is, the roots of the equation

$$\lambda^2 + 2\delta\lambda + \omega^2 = 0 \qquad (3.95)$$

be complex, that is, let the discriminant $\delta^2 - \omega^2$ be negative. Setting $\delta^2 - \omega^2 = -\gamma^2$, where γ is a positive number, we can write the roots of (3.95) in the form $\lambda = -\delta \pm i\gamma$. The eigenvector of the matrix A corresponding to the eigenvalue $\lambda = -\delta + i\gamma$ has the form $q_1 - iq_2$,

$$q_1 = \{1, -\delta\}, \qquad q_2 = \{0, -\gamma\}.$$

We shall take q_1 and q_2 as basis vectors of a new oblique coordinate system y^1, y^2. Then, the transition from the y^1, y^2-system to the x^1, x^2-system will be expressed by the formulas

$$
\begin{aligned}
x^1 &= y^1, \\
x^2 &= -\delta y^1 - \gamma y^2.
\end{aligned}
\tag{3.96}
$$

Using these formulas, we find that in the new coordinates the system of equations (3.94) can be written in the form

$$
\begin{aligned}
\dot{y}^1 &= -\delta y^1 - \gamma y^2, \\
\dot{y}^2 &= \gamma y^1 - \delta y^2 - \frac{1}{\gamma} u,
\end{aligned}
$$

or otherwise, in the form

$$
\begin{aligned}
\dot{y}^1 &= -\delta y^1 - \gamma y^2 + v^1, \\
\dot{y}^2 &= +\gamma y^1 - \delta y^2 + v^2,
\end{aligned}
$$

where the control point (v^1, v^2) may vary within the limits of the polyhedron V which is the interval $[-1/\gamma, 1/\gamma]$ of the y^2-axis.

According to what has been said on p. 148, the points

$$
h_1 = \left(\frac{1}{\gamma^2 + \delta^2}, -\frac{\delta}{\gamma(\gamma^2 + \delta^2)} \right), \qquad
h_2 = \left(-\frac{1}{\gamma^2 + \delta^2}, \frac{\delta}{\gamma(\gamma^2 + \delta^2)} \right)
$$

(whose coordinates are given in the y^1, y^2-system) correspond to the vertices

$$
e_1 = \left(0, -\frac{1}{\gamma} \right), \qquad e_2 = \left(0, \frac{1}{\gamma} \right)
$$

of the polyhedron V, and each of the angles α_1, α_2, corresponding to these vertices, is equal to π. Now, using the arguments given in § 10.38, it is easy to construct the synthesis of optimal controls in the (y^1, y^2)-plane. The pieces of the phase trajectories are semicircles for $\delta = 0$ (Figure 155) and arcs of logarithmic spirals for $\delta > 0$ (Figure 156). Returning to the coordinates x^1, x^2 from the coordinates y^1, y^2 [by (3.96)], the synthesis "portrait" is affinely distorted (Figures 158, 159).

Now, let the eigenvalues of the matrix A, that is, the roots of (3.95), be real and distinct. Let us denote these roots by λ_1 and λ_2 (they are negative), and let q_1 and q_2 denote the corresponding eigenvectors (Figure 159; compare the trajectories in Figure 138):

$$
q_1 = \{-1, -\lambda_1\}, \qquad q_2 = \{-1, -\lambda_2\}.
$$

Let us take q_1 and q_2 as basis vectors of a new oblique coordinate system y^1, y^2. Then the transition from the y^1, y^2-system to the x^1, x^2-system can be expressed by

$$
\begin{aligned}
x^1 &= -y^1 - y^2, \\
x^2 &= -\lambda_1 y^1 - \lambda_2 y^2.
\end{aligned}
\tag{3.97}
$$

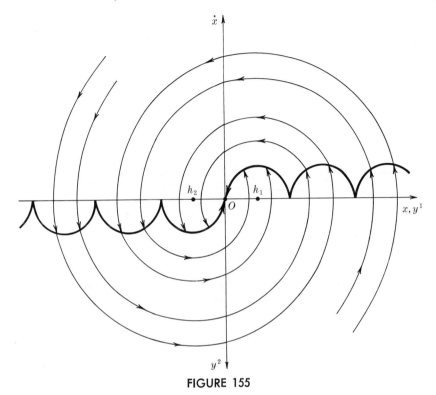

FIGURE 155

In the new coordinates, (3.94) takes the form

$$\dot{y}^1 = \lambda_1 y^1 + \frac{1}{\lambda_1 - \lambda_2} u,$$

$$\dot{y}^2 = \lambda_2 y^2 + \frac{1}{\lambda_2 - \lambda_1} u,$$

or, otherwise, the form

$$\dot{y}^1 = \lambda_1 y^1 + v^1,$$
$$\dot{y}^2 = \lambda_2 y^2 + v^2,$$

where the control point (v^1, v^2) may vary within the limits of the polyhedron V which is a segment with the endpoints

$$e_1 = \left(\frac{1}{\lambda_1 - \lambda_2}, -\frac{1}{\lambda_1 - \lambda_2} \right),$$

$$e_2 = \left(-\frac{1}{\lambda_1 - \lambda_2}, \frac{1}{\lambda_1 - \lambda_2} \right).$$

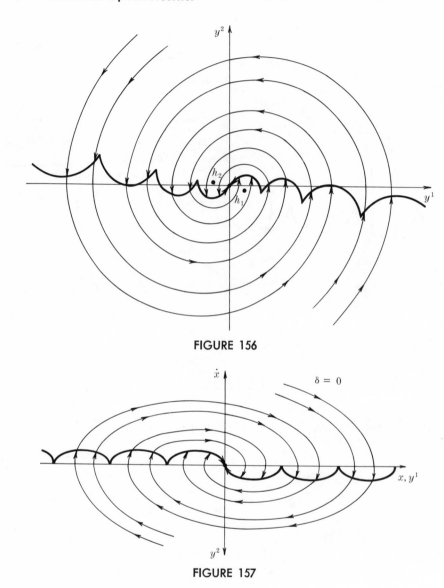

FIGURE 156

FIGURE 157

Now, using the arguments of § 10.39, it is not difficult to construct the synthesis of optimal controls in the y^1, y^2-plane (Figure 160). In returning to the x^1, x^2 coordinates, the synthesis "portrait" is affinely distorted (Figure 161).

Finally, if the eigenvalues coincide ($\lambda_1 = \lambda_2 = -\delta = -\omega$), then it is convenient to use the eigenvector $q_1 = \{1, \lambda\}$ and the vector $q_2 = \{0, 1\}$ as auxiliary basis vectors.

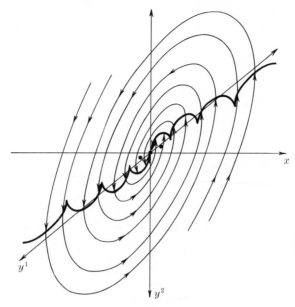

FIGURE 158

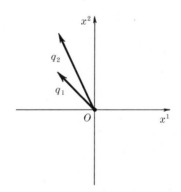

FIGURE 159

The oblique coordinate system y^1, y^2, defined by the basis vectors q_1, q_2, is related to the system x^1, x^2 by

$$x^1 = y^1,$$
$$x^2 = \lambda y^1 + y^2.$$

In the new coordinates, (3.94) can be written in the form

$$\dot{y}^1 = \lambda y^1 + y^2 + v^1,$$
$$\dot{y}^2 = \lambda y^2 + v^2,$$

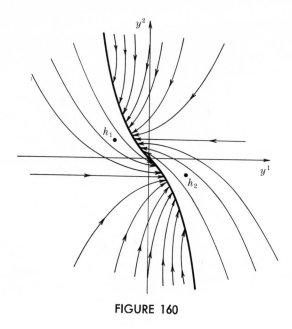

FIGURE 160

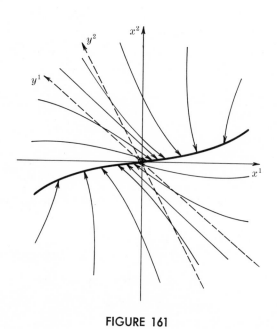

FIGURE 161

where the control point (v^1, v^2) varies within the bounds of the polyhedron V which in this case is the segment $[-1, 1]$ of the y^2-axis. The synthesis is constructed in a manner analogous to that of the previous case. We note that for $\lambda = 0$ (that is, $\delta = \omega = 0$) (3.94) takes the form

$$\dot{x}^1 = x^2,$$
$$\dot{x}^2 = u.$$

The synthesis of optimal controls for this system was considered on pp. 24–30.

chapter 4 Fundamentals of the Method of Dynamic Programming and Sufficient Conditions for Optimality

§ 11. ESTIMATE OF THE TRANSITION TIME OF A PROCESS

41. The Idea of the Method.   In this section we consider the function $\omega(x)$, introduced on p. 14, from an entirely different point of view. Namely, we now take (1.12) [or (1.18)] as a primitive relation and shall obtain therefrom an estimate of the transition time of the process. The basic idea of this method is very simple. We present it in the form of the following lemma which pertains (along with the subsequent lemmas) to the control process (1.2). Moreover, we shall assume, for the sake of simplicity, that the right-hand sides of (1.2) are defined, continuous, and have continuous derivatives $\partial f^i/\partial x^j$ for x in the entire phase space X (and for u in a control region U).

Lemma 4.1.   Assume that a continuously differentiable function $\omega(x) = \omega(x^1, x^2, \ldots, x^n)$ defined on an open set G of the space X satisfies the inequality

$$\sum_{\alpha=1}^{n} \frac{\partial \omega(x)}{\partial x^\alpha} f^\alpha(x, u) \leqslant 1 \tag{4.1}$$

for all $x \in G$, $u \in U$. If $u(t)$, $t_0 \leqslant t \leqslant t_1$, is an admissible control transferring the phase point from the state x_0 to the state x_1 and if the corresponding trajectory $x(t)$ lies entirely in the set G, then the transition time $t_1 - t_0$ from x_0 to x_1 is estimated by the inequality

$$t_1 - t_0 \geqslant \omega(x_1) - \omega(x_0). \tag{4.2}$$

PROOF.   Let $\tau_1, \tau_2, \ldots, \tau_k$ be the points of discontinuity of the control $u(t)$ and let $t_0 < \tau_1 < \tau_2 < \cdots < \tau_k < t_1$; we set $t_0 = \tau_0$, $t_1 = \tau_{k+1}$.

Then the function $u(t)$ is continuous on each of the intervals (τ_0, τ_1), (τ_1, τ_2), $\ldots$, (τ_k, τ_{k+1}). We have

$$\frac{d}{dt}\,\omega(x(t)) = \sum_{\alpha=1}^{n} \frac{\partial\omega(x(t))}{\partial x^\alpha} \cdot \frac{dx^\alpha(t)}{dt} = \sum_{\alpha=1}^{n} \frac{\partial\omega(x(t))}{\partial x^\alpha}\, f^\alpha(x(t),\, u(t)) \leqslant 1$$

[see (4.1)]. Thus, the continuous function $\omega(x(t))$ has a continuous derivative on each of the intervals (τ_i, τ_{i+1}), $i = 0, 1, \ldots, k$, satisfying the inequality

$$\frac{d}{dt}\,\omega(x(t)) \leqslant 1,$$

and therefore,

$$\omega(x(\tau_{i+1})) - \omega(x(\tau_i)) \leqslant \tau_{i+1} - \tau_i.$$

Adding these inequalities for $i = 0, 1, \ldots, k$, we obtain the relation

$$\omega(x(\tau_{k+1})) - \omega(x(\tau_0)) \leqslant \tau_{k+1} - \tau_0,$$

which coincides with the required inequality (4.2).

We note that it can always be assumed that the function ω vanishes at a preassigned point [since the addition of a constant to $\omega(x)$ does not change its properties]. If, for example, $\omega(x)$ satisfies the condition $\omega(x_1) = 0$, then (4.2) can be rewritten in the form

$$t_1 - t_0 \geqslant -\omega(x_0). \qquad (4.3)$$

The lemma we have proved provides a method which is essentially opposite to that of dynamic programming. More precisely, (1.12) [or (1.18)] was derived as a necessary condition for optimality on pp. 14–17, whereas here it appears as a sufficient condition for optimality. As a matter of fact, on pp. 14–17 we fixed a point x_1 and assumed that there exists a continuous and continuously differentiable function $\omega(x)$ which satisfies the condition $\omega(x_1) = 0$ and has the following two properties:

1° For every point x_0 there exists an admissible control $u(t) = u_{x_0}(t)$ transferring the phase point from x_0 to x_1 in the time $-\omega(x_0)$.

2° All these controls are optimal, that is, it is not possible to reach x_1 from x_0 in a time less than $-\omega(x_0)$.

Under these conditions we established the validity of inequality (1.12) [or relation (1.18) which is equivalent to (1.12) by virtue of condition 1°]. In other words, in satisfying condition 1°, (1.12) [or equivalently (1.18)] represents a necessary condition for optimality. On the other hand, Lemma 4.1 shows that under condition 1°, the inequality (1.12) [or relation (1.18)] is a sufficient condition for optimality. In fact, by virtue of condition 1°, the point x_0 can be transferred to x_1 in time $-\omega(x_0)$ with the aid of some control

$u(t) = u_{x_0}(t)$; it is not possible to transfer x_0 to x_1 in a shorter time [see (4.3)], that is, the control $u_{x_0}(t)$ is optimal.

Thus, under condition 1°, Bellman's equation (1.18) is actually a necessary and sufficient condition for optimality. However, as shown by the critique on p. 17, this statement is not of great value because of the unrealistic requirement of continuous differentiability of the function $\omega(x)$.

42. Sufficient Conditions for Optimality in the Form of the Principle of Dynamic Programming.

In this and the next section we shall weaken somewhat the requirements imposed on the function $\omega(x)$. As a result, we will obtain a new necessary and sufficient condition for optimality (in the form of the principle of dynamic programming) which will have a wider range of applicability.

In order to approach the formulation of these weakened requirements let us consider once again the example given on pp. 24–33. In this example, the function $\omega(x)$ did not have continuous derivatives on the entire phase plane X, but everywhere except for the points of the "switching curve" AOB (Figure 26). It is natural to assume in the general case that a set M [a "singular set" of the function $\omega(x)$] is singled out in the set G on which $\omega(x)$ is given, and that $\omega(x)$ *is continuous on the entire set G, and has continuous derivatives only at those points of the set G which do not belong to M.* Indeed, this is the assumption we shall make.

Lemma 4.2. Let G be an open set of the phase space X and let M be a set contained in G. Assume that a continuous function $\omega(x) = \omega(x^1, \ldots, x^n)$, defined on G, has continuous derivatives satisfying (4.1) at the points of G which do not belong to M. Furthermore, let $u(t)$, $t_0 \leqslant t \leqslant t_1$, be an admissible control which transfers the phase point from the state x_0 to the state x_1; moreover, let the corresponding trajectory $x(t)$ lie entirely in G and intersect the set M only a finite number of times. Then the estimate (4.2) remains valid.

PROOF. First of all, it is clear that the estimate (4.2) remains valid if the trajectory $x(t)$ intersects the set M only at the instants t_0, t_1 (or at one of these instants). In fact, the proof of the lemma can be carried out completely in this case, since the function ω is, as before, continuous, and if t is contained within the intervals (τ_i, τ_{i+1}), then the point $\omega(x(t))$ lies outside the set M [so that on these intervals the derivatives $\partial\omega(x(t))/\partial x^\alpha$ exist and are continuous].

Now, let $\theta_1, \theta_2, \ldots, \theta_s$ be all the instants belonging to the interval (t_0, t_1) at which the trajectory $x(t)$ intersects the set M; moreover, let $t_0 < \theta_1 < \theta_2 < \cdots < \theta_s < t_1$. Let us set $t_0 = \theta_0$, $t_1 = \theta_{s+1}$. Then the remark made above is applicable to every interval $[\theta_i, \theta_{i+1}]$, $i = 0, 1, \ldots, s$, and therefore,

$$\theta_{i+1} - \theta_i \geqslant \omega(x(\theta_{i+1})) - \omega(x(\theta_i)).$$

Adding these inequalities for $i = 0, 1, \ldots, s$, we obtain the relation

$$\theta_{s+1} - \theta_0 \geqslant \omega(x(\theta_{s+1})) - \omega(x(\theta_0)),$$

which coincides with the required inequality (4.2).

Lemma 4.3. Let us retain the assumptions made in Lemma 4.2 concerning the function ω. Let $u(t)$, $t_0 \leqslant t \leqslant t_1$, be an admissible control transferring the phase point from the state x_0 to the state x_1 and let the corresponding trajectory $x(t)$ lie entirely in G. Finally, let us assume that there is a point y_0, as close as desired to x_0, such that the trajectory $y(t)$ emanating from the point y_0 and corresponding to the same control $u(t)$, $t_0 \leqslant t \leqslant t_1$, intersects M only a finite number of times. Then the estimate (4.2) remains valid.

PROOF.   Let us choose an arbitrary number $\epsilon > 0$, and let W_0, W_1 be neighborhoods of the points x_0, x_1, respectively, such that

$$|\omega(x) - \omega(x_0)| < \epsilon \qquad \text{for} \qquad x \in W_0,$$
$$|\omega(x) - \omega(x_1)| < \epsilon \qquad \text{for} \qquad x \in W_1.$$

By virtue of the theorem on the continuous dependence of solutions of a system of differential equations on the initial conditions, there exists a neighborhood $W'_0 \subset W_0$ of the point x_0 such that any solution $y(t)$ of the system (1.4) [with the same control $u(t)$, $t_0 \leqslant t \leqslant t_1$], for which $y(t_0) \in W'_0$, is defined on the entire interval $t_0 \leqslant t \leqslant t_1$ and satisfies the relation $y(t_1) \in W_1$.

By the condition of the lemma, there exists a solution $y(t)$, $t_0 \leqslant t \leqslant t_1$, of (1.4) which satisfies the condition $y(t_0) \in W'_0$ and intersects the set M at only a finite number of points. Hence, by Lemma 4.2

$$\omega(y(t_1)) - \omega(y(t_0)) \leqslant t_1 - t_0. \tag{4.4}$$

Furthermore, since $y(t_0) \in W'_0 \subset W_0$, it follows that $y(t_1) \in W_1$. Consequently, by virtue of the choice of the neighborhoods W_0 and W_1, we have

$$|\omega(y(t_0)) - \omega(x_0)| < \epsilon, \qquad |\omega(y(t_1)) - \omega(x_1)| < \epsilon;$$

in particular,

$$\omega(y(t_0)) - \omega(x_0) < \epsilon, \tag{4.5}$$
$$-\omega(y(t_1)) + \omega(x_1) < \epsilon. \tag{4.6}$$

Adding the inequalities (4.4)–(4.6), we obtain

$$\omega(x_1) - \omega(x_0) < (t_1 - t_0) + 2\epsilon.$$

The required inequality (4.2) follows from the arbitrariness of ϵ.

Lemma 4.3 provides the most general conditions under which the estimate (4.2) is valid. However, these conditions are not convenient since they are formulated for each control separately and it is exceedingly difficult to verify the validity of these conditions for all possible controls $u(t)$. It turns out, however, that by imposing some additional conditions on the set M, the conditions formulated in Lemma 4.3 are satisfied for any admissible control $u(t)$. Thus, by imposing these conditions on the set M, the estimate (4.2) proves valid without any restrictions on the control $u(t)$. We shall prove in § 11.44 that this is true, in particular, if M is a *piecewise smooth set* of dimension less than n. We shall not give here a precise definition of piecewise smooth sets (this will be done in the next section); for the time being, let us confine ourselves, for the sake of clarity, to the statement that for $n = 2$ the piecewise smooth sets of dimension less than 2 are curves consisting of separate differentiable pieces (for example, the curve AOB in Figure 26 or the switching curves in the examples given in § 10.38–§ 10.40).

Thus, we have the following statement whose proof, along with an explanation of the formulation, will be given in the next sections.

Fundamental Lemma. Let M be a piecewise smooth set of dimension less than n lying in an open set G of the phase space X. Let us assume that a continuous function $\omega(x) = \omega(x^1, \ldots, x^n)$ is given on the set G, which has continuous derivatives satisfying inequality (4.1) at all points of G not belonging to M. If $u(t)$, $t_0 \leqslant t \leqslant t_1$, is an admissible control transferring the phase point from the state x_0 to the state x_1 and if the corresponding trajectory $x(t)$ lies entirely in the set G, then the estimate of the transition time (4.2) is valid.

We now can formulate and prove a theorem which represents a necessary and sufficient condition for optimality in the form of the method of dynamic programming. The weakening of the requirement of continuous differentiability of $\omega(x)$ makes the range of applicability of this theorem very wide (this will be clarified in § 12.45 and § 12.46 in the proof of the sufficient condition for optimality in the form of the maximum principle). In this connection, Theorem 4.4 may be considered as a basis of the method of dynamic programming. For the sake of simplicity, we shall formulate this theorem only under the assumption that the set G coincides with the entire phase space X (the changes which occur in the formulation of this theorem if the optimal trajectories do not fill out the entire phase space are quite obvious, see Figures 137, 149, and 151).

Theorem 4.4. Let M be a piecewise smooth set of dimension less than n lying in the phase space X and let $\omega(x) = \omega(x^1, \ldots, x^n)$ be a continuous function defined on X having a continuous derivative at the points not belonging to the set M. Furthermore, let $\omega(x_1) = 0$ for a point $x_1 \in X$. Assume that for every point $x_0 \in X$ different from x_1 there exists an admissible control $u(t) = u_{x_0}(t)$ which transfers the phase point from the state x_0 to the state x_1 in the time $-\omega(x_0)$. In order that all the controls $u_{x_0}(t)$ be optimal, it is necessary and sufficient that the function $\omega(x)$ satisfy Bellman's equation (1.18) [or, equivalently, the inequality (1.12)] at all points x which do not belong to the set M.

PROOF.   The necessity of this condition is proved in the same way as on pp. 14–17 (the proof given there is applicable to every point $x = x_0$ at which the derivatives $\partial \omega / \partial x^i$ exist and are continuous). The sufficiency follows from the fundamental lemma. In fact, by the condition of the theorem, the point x_0 can be transferred to x_1 in the time $-\omega(x_0)$ with the aid of a control $u(t) = u_{x_0}(t)$; a transfer from x_0 to x_1 in a shorter time is not possible [see (4.3)], that is, the control $u_{x_0}(t)$ is optimal.

43. Piecewise Smooth Sets. In this section we shall give the precise definition of piecewise smooth sets and in the next section we shall prove the fundamental lemma formulated in § 11.42. A number of concepts from the theory of smooth manifolds are used in the proof of the fundamental lemma. To familiarize oneself with the necessary concepts and theorems of this theory, the reader is referred to the book of Pontryagin [7] (it suffices to read the first 20 pages of this book). A reader unacquainted with the above material is advised to skip the proof of the fundamental lemma (§ 14.44).

Thus, let us proceed to the definition of the concept of a *piecewise smooth set*. Let K be a bounded s-dimensional convex polyhedron ($s \leqslant n$) considered together with its boundary (that is, a closed polyhedron) in a vector space Ξ of the variables $\xi^1, \xi^2, \ldots, \xi^s$. Assume that n continuously differentiable functions

$$\varphi^i(\xi^1, \xi^2, \ldots, \xi^s), \qquad i = 1, 2, \ldots, n, \tag{4.7}$$

are defined in an open set in Ξ containing the polyhedron K and that these functions have the property that the functional matrix

$$\left(\frac{\partial \varphi^i}{\partial \xi^j} \right) \qquad (i = 1, \ldots, n; j = 1, \ldots, s)$$

has rank s at every point $\xi \in K$. The functions (4.7) induce a *smooth mapping* φ of the polyhedron K into the space X through the formulas

$$x^i = \varphi^i(\xi^1, \ldots, \xi^s), \qquad i = 1, \ldots, n. \tag{4.8}$$

If this mapping is one-to-one (that is, different points of K are transformed into different points of X) then the image $L = \varphi(K)$ of K is called a *curvilinear s-dimensional polyhedron in X*. Evidently, a curvilinear polyhedron is a closed bounded set in X.

Now, let G be an open set of the phase space X. Any set $M \subset G$ which can be represented as a union of a finite or infinite number of curvilinear polyhedra arranged in such a way that any closed bounded set lying in G intersects only a finite number of these polyhedra will be called a *piecewise smooth set in G*. (The polyhedra may "cluster" at the boundary of the set G.) If there is at least one k-dimensional polyhedron among the curvilinear polyhedra whose union is a piecewise smooth set M, and if the dimensionality of the remaining polyhedra is $\leqslant k$, then the piecewise smooth set is called *k-dimensional*. In particular, *any smooth surface of dimension less than n which is closed in G is a piecewise smooth set in G* (since it can be decomposed into curvilinear polyhedra) [3]. It is obvious that a piecewise smooth set in G of dimension less than n does not contain interior points.

44. Proof of the Fundamental Lemma.   Let $L = \varphi(K)$ be a curvilinear polyhedron [see (4.8)] lying in the region G, and let $x(t)$, $t_0 \leqslant t \leqslant t_1$, be a phase trajectory in G corresponding to the admissible control $u(t)$, $t_0 \leqslant t \leqslant t_1$, that is, it is a solution of (1.5). One says that the phase trajectory $x(t)$ is in general position with a polyhedron L of dimension less than $n - 1$ if this trajectory does not intersect the polyhedron; the trajectory $x(t)$ is in general position with an $(n - 1)$-dimensional polyhedron L if the following conditions are satisfied:

(1)   The trajectory $x(t)$ does not intersect the boundary of the polyhedron L;

(2)   Let $\tau_1, \tau_2, \ldots, \tau_{k-1}$ be all the points of discontinuity of the control $u(t)$; then none of the points

$$x(t_0), \ x(\tau_1), \ x(\tau_2), \ \ldots, \ x(\tau_{k-1}), \ x(t_1)$$

belong to the polyhedron L;

(3)   The trajectory $x(t)$ is not tangent to the polyhedron L at any of its points, that is, if $x(t') \in L(t_0 < t' < t_1)$, then the vector $f(x(t'), u(t'))$ does not lie in the tangent plane of the polyhedron L drawn at the point $x(t')$ (in particular, this vector is not zero).

If the phase trajectory $x(t)$ is in general position with an $(n - 1)$-dimensional polyhedron L, then each of their common points is an isolated point on the trajectory $x(t)$, and therefore [since the trajectory $x(t)$, $t_0 \leqslant t \leqslant t_1$, is a closed bounded set], there exists only a finite number of points at which the trajectory $x(t)$ intersects the polyhedron L.

Lemma 4.5.   Let $u(t)$, $t_0 \leqslant t \leqslant t_1$, be an admissible control transferring the phase point from the state x_0 to the state x_1 and let the corresponding trajectory $x(t)$ lie entirely in the set G. Furthermore, let L be a curvilinear polyhedron in G of dimension $\leqslant n - 1$. Then for any neighborhood W_0 of x_0 there exists an open set $V \subset W_0$ such that for any point $y \in V$ the solution $y(t)$ of (1.5) with initial condition $y(t_0) = y_0$ is defined on the entire interval $t_0 \leqslant t \leqslant t_1$ and is in general position with the polyhedron L.

PROOF.   First of all, we can assume (decreasing if necessary the neighborhood W_0) that for any point $y_0 \in W_0$, the solution $y(t)$ of (1.5) with initial condition $y(t_0) = y_0$ is defined on the entire interval $t_0 \leqslant t \leqslant t_1$.

At first, let the polyhedron L be of dimension $\leqslant n - 2$. We choose in G an open manifold N of the same dimensionality as L containing the polyhedron L [such a manifold exists since the mapping (4.8) which produces the curvilinear polyhedron L is defined not only on the polyhedron K but also on a neighborhood of K].

The *direct product* $N \times [t_0, t_1]$ of the manifold N and the interval $[t_0, t_1]$ (see [7], p. 14) is a manifold with an edge having dimension $\leqslant n - 1$. Let P denote the set of all points $(x', t') \in N \times [t_0, t_1]$ for which the solution $x(t; x', t')$ of (1.5) with initial condition $x(t') = x'$ is defined on the interval $t_0 \leqslant t \leqslant t'$. The set P is an open subset (that is, a submanifold) of the manifold $N \times [t_0, t_1]$. For each point $(x', t') \in P$, we consider the trajectory $x(t; x', t')$ [that is, the solution of (1.5) with initial condition $x(t') = x'$] and set $x(t_0; x', t') = \psi(x', t')$. We obtain a mapping ψ (obviously continuous) of the manifold P into the region G.

We shall show that the image $\psi(P)$ of this mapping is a *set of the first category* in G (that is, it is representable as a union of not more than a denumerable number of nowhere dense sets). For this purpose we set $\tau_0 = t_0$, $\tau_k = t_1$ [recall that $\tau_1, \tau_2, \ldots, \tau_{k-1}$ are the points of discontinuity of the control $u(t)$] and let P_i, $i = 1, 2, \ldots, k$, denote the set of all points $(x', t') \in P$ for which $\tau_{i-1} \leqslant t' \leqslant \tau_i$. It is clear that $P = P_1 \cup P_2 \cup \cdots \cup P_k$, and therefore

$$\psi(P) = \psi(P_1) \cup \psi(P_2) \cup \cdots \cup \psi(P_k).$$

Thus, it suffices to prove that each of the sets $\psi(P_1)$, $\psi(P_2)$, $\ldots$, $\psi(P_k)$ in G is of the first category. This proof is identical for all the sets $\psi(P_1)$, $\ldots$, $\psi(P_k)$. We shall carry it out for $\psi(P_k)$.

Since for $\tau_{k-1} \leqslant t \leqslant \tau_k$ the right-hand side of (1.5) depends continuously on $x^1, \ldots, x^n, t$ and is continuously differentiable with respect to $x^1, \ldots, x^n$, the point $x(t; x', t')$ is continuously differentiable with respect to x', t' for $(x', t') \in P_k$, $\tau_{k-1} \leqslant t \leqslant t'$ (by virtue of the theorem on the differentiability of solutions with respect to initial values, see [8], Theorem 17). Therefore the mapping $(x', t') \to x(\tau_{k-1}; x', t')$ is a smooth (class 1) mapping of the manifold P_k into G. By virtue of the same theorem on the differentiability with respect to initial values applied to (1.5) for $\tau_{k-2} \leqslant t \leqslant \tau_{k-1}$, the point $x(\tau_{k-2}; x', t') = x(\tau_{k-2}; x(\tau_{k-1}; x', t'), \tau_{k-1})$ depends smoothly on $x(\tau_{k-1}; x', t')$, that is, depends smoothly on $(x', t') \in P_k$ by virtue of what has been proved above. Then, considering the intervals $\tau_{k-3} \leqslant t \leqslant \tau_{k-2}$, $\ldots$, $\tau_0 \leqslant t \leqslant \tau_1$, we finally obtain that the point $x(\tau_0; x', t')$ [or equivalently, $\psi(x', t')$] depends smoothly on x', t' for $(x', t') \in P_k$. In other words, the mapping ψ considered on P_k is a smooth (class 1) mapping. Consequently, the set $\psi(P_k)$ is of the first category in G (see [7], Theorem 1, p. 15).

Thus, the set $\psi(P)$ is of the first category in G. Therefore there exists a point y_0 in W_0 such that $y_0 \notin \psi(P)$. Let us consider the solution $y(t)$ with initial condition $y(t_0) = y_0$. By virtue of the choice of the neighborhood W_0, this solution is defined on the entire interval $t_0 \leqslant t \leqslant t_1$. Furthermore, the solution $y(t)$ does not intersect the manifold N. In fact, if there were a t', $t_0 \leqslant t' \leqslant t_1$, such that $y(t') \in N$, then we would have $(y(t'), t') \in P$ [since the solution $y(t)$ is defined on the interval $t_0 \leqslant t \leqslant t'$]. By definition, the solution $x(t; y(t'), t')$ satisfies the initial condition $x(t') = y(t')$ and therefore coincides (by virtue of the uniqueness theorem) with the solution $y(t)$. But then $y_0 = y(t_0) = x(t_0; y(t'), t') = \psi(y(t'), t') \in \psi(P)$, which contradicts the choice of the point y_0.

Thus, the trajectory $y(t)$ with initial condition $y(t_0) = y_0$ is defined on the entire interval $t_0 \leqslant t \leqslant t_1$, does not intersect the manifold N, and hence does not intersect L. It follows from the theorem on the continuous dependence of a solution on initial values (since L is closed and bounded) that there exists a neighborhood $W_0' \subset W_0$ of the point y_0 such that any solution $x(t)$ of (1.5) for which $x(t_0) \in W_0'$ does not intersect L (these solutions are defined on the entire interval $t_0 \leqslant t \leqslant t_1$ by virtue of the inclusion $W_0' \subset W_0$). This completes the consideration of the case in which the polyhedron has dimension less than $n - 1$.

Let us now consider the case in which L is an $(n - 1)$-dimensional polyhedron. Then the above arguments are applicable to any face of the polyhedron L. Since L has only a finite number of faces, there exists an open set $W_0'' \subset W_0$ such that the solutions $x(t)$ satisfying the condition $x(t_0) \in W_0''$ are defined on the entire interval $t_0 \leqslant t \leqslant t_1$ and do not intersect the boundary of the polyhedron L.

Thus, condition (1) in the definition of general position is satisfied for any open set $V \subset W_0''$.

Let us now consider condition (2). We choose in G an open $(n - 1)$-dimensional manifold N containing the polyhedron L. Let N_i ($i = 0, 1, \ldots, k$) denote the set of points $x' \in N$ for which the solution $x_i(t; x')$ of (1.5) with initial condition $x(\tau_i) = x'$ is defined on the interval $\tau_0 \leqslant t \leqslant \tau_i$. Then N_i is an open subset (that is, a submanifold) of the manifold N. For each point $x' \in N_i$, we shall consider the solution $x_i(t; x')$ [that is, the solution of (1.5) with initial condition $x(\tau_i) = x'$] and set $\psi_i(x') = x_i(\tau_0; x')$. We obtain a mapping ψ_i (obviously continuous) of N_i into G.

As before, one establishes that ψ_i is a smooth (class 1) mapping of N_i into G. Since N (and consequently N_i) is $(n - 1)$-dimensional, the image $\psi_i(N_i)$ is a set of the first category in G. Consequently,

$$\psi_0(N_0) \cup \psi_1(N_1) \cup \cdots \cup \psi_k(N_k)$$

is a set of the first category in G. Therefore, there exists a point y_0 in W_0'' such that $y_0 \notin \psi_0(N_0) \cup \psi_1(N_1) \cup \cdots \cup \psi_k(N_k)$. Let us consider the solution $y(t)$ with initial condition $y(t_0) = y_0$. This solution is defined on the entire interval $t_0 \leqslant t \leqslant t_1$ and, as can be readily seen, satisfies the condition

$y(\tau_i) \notin N$, $i = 0, 1, \ldots, k$. It follows from the theorem on continuous dependence of a solution on initial values that there exists a neighborhood $W_0''' \subset W_0''$ of the point y_0 such that the relations $x(\tau_i) \notin N$, $i = 0, 1, \ldots, k$, and consequently the relations $x(\tau_i) \notin L$, $i = 0, 1, \ldots, k$ are satisfied for any solution $x(t)$ of (1.5) for which $x(t_0) \in W_0'''$ (these solutions are defined on the entire interval $t_0 \leqslant t \leqslant t_1$).

Thus, conditions (1) and (2) in the definition of general position are satisfied for any open set $V \subset W_0'''$.

Finally, let us consider condition (3). The direct product $N \times [t_0, t_1]$ is an n-dimensional manifold with an edge. Let Q denote the set of points $(x', t') \in N \times [t_0, t_1]$ for which the solution $x(t; x', t')$ of (1.5) with initial condition $x(t') = x'$ is defined on the interval $t_0 \leqslant t \leqslant t'$. The set Q is a submanifold of the manifold $N \times [t_0, t_1]$. Let us divide the manifold Q into $Q_1, Q_2, \ldots, Q_k$ in which Q_i is the totality of all points $(x', t') \in Q$ for which $\tau_{i-1} \leqslant t' \leqslant \tau_i$. Furthermore, let us define a mapping ψ of the manifold Q into the region G by setting $\psi(x', t') = x(t_0; x', t')$. As before, ψ, considered on Q_i, is a smooth (class 1) mapping.

Let us assume that the trajectory $y(t)$ with initial condition $y(t_0) = y_0 \in W_0'''$ is tangent to the manifold N at the point x' at time $t'(t_0 \leqslant t' \leqslant t_1)$. [The time t' is distinct from $\tau_0, \tau_1, \ldots, \tau_k$ by virtue of condition (2) which holds for $y_0 \in W_0'''$.] Then $x(t' + dt; x', t') = x' + dx$, where dx is a vector tangent to the manifold N and $dt \neq 0$. In other words, $x(t' + dt; x', t') = x(t' + dt; x' + dx, t' + dt)$; hence, by virtue of the uniqueness theorem $x(t; x', t') = x(t; x' + dx, t' + dt)$. In particular, for $t = t_0$ we obtain $\psi(x', t') = \psi(x' + dx, t' + dt)$. This means that a nonzero tangent vector (dx, dt) of the manifold $N \times [t_0, t_1]$ at the point (x', t') is transformed into zero under the tangent mapping ψ, that is, the tangent mapping degenerates at the point (x', t'). In other words, the point (x', t') is not a regular point of the mapping ψ, and therefore the point $\psi(x', t') = y_0$ belongs to the image of the set of irregular points (the definition of regular and irregular points is given in [7], p. 10).

Thus, if a trajectory emanating from the point $y_0 \in W_0'''$ is tangent to the manifold N, then y_0 belongs to the image of the set of irregular points under the mapping ψ. But under a smooth (class 1) mapping of an n-dimensional manifold into an n-dimensional manifold, the image of the set of irregular points is of the first category (see for example [4]). Therefore, there exists a point y_0 in W_0''' which does not belong to the image of the set of irregular points. The trajectory $y(t)$ emanating from this point y_0 is not tangent to the manifold N, that is, it is not tangent to the polyhedron L and therefore is in general position with L. In particular, the trajectory $y(t)$ intersects L at only a finite number of points at which $y(t)$ is not tangent to L. Hence, it is not difficult to deduce that there exists a neighborhood $V \subset W_0'''$ of the point y_0 such that for $x(t_0) \in V$ the trajectory $x(t)$ [defined on the interval $t_0 \leqslant t \leqslant t_1$ and satisfying the conditions (1) and (2)] also intersects

L at a finite number of points at which it is not tangent to L. In other words, for $x(t_0) \in V$, the trajectory $x(t)$ satisfies the conditions (1), (2), and (3).

Thus Lemma 4.5 has been completely proved.

Lemma 4.6.   Let M be a piecewise smooth set in G of dimension $\leqslant n - 1$. Furthermore, let $u(t)$, $t_0 \leqslant t \leqslant t_1$, be an admissible control transferring the phase point from the state x_0 to the state x_1 and let the corresponding trajectory $x(t)$, $t_0 \leqslant t \leqslant t_1$, lie entirely in G. Then, in any neighborhood W_0 of the point x_0, there exists a point y_0 such that the trajectory $y(t)$, $t_0 \leqslant t \leqslant t_1$, emanating from y_0 and corresponding to the control $u(t)$ lies entirely in G and intersects M at only a finite number of points [that is, there exists only a finite number of times t, $t_0 \leqslant t \leqslant t_1$, for which $y(t) \in M$].

PROOF.   The trajectory $x(t)$, $t_0 \leqslant t \leqslant t_1$, lies entirely in G and is a closed bounded set. Therefore there exists a neighborhood $W \subset G$ of this trajectory which intersects only a finite number of the curvilinear polyhedra comprising the set M. Let us number these polyhedra L_1, L_2, . . . , L_ν.

We shall consider (1.5) to be obtained by replacing u in the right-hand side of (1.3) by the control $u(t)$ mentioned in the formulation of the lemma. Then $x(t)$ is the solution of (1.5) defined for $t_0 \leqslant t \leqslant t_1$ and satisfying the initial condition $x(t_0) = x_0$. By virtue of the theorem on continuous dependence of a solution on initial values, there exists a neighborhood $W_0' \subset W_0$ of the point x_0 such that any solution $y(t)$ of (1.5) for which $y(t_0) \in W_0'$ is defined on the entire interval $t_0 \leqslant t \leqslant t_1$ and lies entirely in W.

By virtue of Lemma 4.5, there exists a neighborhood $W_0^{(1)} \subset W_0'$ such that any solution $y(t)$ for which $y(t_0) \in W_0^{(1)}$ is in general position with the polyhedron L_1. By the same lemma, there exists an open set $W_0^{(2)} \subset W_0^{(1)}$ such that for $y(t_0) \in W_0^{(2)}$ the solution $y(t)$ is in general position with the polyhedron L_2. Proceeding in the same manner, we obtain open sets $W_0^{(\nu)} \subset \cdots \subset W_0^{(2)} \subset W_0^{(1)} \subset W_0' \subset W_0$. For $y(t_0) \in W_0^{(\nu)}$, the solution $y(t)$ is in general position with L_1, L_2, . . . , L_ν. Moreover, it lies entirely in W [since $y(t_0) \in W_0'$] and therefore does not intersect any other polyhedra. Thus, for $y_0 = y(t_0) \in W_0^{(\nu)} \subset W_0$, the solution $y(t)$ intersects M at only a finite number of points.

Lemma 4.6 has been proved.

The validity of the fundamental lemma of the previous section follows directly from Lemma 4.6 and Lemma 4.3.

§ 12. SUFFICIENT CONDITION FOR OPTIMALITY IN THE FORM OF THE MAXIMUM PRINCIPLE

45. Regular Synthesis and Formulation of the Sufficient Condition.
First of all, we shall introduce the concept of *regular synthesis* for (1.3) for which the continuity of the derivatives $\partial f^i / \partial x^j$, $\partial f^i / \partial u^k$ will not be assumed.

Suppose that a piecewise smooth set N of dimension $\leqslant n - 1$, piecewise smooth sets

$$P^0 \subset P^1 \subset P^2 \subset \cdots \subset P^{n-1} \subset P^n = G, \qquad (4.9)$$

and a function $v(x)$ defined in G with values in U are given. We will say that the sets (4.9) and the function $v(x)$ realize a *regular synthesis* for (1.3) in the region G if the following conditions are satisfied.

(A)   The set P^0 contains the point $a = x_1$ but does not have limiting points in the open set G. Each component of the set $P^i - (P^{i-1} \cup N)$ ($i = 1$, $2, \ldots, n$) is a smooth i-dimensional manifold in G; these components will be called *i-dimensional cells*. The points of the set P^0 will be called zero-dimensional cells. The function $v(x)$ is continuous and continuously differentiable on each cell and can be extended as a continuously differentiable function into a neighborhood of the cell.

(B)   All cells are grouped into cells of the first and second kind. All n-dimensional cells are cells of the first kind, all zero-dimensional cells are cells of the second kind.

(C)   If σ is an i-dimensional cell of the first kind, then each point of this cell has a unique trajectory of the equation

$$\dot{x} = f(x, v(x)) \qquad (4.10)$$

passing through it. There exists an $(i - 1)$-dimensional cell $\Pi(\sigma)$ such that each trajectory of the system (4.10) traversing the cell σ leaves this cell after a finite time, arrives at the cell $\Pi(\sigma)$ at a nonzero angle, and approaches the latter with a nonzero phase velocity. If σ is a one-dimensional cell of the first kind, then it is a piece of a phase trajectory of (4.10) approaching a zero-dimensional cell $\Pi(\sigma)$ with a nonzero phase velocity. If σ is an i-dimensional cell of the second kind distinct from the point a, then there exists an $(i + 1)$-dimensional cell $\Sigma(\sigma)$ of the first kind such that from any point of the cell σ there emanates a unique trajectory of (4.10) traversing the cell $\Sigma(\sigma)$; moreover, the function $v(x)$ is continuous and continuously differentiable on $\sigma \cup \Sigma(\sigma)$.

(D)   The above conditions make it possible to continue the trajectories of (4.10) from cell to cell: from the cell σ to the cell $\Pi(\sigma)$ if $\Pi(\sigma)$ is of the first kind, and from the cell σ to the cell $\Sigma(\Pi(\sigma))$ if $\Pi(\sigma)$ is of the second kind. It is required that each such trajectory traverse only a finite number of cells (that is, each such trajectory "pierces" only a finite number of cells of the second kind). Moreover, each such trajectory terminates at the point a. The above trajectories will be called *"distinguished"* trajectories. Thus, a single distinguished trajectory (leading to the point a) emanates from every point of the set $G - N$. It is also required that a (possibly nonunique) trajectory of system

(4.10) leading to the point a emanates from every point of the set N. These will also be called "distinguished" trajectories.

(E)   All distinguished trajectories satisfy the maximum principle.

(F)   The value of the transition time from the point x_0 to the point a, calculated along distinguished trajectories (terminating at the point a), is a continuous function of the initial point x_0. (In particular, if several distinguished trajectories emanate from the point $x_0 \in N$, then the value of the transition time is the same for these trajectories.)

All the examples of the synthesis of optimal controls given in § 10.38–§ 10.40 are special cases of regular synthesis. We shall verify this on the example discussed on pp. 144–156. In this case, we take $G = P^2$ to be the region of controllability of the system, that is, the entire region in which the synthesis of optimal controls is realized. All curves A_iO, B_iA_i, $B_{i-1}A_i$, B_iC_i, $C_{i-1}B_i$, . . . , which are the sides of the curvilinear quadrangles constructed in realizing the synthesis, form a piecewise smooth set P^1 (Figure 162). Finally, the set P^0 consists of all the points O ($= a$), A_i, B_i, C_i, ... Of course, the function which realizes the synthesis of optimal controls will serve as $v(x)$ [that is, $v(x)$ takes on the value e_i in the interior of the "angle"

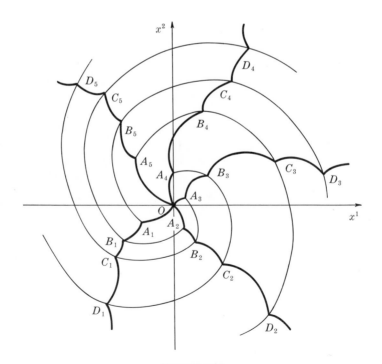

FIGURE 162

between the curves $OA_{i+1}B_{i+1}C_{i+1}$ . . . and $OA_iB_iC_i$ . . . and on the arc A_iO].

The validity of conditions (A)–(F) for a regular synthesis can be verified without difficulty. Among one-dimensional cells, the arcs A_iO and also the arcs $B_{i-1}A_i$, $C_{i-1}B_i$, . . . will be considered as cells of the first kind; the remaining arcs (that is, A_iB_i, B_iC_i, . . .) will be considered as cells of the second kind. The "curvilinear quadrangles" are the two-dimensional cells. We note, for example, that for the cell $\sigma = OA_iB_iA_{i+1}$, the cell $\Pi(\sigma)$ coincides with the arc $A_{i+1}O$, and for the cell $\sigma' = A_iB_i$, the cell $\Sigma(\sigma')$ coincides with $OA_iB_iA_{i+1}$. Thus, conditions (A)–(F) are readily verified; the continuity of the transition time [condition (F)] is also clear.

The validity of the conditions of regular synthesis is easily verified in other examples. Thus, in the example depicted in Figure 26, there are only two one-dimensional cells AO and BO, one zero-dimensional cell O $(= a)$, and two two-dimensional cells.

Conditions (A)–(F) do not impose essential limitations on (1.2) but rather postulate properties which usually hold in realizing a synthesis. It is very important that the fulfillment of these conditions is sufficient for all the distinguished trajectories to be optimal. In fact, we have the following theorem which makes it possible to assert that a synthesis, realized on the basis of the maximum principle, as a rule [that is, if the conditions (A)–(F) are satisfied], leads to optimal trajectories. In this sense the maximum principle is a sufficient condition for optimality.

Theorem 4.7. *If a regular synthesis is realized in G for* (1.3) (*assuming the existence of continuous derivatives $\partial f^i/\partial x^j$, $\partial f^i/\partial u^k$), then all distinguished trajectories are optimal* (*in G*).

46. Proof of Sufficiency. We turn now to the proof of Theorem 4.7. Let $\omega(x)$ denote the duration of the motion of the phase point from x along a distinguished trajectory leading from the point x to the point a. The set $P^{n-1} \cup N$ will be denoted by M. We shall prove that $\omega(x)$ is Bellman's function with singular set M. Then Theorem 4.7 will follow directly from Theorem 4.4. Thus, it is only necessary to prove that the function $\omega(x)$ is differentiable on the set $G - M$ and that it satisfies Bellman's equation (1.18).

Let x be an arbitrary point belonging to an n-dimensional cell σ. Let us select an arbitrary number t_0 and let $t_0 + \theta_1(x)$ be the instant at which the trajectory of (4.10) emanating from the point x at time t_0 reaches the cell $\Pi(\sigma)$ [that is, $\theta_1(x)$ is the time of motion from x to $\Pi(\sigma)$]. The point at which this trajectory "lands" on the cell $\Pi(\sigma)$ will be denoted by $\xi_1(x)$. From the general theorems on the differentiability of a solution with respect to parameters it follows that $\xi_1(x)$ and $\theta_1(x)$ are continuously differentiable functions of x. Indeed, let x_0 be an arbitrary point of the n-dimensional cell σ under consideration. Let us reverse the course of time, that is, let us consider the system

$$\dot{y} = -f(y, v(y)) \tag{4.11}$$

on the cell σ. The trajectories of this system (in the cell σ) coincide with the trajectories of (4.10) but run in the opposite direction. Let $y(t, \xi)$ denote the solution of (4.11) with initial condition $y(0, \xi) = \xi$ for any point $\xi \in \Pi(\sigma)$ [close to $\xi_1(x_0)$]. Then $y(t, \xi)$ is continuously differentiable jointly in the variables $t, \xi[t > 0, \xi \in \Pi(\sigma)]$. Obviously, we have

$$y(\theta_1(x_0), \xi_1(x_0)) = x_0. \tag{4.12}$$

It is not difficult to see that the functional determinant*

$$\frac{D(y(t, \xi))}{D(t, \xi)} \bigg|_{t = \theta_1(x_0), \xi = \xi_1(x_0)}$$

is different from zero. Indeed, for $t = 0$, $\xi = \xi_1(x_0)$ the functional determinant is different from zero since, by virtue of condition (C), the trajectory $x(t)$ of (4.10) emanating from x_0 approaches the cell $\Pi(\sigma)$ at a nonzero angle. Consequently, the functional determinant differs from zero also for $t = \theta_1(x_0)$, $\xi = \xi_1(x_0)$ since the variational system of equations is linear.

Therefore, the equation $y(t, \xi) = x$ is uniquely solvable when x is close to x_0 [see (4.12)]:

$$\xi = \xi_1(x), \, t = \theta_1(x);$$

moreover, the functions $\xi_1(x)$ and $\theta_1(x)$ are continuously differentiable with respect to x.

Next, the trajectory is continued from the point $\xi_1(x)$ to the cell $\Pi(\sigma)$ or $\Sigma(\Pi(\sigma))$; moreover, it is established in an analogous manner that the point $\xi_2(x)$ at which this trajectory leaves $\Pi(\sigma)$ [or $\Sigma(\Pi(\sigma))$] and the time $\theta_2(x)$ of the motion within this cell depend differentiably on $\xi_1(x)$ and hence on x. Proceeding in this manner, we find that the total time $-\omega(x) = \theta_1(x) + \theta_2(x) + \cdots$ of the motion (along the distinguished trajectory) from x to a is (in the interior of the cell σ) a continuously differentiable function of the point x.

Thus, $\omega(x)$ is continuously differentiable on $G - M$.

It remains to establish that $\omega(x)$ satisfies Bellman's equation (1.18) on $G - M$. Let $x_0 \in G - M$; let $x(t)$ denote the distinguished trajectory emanating from x_0 at time t_0, and let t_1 be the time at which this trajectory reaches the point a. Let us consider the set S consisting of all points satisfying the condition

$$\omega(x) = \omega(x_0).$$

---

* The point $y(t, \xi)$ (in the cell σ) has n coordinates; the point ξ [in the cell $\Pi(\sigma)$] has $n - 1$ coordinates.

Then near x_0, the set S is a smooth hypersurface in G with normal vector

$$\text{grad } \omega(x_0) = \left(\frac{\partial \omega(x_0)}{\partial x^1}, \cdots, \frac{\partial \omega(x_0)}{\partial x^n} \right).$$

This vector is different from zero since

$$\sum_{\alpha=1}^{n} \frac{\partial \omega(x_0)}{\partial x^\alpha} f^\alpha(x_0, v(x_0)) = \frac{d\omega(x(t))}{dt}\bigg|_{x(t)=x_0} = 1. \tag{4.13}$$

According to condition (F), the trajectory $x(t)$ satisfies the maximum principle. Let $\psi(t) = (\psi_1(t), \ldots, \psi_n(t))$ denote the vector function which corresponds to $x(t)$ by virtue of the maximum principle. We shall show that $\psi(t_0)$ is orthogonal to the surface S at x_0, that is,

$$\psi(t_0) = \lambda \text{ grad } \omega(x_0)$$

or, equivalently,

$$\psi_\alpha(t_0) = \lambda \frac{\partial \omega(x_0)}{\partial x^\alpha}, \qquad \alpha = 1, 2, \ldots, n. \tag{4.14}$$

Let us suppose that (4.14) has been established. Then, by the maximum principle, we have

$$H = \sum_{\alpha=1}^{n} \psi_\alpha(t_0) f^\alpha(x_0, v(x_0)) = \lambda \sum_{\alpha=1}^{n} \frac{\partial \omega(x_0)}{\partial x^\alpha} f^\alpha(x, v(x_0)) = \lambda$$

[see (4.13)]. We conclude, from the relation $H \geqslant 0$ contained in the maximum principle, that $\lambda \geqslant 0$. In addition, $\lambda \neq 0$ [since otherwise we would have $\psi(t_0) = 0$, see (4.14)]. Thus, $\lambda > 0$. Furthermore, we find from the maximum principle that

$$H(\psi(t_0), x_0, v(x_0)) \geqslant H(\psi(t_0), x_0, u) \qquad \text{for any } u \in U;$$

hence [from (4.13), (4.14), and the relation $\lambda > 0$], we obtain

$$1 = \sum_{\alpha=1}^{n} \frac{\partial \omega(x_0)}{\partial x^\alpha} f^\alpha(x_0, v(x_0)) = \frac{1}{\lambda} \sum_{\alpha=1}^{n} \psi_\alpha(t_0) f^\alpha(x_0, v(x_0))$$

$$= \frac{1}{\lambda} H(\psi(t_0), x(t_0), v(x_0)) \geqslant \frac{1}{\lambda} H(\psi(t_0), x(t_0), u)$$

$$= \frac{1}{\lambda} \sum_{\alpha=1}^{n} \psi_\alpha(t_0) f^\alpha(x_0, u) = \sum_{\alpha=1}^{n} \frac{\partial \omega(x_0)}{\partial x^\alpha} f^\alpha(x_0, u)$$

for any $u \in U$. Thus, relation (1.18) is satisfied in $G - M$.

It remains to establish the validity of (4.14). Let $\sigma_1, \sigma_2, \ldots, \sigma_q$ be those cells of the first kind through which the trajectory $x(t)$ passes consecutively so that $x_0 \in \sigma_1$, and σ_q is the one-dimensional cell abutting the point a. Let us set $t_0 = \tau_0$, $t_1 = \tau_q$, and let $\tau_1, \tau_2, \ldots, \tau_{q-1}$ denote the switching times (that is, the times of crossing from one cell to another) so that the trajectory $x(t)$ traverses the cell σ_i $(i = 1, \ldots, q)$ on the interval $\tau_{i-1} < t < \tau_i$. For every two adjacent cells σ_i and σ_{i+1} in the sequence $\sigma_1, \sigma_2, \ldots, \sigma_q$, one of the following two cases is possible [see conditions (C) and (D) in the definition of regular synthesis]:

(a)   Both cells σ_i and σ_{i+1} are k-dimensional and then $\sigma_{i+1} = \Sigma(\Pi(\sigma_i))$; in this case, at the "switching time" τ_i, the trajectory $x(t)$ pierces $\Pi(\sigma_i)$ which is a $(k-1)$-dimensional cell of the second kind.

(b)   The cell σ_i is k-dimensional and the cell σ_{i+1} is $(k-1)$-dimensional and coincides with the cell $\Pi(\sigma_i)$.

In both cases, the "switching point" $x(\sigma_i)$ is an interior point of $\Pi(\sigma_i)$; moreover, a unique trajectory of (4.10) traversing the cell σ_{i+1} emanates from each point $\Pi(\sigma_i)$. Therefore the trajectory $x^*(t)$ of (4.10) emanating from any interior point x_0^* of the cell σ_1 at time t_0 traverses the same sequence of cells $\sigma_1, \sigma_2, \ldots, \sigma_q$ and arrives at the point a. We shall assume that x_0^* lies on the hypersurface S (sufficiently close to x_0) so that the time of motion along the trajectory $x^*(t)$ from x_0^* to a coincides with the time of motion along the trajectory $x(t)$. In other words, both trajectories $x(t)$ and $x^*(t)$, emanating from the points x_0 and x_0^* at time t_0, arrive at the point a at the same time $t_1 = -\omega(x_0) = -\omega(x_0^*)$.

As we have seen above, the "switching times" $\tau_0^* = t_0, \tau_1^*, \tau_2^*, \ldots,$ τ_{q-1}^* for the trajectory $x^*(t)$ and the corresponding "switching points" $x^*(\tau_1^*), x^*(\tau_2^*), \ldots, x^*(\tau_{q-1}^*)$, being interior points of the cells $\Pi(\sigma_1), \Pi(\sigma_2),$ $\ldots, \Pi(\sigma_{q-1})$, depend differentiably on the point $x_0^* \in \sigma_1$.

If x_0^* is sufficiently close to x_0, then the inequalities $\tau_{i-1} < \tau_i^*$ and $\tau_{i-1}^* < \tau_i$ are satisfied (since $\tau_{i-1} < \tau_i$). Let δ_i denote the time interval between τ_i and τ_i^*, $i = 1, 2, 3, \ldots, q-1$. Since each of the numbers τ_i and τ_i^* is less than each of the numbers τ_{i+1} and τ_{i+1}^* (provided the point x_0^* is sufficiently close to x_0), the entire interval δ_i lies to the left of the interval δ_{i+1}. Let Δ_1 denote the interval from time t_0 to the left-hand endpoint of the interval δ_1, let Δ_i $(i = 2, 3, \ldots, q-1)$ denote the interval from the right-hand endpoint of the interval δ_{i-1} to the left-hand endpoint of the interval δ_i, and let Δ_q denote the interval from the right-hand endpoint of the interval δ_{q-1} to the point t_1. Thus, the intervals

$$\Delta_1, \delta_1, \Delta_2, \delta_2, \Delta_3, \ldots, \delta_{q-1}, \Delta_q$$

abut on the axis. Moreover, on the time interval Δ_i both phase points $x(t)$ and $x^*(t)$ are in σ_i, and on the time interval δ_i one of these points is in σ_i and the other is in σ_{i+1}.

Let ϵ denote the distance between x_0 and x_0^* (we shall assume that ϵ is sufficiently small). It follows easily from the fact that the "switching times" $\tau_1^*, \tau_2^*, \ldots, \tau_{q-1}^*$ and the corresponding "switching points" $x^*(\tau_1^*)$, $x^*(\tau_2^*), \ldots, x^*(\tau_{q-1}^*)$ depend differentiably on the point $x_0^* \in \sigma_1$ that there exists a positive constant C such that the length of each of the intervals δ_i does not exceed $C\epsilon$, and the distance between the trajectories $x(t)$ and $x^*(t)$ is of order ϵ:

$$|x(t) - x^*(t)| \leqslant C\epsilon \qquad (t_0 \leqslant t \leqslant t_1). \tag{4.15}$$

As before, let $\psi(t) = (\psi_1(t), \ldots, \psi_n(t))$ denote the vector function corresponding to the trajectory $x(t)$ by virtue of the maximum principle. We have [due to the relation $x(t_1) = x^*(t_1) = a$]

$$-\sum_{\alpha=1}^{n} [x^\alpha(t_0) - x^{*\alpha}(t_0)]\psi_\alpha(t_0)$$

$$= -\sum_{\alpha=1}^{n} [x^\alpha(t_0) - x^{*\alpha}(t_0)]\psi_\alpha(t_0) + \sum_{\alpha=1}^{n} [x^\alpha(t_1) - x^{*\alpha}(t_1)]\psi_\alpha(t_1)$$

$$= +\int_{t_0}^{t_1} \frac{d}{dt}\left(\sum_{\alpha=1}^{n} (x^\alpha(t) - x^{*\alpha}(t))\psi_\alpha(t)\right) dt$$

$$= +\int_{t_0}^{t_1} \left(\sum_{\alpha=1}^{n} \psi_\alpha(t)\frac{d}{dt}(x^\alpha(t) - x^{*\alpha}(t)) + \sum_{\alpha=1}^{n} (x^\alpha(t) - x^{*\alpha}(t))\frac{d\psi_\alpha(t)}{dt}\right) dt$$

$$= +\int_{t_0}^{t_1} \left[\sum_{\alpha=1}^{n} \psi_\alpha(t)(f^\alpha(x(t), v(x(t))) - f^\alpha(x^*(t), v(x^*(t))))\right.$$

$$\left. -\sum_{\alpha=1}^{n} (x^\alpha(t) - x^{*\alpha}(t))\frac{\partial H(\psi(t), x(t), v(x(t)))}{\partial x^\alpha}\right] dt$$

$$= \int_{t_0}^{t_1} \left[H(\psi(t), x(t), v(x(t))) - H(\psi(t), x^*(t), v(x^*(t)))\right.$$

$$\left. -\sum_{\alpha=1}^{n} (x^\alpha(t) - x^{*\alpha}(t))\frac{\partial H(\psi(t), x(t), v(x(t)))}{\partial x^\alpha}\right] dt.$$

Thus

$$-\sum_{\alpha=1}^{n} [x^\alpha(t_0) - x^{*\alpha}(t_0)]\psi_\alpha(t_0) = \int_{t_0}^{t_1} F(t)\, dt, \tag{4.16}$$

where

$$F(t) = H(\psi(t), x(t), v(x(t))) - H(\psi(t), x^*(t), v(x^*(t)))$$
$$- \sum_{\alpha=1}^{n} (x^\alpha(t) - x^{*\alpha}(t)) \frac{\partial H(\psi(t), x(t), v(x(t)))}{\partial x^\alpha}$$

$$= H(\psi(t), x(t), v(x(t))) - \left\{ H(\psi(t), x(t), v(x^*(t))) \right.$$
$$\left. + \sum_{\alpha=1}^{n} (x^{*\alpha}(t) - x^\alpha(t)) \frac{\partial H(\psi(t), \xi, v(x^*(t)))}{\partial x^\alpha} \right\}$$
$$- \sum_{\alpha=1}^{n} (x^\alpha(t) - x^{*\alpha}(t)) \frac{\partial H(\psi(t), x(t), v(x(t)))}{\partial x^\alpha}$$

$$= -H(\psi(t), x(t), v(x^*(t))) + H(\psi(t), x(t), v(x(t)))$$
$$- \sum_{\alpha=1}^{n} (x^\alpha(t) - x^{*\alpha}(t)) \left[\frac{\partial H(\psi(t), x(t), v(x(t)))}{\partial x^\alpha} \right.$$
$$\left. - \frac{\partial H(\psi(t), \xi, v(x^*(t)))}{\partial x^\alpha} \right]$$

[ξ is a point of the segment joining $x(t)$ and $x^*(t)$].

Hence, we obtain

$$F(t) \geqslant - \sum_{\alpha=1}^{n} (x^\alpha(t) - x^{*\alpha}(t)) \left[\frac{\partial H(\psi(t), x(t), v(x(t)))}{\partial x^\alpha} \right.$$
$$\left. - \frac{\partial H(\psi(t), \xi, v(x^*(t)))}{\partial x^\alpha} \right]$$

[since $H(\psi(t), x(t), v(x(t))) \geqslant H(\psi(t), x(t), u)$ for every $u \in U$].

Thus

$$F(t) \geqslant G(t), \tag{4.17}$$

where

$$G(t) = - \sum_{\alpha=1}^{n} (x^\alpha(t) - x^{*\alpha}(t)) \left[\frac{\partial H(\psi(t), x(t), v(x(t)))}{\partial x^\alpha} \right.$$
$$\left. - \frac{\partial H(\psi(t), \xi, v(x^*(t)))}{\partial x^\alpha} \right]. \tag{4.18}$$

Now, if the point t belongs to one of the intervals Δ_i, then the points $x(t)$ and $x^*(t)$ belong to one and the same cell σ_i on which the function $v(x)$ is continuously differentiable. In addition, the trajectory $x^*(t)$ lies in a small neighborhood of the trajectory $x(t)$ [see (4.15)], which is a bounded closed set. Consequently, the estimate

$$|v(x^*(t)) - v(x(t))| \leqslant C'|x^*(t) - x(t)| \leqslant CC'\epsilon \qquad (t \in \Delta_i)$$

is valid. Similarly,

$$|\xi - x(t)| \leqslant |x^*(t) - x(t)| \leqslant C\epsilon.$$

From the continuity of the function $\partial H/\partial x^\alpha$ it follows that (for $t \in \Delta_i$) the difference

$$\frac{\partial H(\psi(t), x(t), v(x(t)))}{\partial x^\alpha} - \frac{\partial H(\psi(t), \xi, v(x^*(t)))}{\partial x^\alpha} \qquad (4.19)$$

is an infinitesimal of order ϵ (that is, this difference tends to zero, in fact, uniformly in t, as $\epsilon \to 0$). Finally, considering the estimate (4.15), we obtain, from (4.18),

$$\lim_{\epsilon \to 0} \frac{G(t)}{\epsilon} = 0 \qquad \text{(uniformly in } t \in \Delta_i);$$

hence we find

$$\lim_{\epsilon \to 0} \frac{1}{\epsilon} \int_{\Delta_i} G(t) \, dt = 0, \qquad i = 1, 2, \ldots, q. \qquad (4.20)$$

If the point t belongs to one of the intervals δ_i, then we cannot assert that the difference (4.19) is an infinitesimal of order ϵ [since the points $x(t)$ and $x^*(t)$ belong to different cells and the function $v(x)$ may undergo discontinuities during the transition from one cell to another]. However, the difference (4.19) remains bounded for all t [by virtue of the boundedness of the trajectory $x(t)$]. Therefore, by (4.15), we have

$$\lim_{\epsilon \to 0} G(t) = 0 \qquad \text{(uniformly in } t).$$

Since the length of the interval δ_i does not exceed $C\epsilon$, we obtain

$$\lim_{\epsilon \to 0} \frac{1}{\epsilon} \int_{\delta_i} G(t) \, dt = 0, \qquad i = 1, 2, \ldots, q - 1. \qquad (4.21)$$

Adding relations (4.20) and (4.21), we find

$$\lim_{\epsilon \to 0} \frac{1}{\epsilon} \int_{t_0}^{t_1} G(t) \, dt = 0. \qquad (4.22)$$

Now let the point x_0^* approach the point x_0 on the surface S along a curve which is tangent at x_0 to a vector $p = (p^1, p^2, \ldots, p^n)$. In other words,

$$\lim_{\epsilon \to 0} \frac{x^*(t_0) - x(t_0)}{\epsilon} = \lim_{\epsilon \to 0} \frac{x_0^* - x_0}{\epsilon} = p.$$

Then, by (4.16), (4.17), and (4.22), we have

$$\sum_{\alpha=1}^{n} p^{\alpha}\psi_{\alpha}(t_0) = \lim_{\epsilon \to 0} \sum_{\alpha=1}^{n} \frac{1}{\epsilon}(x^{*\alpha}(t_0) - x^{\alpha}(t_0))\psi_{\alpha}(t_0)$$

$$= \lim_{\epsilon \to 0} \frac{1}{\epsilon} \int_{t_0}^{t_1} F(t)\, dt \geq \lim_{\epsilon \to 0} \frac{1}{\epsilon} \int_{t_0}^{t_1} G(t)\, dt = 0.$$

Since the relation $\displaystyle\sum_{\alpha=1}^{n} p^{\alpha}\psi_{\alpha}(t_0) \geq 0$ holds for any vector p tangent to S,

$$\sum_{\alpha=1}^{n} p^{\alpha}\psi_{\alpha}(t_0) = 0$$

for any vector p tangent to the hyperplane S, from which (4.14) follows. Thus, Theorem 4.7 has been completely proved.

§ 13. EXAMPLES OF THE SYNTHESIS OF OPTIMAL CONTROLS IN NONLINEAR SYSTEMS OF THE SECOND ORDER

47. First Example.   The sufficient conditions for optimality obtained above play an important role for the following reason. In a number of cases, the maximum principle allows us to single out uniquely the trajectories which could be optimal. Are these trajectories actually optimal? To answer this question in the case in which (1.2) is linear, one may use the fact that the maximum principle for linear systems is not only a necessary but also a sufficient condition for optimality. One can also use the existence theorem for optimal controls: since optimal trajectories do exist and since the maximum principle uniquely defines a trajectory which may be optimal, it follows that this trajectory is, indeed, the (unique) optimal trajectory joining two given points. However, Theorem 3.3 and the existence theorem have been proved only for linear systems (1.2). Therefore, for nonlinear systems (even the simplest ones) we have no assurance that the trajectories synthesized on the basis of the maximum principle are actually optimal. The way out of this situation is indicated in Theorem 4.7.

As examples, we consider, in this section and in § 13.50, two nonlinear systems.*

Let us consider a controlled object whose behavior is described by the second-order differential equation

$$\ddot{x} = f(x, \dot{x}, u), \tag{4.23}$$

---

* E. Ya. Roitenberg participated in preparing these examples.

where u is a real control parameter subject to the conditions

$$-1 \leqslant u \leqslant 1. \tag{4.24}$$

In phase coordinates $x^1 = x$, $x^2 = \dot{x}$, (4.23) can be written in normal form as the system

$$\begin{aligned} \dot{x}^1 &= x^2, \\ \dot{x}^2 &= f(x^1, x^2, u). \end{aligned} \tag{4.25}$$

Let us impose some restrictions on the function f. Namely, we shall assume that this function is continuously differentiable with respect to all its arguments and satisfies the inequalities

$$f(x^1, x^2, +1) > 0, \qquad f(x^1, x^2, -1) < 0 \qquad \text{for all} \qquad x^1, x^2, \tag{4.26}$$

$$\frac{\partial f(x^1, x^2, u)}{\partial u} > 0, \qquad \frac{\partial f(x^1, x^2, u)}{\partial x^1} \geqslant 0 \qquad \text{for all} \qquad x^1, x^2, u. \tag{4.27}$$

For the controlled object (4.23), we shall consider the time-optimal problem of reaching $x = 0$ with zero velocity from a given initial state; in other words, we shall consider the time-optimal problem of reaching the origin for the object (4.25).

An object governed by the equation $\ddot{x} = u$ (compare pp. 24–30) is an example of a linear object satisfying all of the imposed conditions. Nonlinear objects which "differ little" from the above linear object also satisfy the imposed conditions.

We can consider the following "quasi-practical" situation which leads to the problem stated above. An emergency has occurred on a space station in a circular orbit about the Earth to which it is required to render help in the shortest time. For this purpose, a rocket is launched from the Earth along a straight line toward the space stations (considering the station's displacement along the orbit, Figure 163). Let us introduce a coordinate on

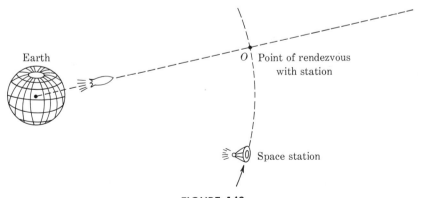

Earth

O Point of rendezvous with station

Space station

FIGURE 163

the above straight line taking the origin $x = 0$ to be the assumed point of rendezvous with the station. Then the equation of motion of the rocket can be written (neglecting the change of the rocket's mass due to fuel consumption) in the form

$$m\ddot{x} = \varphi(\dot{x}, u) - \frac{k}{(x + r)^2},$$

where the first term on the right-hand side is the thrust and the second term is the force of gravity (r is the distance of the station from the center of the Earth). Conditions (4.26) are satisfied quite naturally: for $u = +1$ ("full speed ahead") the right-hand side f is positive, while for $u = -1$ (firing the retrorockets) it is negative. Similarly, the first inequality in (4.27) is also satisfied. Finally, the validity of the second inequality in (4.27) for $x > -r$ (that is, in the domain of the rocket's motion) can be directly verified. Thus, conditions (4.26) and (4.27) are satisfied. These conditions are also satisfied for a rocket approaching a space station prior to returning to earth (Figure 164).

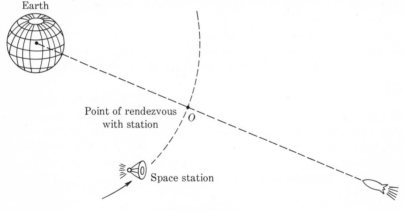

FIGURE 164

48. Description of the Synthesis.   Let us now proceed to the solution of the time-optimal problem we have posed for the object (4.24)–(4.27). Let α_a^+ denote the semitrajectory of the system

$$\begin{aligned} \dot{x}^1 &= x^2, \\ \dot{x}^2 &= f(x^1, x^2, 1) \end{aligned} \tag{4.28}$$

reaching the point $(a, 0)$, and let α_b^- denote the semitrajectory of the system

$$\begin{aligned} \dot{x}^1 &= x^2, \\ \dot{x}^2 &= f(x^1, x^2, -1) \end{aligned} \tag{4.29}$$

reaching the point $(b, 0)$.

Let us find the position and the approximate shape of the semitrajectories α_a^+ and α_b^-. If we move along the curve α_a^+ from the point $(a, 0)$, that is, in the direction opposite to the motion along the semitrajectory α_a^+, then, as seen from the second equation in (4.28) and the first inequality in (4.26), the coordinate x^2 decreases monotonically. Therefore, departing indefinitely from the point $(a, 0)$ along the curve α_a^+, x^2 will approach a limiting value which is equal to either a negative number or $-\infty$. Moreover, the quantity x^1 will increase without bound (Figure 165). In particular, the semitrajectory α_0^+ lies entirely in the fourth quadrant.

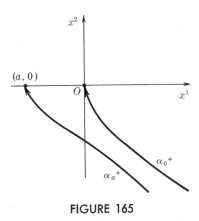

FIGURE 165

All the semitrajectories α_a^+ corresponding to negative values of a fill out an open set on the phase plane which will be denoted by U^+. The boundary of the region U^+ consists of the negative portion of the x^1-axis, the semitrajectory α_0^+, and a curve Δ^+ which may extend to infinity [compare Figure 166(a), (b)]. The portion of the plane contained between α_0^+ and the positive part of the x^1-axis will be denoted by V^-.

Similarly, moving from the point $(b, 0)$ along the curve α_b^-, the quantity

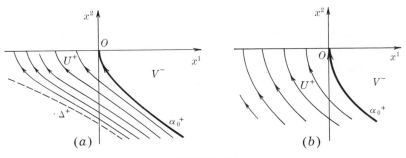

FIGURE 166

x^2 will increase monotonically, which follows from the second equation in (4.29) and the second inequality in (4.26). The value of the coordinate x^2 will approach a limiting value which is equal to either a positive number or $+\infty$, while the coordinate x^1 will decrease monotonically to $-\infty$. In particular, the semitrajectory α_0^- lies in the second quadrant. All semitrajectories α_b^- corresponding to positive values of b fill out an open set on the phase plane which will be denoted by U^-. The boundary of the region U^- consists of the positive portion of the x^1-axis, the semitrajectory α_0^-, and a curve Δ^- which may extend to infinity. The portion of the plane enclosed between α_0^- and the negative portion of the x^1-axis will be denoted by V^+.

The portion of the phase plane contained between the curves Δ^- and Δ^+ will be denoted by G (if Δ^- and Δ^+ extend to infinity, then the region G coincides with the entire plane). The unbounded curve formed by the semitrajectories α_0^+ and α_0^- will be denoted by Γ. The x^1-axis and the curve Γ intersect at the origin and divide the region G into four parts: U^+, U^-, V^+, V^- (Figure 167).

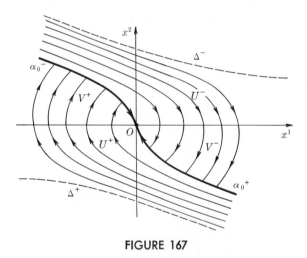

FIGURE 167

Theorem 4.8. For the controlled object (4.25), (4.24) *satisfying conditions* (4.26), (4.27), *time-optimal motion to the origin is possible from any point belonging to the region G. It is altogether impossible to reach the origin from points not lying in this region (such points exist if at least one of the curves* Δ^-, Δ^+ *does not extend to infinity). The synthesis of optimal controls is realized in the region G as follows: at points lying above the curve* Γ *and on the semitrajectory* α_0^-, *we set* $u = -1$; *at points lying below the curve* Γ *and on the semitrajectory* α_0^+, *we set* $u = +1$. *Moreover, no more than one switching occurs on each optimal trajectory.*

49. Proof.   Let the point x_0 belong to the region G and let it lie below the curve Γ (Figure 168). We shall show that the semitrajectory β of (4.28) emanating from the point x_0 intersects the semitrajectory α_0^- at only one point.

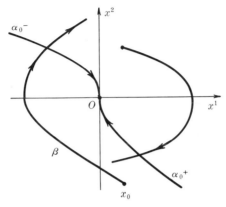

FIGURE 168

First, let the x^2 coordinate of the point x_0 be positive. It follows from the second equation in (4.28) and the first inequality in (4.26) that x^2 will increase monotonically. Therefore, by virtue of the first equation in (4.28), the x^1 coordinate increases monotonically and without bound on the semitrajectory β. Hence it follows that the semitrajectory β must intersect α_0^-. Moving along the semitrajectory β the value of the phase coordinate x^2 increases, while moving along α_0^- it decreases. The value of x^1 in the half-plane $x^2 > 0$ increases on both trajectories β and α_0^-. Consequently, there is a unique point of intersection of the semitrajectories β and α_0^-.

Now, let the x^2 coordinate of the point x_0 be negative. The curves α_0^+ and β do not intersect, being two phase trajectories of the same system (4.28).

It follows from the second equation in (4.28) and the first condition in (4.26) that the x^2 coordinate increases monotonically on the trajectory β, so that the semitrajectory β intersects the x^1-axis at some point on the negative semiaxis. During the subsequent motion, the x^2 coordinate becomes positive and, by virtue of what has been proved above, there exists a single point of intersection of the semitrajectories β and α_0^-.

It is proved in an analogous manner that any semitrajectory of (4.29) originating in the region G above the curve Γ intersects the semitrajectory α_0^+ at only one point.

Thus, in order to reach the origin from a point x_0 in G lying below the curve Γ, one first moves along a trajectory of (4.25) with $u = +1$ up to the semitrajectory α_0^-. A switching occurs at the instant the trajectory α_0^- is reached, and the motion continues along a trajectory of (4.25) with $u = -1$, that is, along the trajectory α_0^-, until the origin is reached. The trajectory thus obtained will be denoted by ξ_{x_0}. If, however, the point x_0 in G lies above the curve Γ, then to obtain the trajectory ξ_{x_0}, it is necessary to move first from x_0 with $u = -1$ in order to reach the semitrajectory α_0^+, and then to move with $u = +1$ along the semitrajectory α_0^+ until reaching the origin. If the point x_0 lies on the curve Γ, then the motion from x_0 to the origin must proceed without switching, that is, along the semitrajectory α_0^+ or α_0^-.

We have obtained a family of trajectories ξ_{x_0} of (4.25) which fill out the entire region G (Figure 167). Only one trajectory of the above family emanates from each point of G. We shall show that all these trajectories satisfy the maximum principle.

The function H corresponding to (4.25) has the form

$$H = \psi_1 x^2 + \psi_2 f(x^1, x^2, u).$$

From the first inequality in (4.27) it is seen that the function $f(x^1, x^2, u)$ increases monotonically in u, and therefore, as u varies on the interval $[-1, 1]$, this function attains its minimum at $u = -1$ and its maximum at $u = +1$. Therefore, the maximum condition on the function H contained in the maximum principle leads to the following relation:

$$u = \text{sign } \psi_2 \qquad (\text{if } \psi_2 \neq 0).$$

The system of equations for the auxiliary variables has the form

$$\dot{\psi}_1 = -\frac{\partial H}{\partial x^1} = -\psi_2 \frac{\partial f}{\partial x^1},$$

$$\dot{\psi}_2 = -\frac{\partial H}{\partial x^2} = -\psi_1 - \psi_2 \frac{\partial f}{\partial x^2}. \tag{4.30}$$

Let c be a point on the curve α_0^+. Let $x(t)$ denote the trajectory which at $t = 0$ passes through the point c, for $t < 0$ satisfies (4.29), and for $t > 0$ satisfies (4.28). Thus, the trajectory $x(t)$ reaches the origin at time $t = \tau > 0$; we shall consider the motion $x(t)$ for $t \leqslant \tau$. The control corresponding to this motion (Figure 169) is given by

$$\begin{aligned} u(t) &= -1 \qquad \text{for} \qquad t < 0, \\ u(t) &= +1 \qquad \text{for} \qquad t > 0. \end{aligned} \tag{4.31}$$

Let us substitute $x(t)$ for x and the corresponding control (4.31) for u in the right-hand sides of (4.30); furthermore, let $\psi(t)$ denote the solution of this system with initial conditions

$$\psi_1(0) = -1, \qquad \psi_2(0) = 0. \tag{4.32}$$

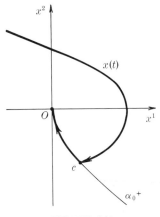

FIGURE 169

We shall prove that the functions $\psi(t)$, $x(t)$, $u(t)$ satisfy the maximum principle, that is, that $u(t) = \text{sign } \psi_2(t)$ for all $t \leqslant \tau$. In other words, we shall prove that

$$\begin{aligned}
\psi_2(t) &< 0 \qquad \text{for} \qquad t < 0, \\
\psi_2(t) &> 0 \qquad \text{for} \qquad t > 0.
\end{aligned} \qquad (4.33)$$

We note that by (4.30) and (4.32) we have

$$\psi_2(0) = -\psi_1(0) - \psi_2(0) \frac{\partial f}{\partial x^2} = 1, \qquad (4.34)$$

and therefore $\psi_2(t)$ is different from zero in some neighborhood of the point $t = 0$. In addition, it follows from (4.34) that the relations $\psi_2(t) < 0$ for $t < 0$, and $\psi_2(t) > 0$ for $t > 0$ are satisfied near the point $t = 0$, which is in complete agreement with the inequalities (4.33) being proved.

Let us now suppose that the inequalities (4.33) are not satisfied at some instant of time. Then the function $\psi_2(t)$ must vanish at an instant of time different from $t = 0$. Let θ be a root of the function $\psi_2(t)$ closest to 0. Then

$$\psi_2(\theta) = \psi_2(0) = 0 \qquad (4.35)$$

and the function $\psi_2(t)$ maintains a fixed sign between 0 and θ.

We shall prove that $\psi_1(t)$ also has a fixed (namely, negative) sign between 0 and θ. If $\theta < 0$, then it follows from (4.34) that $\psi_2(t) < 0$ on the interval $(\theta, 0)$ and therefore, $\psi_1(t) \geqslant 0$ on this interval [see (4.30) and (4.27)]. Now it follows from (4.32) that $\psi_1(t) > 0$ on the interval $(\theta, 0)$. If, however, $\theta > 0$, then it follows from (4.34) that $\psi_2(t) > 0$ on the interval $(0, \theta)$ and therefore $\psi_1(t) \leqslant 0$ on this interval [see (4.30) and (4.27)]. Thus, also in this case $\psi_1(t) < 0$ on the interval $(0, \theta)$.

By virtue of (4.35) and the second equation in (4.30), we have [for any function $\eta(t)$]

$$0 = \psi_2(t)\eta(t) \Big|_0^\theta = \int_0^\theta \frac{d}{dt} (\psi_2(t)\eta(t))\, dt = \int_0^\theta (\dot\psi_2(t)\eta(t) + \psi_2(t)\dot\eta(t))\, dt$$

$$= \int_0^\theta \left(-\psi_1(t)\eta(t) - \psi_2(t)\frac{\partial f}{\partial x^2}\eta(t) + \psi_2(t)\dot\eta(t)\right) dt. \tag{4.36}$$

If we choose the function $\eta(t)$ in such a way that $\dot\eta(t) - \eta(t)(\partial f/\partial x^2) \equiv 0$, that is, setting $\eta(t) = e^{\int \frac{\partial f}{\partial x^2}\, dt}$, then the second and third terms under the integral in (4.38) cancel, and we obtain

$$0 = -\int_0^\theta \psi_1(t)\eta(t)\, dt.$$

But this contradicts the fact that each of the functions $\psi_1(t)$, $\eta(t)$ has a fixed sign between 0 and θ. The contradiction obtained proves that the relations (4.33) are satisfied, that is, that $\psi(t)$, $x(t)$, $u(t)$ satisfy the maximum principle.

Similarly one can prove that a trajectory passing through a point c' of the curve α_0^- (Figure 170) satisfies the maximum principle. Accordingly, it has been proved that all the trajectories ξ_{x_0} satisfy the maximum principle.

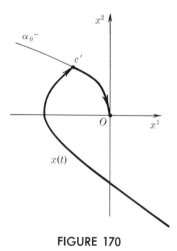

FIGURE 170

Now, let us consider the semitrajectories α_0^+ and α_0^- as one-dimensional cells of the first kind and the two regions into which the region G is divided by the curve Γ as two-dimensional cells of the first kind. The origin is the only zero-dimensional cell. Finally, let us set $P^0 = 0$, $P^1 = \Gamma$, $P^2 = G$ and assume that N is an empty set. Then all the conditions of regular synthesis are satisfied for the function $v(x)$ which is equal to -1 on the cell α_0^- and above the curve Γ, and to $+1$ on the cell α_0^+ and below the curve Γ. We need only verify the fact that trajectories traversing the two-dimensional cells approach α_0^- and α_0^+ at nonzero angles. But by virtue of (4.25), the vector $\{x^2; f(x^1, x^2, +1)\}$ is tangent to the trajectory α_0^+ at the point $c = (x^1, x^2)$ of α_0^+, and the trajectory approaching this point from a two-dimensional cell is tangent to the vector $\{x^2; f(x^1, x^2, -1)\}$. Since $x^2 \neq 0$ at points of the semitrajectory α_0^+, it follows from (4.26) that the above vectors are not collinear. This, in fact, means that the trajectories which traverse the two-dimensional cells approach α_0^+ at nonzero angles. Analogous arguments are applicable to the curve α_0^-.

Thus, all the conditions of regular synthesis in the region G are satisfied and, according to Theorem 4.7, all the trajectories ξ_{x_0} are optimal in G. It remains to establish, in order to complete the proof of the theorem, that it is altogether impossible to reach the origin from points not belonging to G (if there are such points).

Suppose, for example, that p is a point on the boundary curve Δ^- of the region G (Figure 171). Let us take a point b on the x^1 semiaxis and

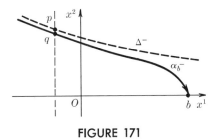

FIGURE 171

let us consider the trajectory α_b^- of (4.29) which reaches this point. The trajectory α_b^- intersects the straight line passing through p parallel to the x^2-axis at a point q. If the point b moves to the right along the x^1-axis, then the point q will rise but not reach the point p (since p does not belong to G). Denoting the ordinate of the point p by p^2, we find that the x^2 coordinate takes on positive values less than p^2 on the entire interval (q, b) of the trajectory α_b^-. Therefore, by virtue of the first equation in (4.25), the relation $\dot{x}^1 < p^2$ holds on the entire interval (q, b). Consequently, the duration of the motion over the interval (q, b) can be made as large as desired if the point b is taken sufficiently far. Hence, we can conclude that when the initial phase state approaches the point p, the duration of the optimal motion (in the region G) from this state to the origin increases without bound. An analogous conclusion is obtained also for any point lying on the boundary curve Δ^+.

Now, suppose that it is possible to reach the origin in a finite time from a point not belonging to G along a trajectory ξ. Then we could find the last (in time) point of intersection of the trajectory ξ with the boundary of the region G. Let us denote this point by p. Then, it would be possible to reach the origin in a finite time from the point p moving within G, which however contradicts the unbounded increase of duration of the optimal motion as the initial state approaches the point p. The contradiction obtained proves our assertion.

50. Second Example.   Let us consider the object

$$\begin{aligned} \dot{x}^1 &= u^1 x^2, \\ \dot{x}^2 &= u^2 \end{aligned} \tag{4.37}$$

with the control region defined by the inequalities

$$-1 \leqslant u^1 \leqslant 1, \qquad -1 \leqslant u^2 \leqslant 1, \tag{4.38}$$

for which we again consider the time-optimal problem of reaching the origin. Without giving the calculations (which can be easily carried out), we shall indicate the final result, that is, we shall describe the regular synthesis for the object (4.37), (4.38).

Two parabolas $x^1 = \pm\frac{1}{2}(x^2)^2$ constitute the set P^1; the axis $x^2 = 0$ is taken as N. The set $M = N \cup P^1$ divides the plane P^2 of the variables x^1, x^2 into six regions: V_1, V_2, V_3, V_4, V_5, V_6 (Figure 172), and the point

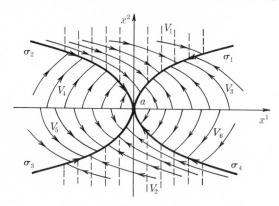

FIGURE 172

$a = (0, 0)$ divides the set P^1 into four branches (cells) extending to infinity, namely, $\sigma_1, \sigma_2, \sigma_3, \sigma_4$ also shown in Figure 172. We set $v(x) = (v^1(x), v^2(x))$, where

$$v^1(x) = \begin{cases} +1 & \text{for} & x \in V_1,\, V_2,\, V_4,\, V_6,\, \sigma_2,\, \sigma_4, \\ -1 & \text{for} & x \in V_3,\, V_5,\, \sigma_1,\, \sigma_3; \end{cases}$$

$$v^2(x) = \begin{cases} +1 & \text{for} & x \in V_2,\, V_3,\, V_4,\, \sigma_3,\, \sigma_4, \\ -1 & \text{for} & x \in V_1,\, V_5,\, V_6,\, \sigma_1,\, \sigma_2. \end{cases}$$

This defines the regular synthesis for the object (4.37), (4.38). The optimal trajectories consist of pieces of the parabolas

$$x^1 = \pm \tfrac{1}{2}(x^2)^2 + \text{const};$$

they are shown in Figure 172. We note that two trajectories emanate from each point of the set N in this regular synthesis (this fact does not prevent the application of Theorem 4.7).

It should be noted that Theorem 4.7 does not exclude the existence of other optimal trajectories (besides the distinguished ones). In the example considered in § 13.47–§ 13.49 there were no other optimal trajectories (except the distinguished ones) satisfying the maximum principle. In this example, an infinitely large number of optimal trajectories emanates from every point of the cells V_1 and V_2. Namely, in V_1 (or V_2) we must set $u^2 = -1$ (respectively, $u^2 = +1$), and any piecewise continuous function satisfying the conditions (4.38) may be used as u^1. At the instant of time at which the phase point reaches P^1, a switching occurs, and the subsequent motion continues over the set P^1. All the trajectories thus obtained, as can be easily seen, have the same duration of motion from the point x_0 to a, and are all optimal. For example, in the cells V_1 and V_2, one can take $u^2 = \pm 1$, $u^1 = 0$ (dotted lines in Figure 172).

As shown by the above examples, the main difficulty of calculation lies in realizing the synthesis on the basis of the maximum principle. If the synthesis already has been realized, then, as a rule, conditions (A)–(F) are automatically satisfied. Thus, the maximum principle (a necessary condition for optimality), which, as a rule, makes it possible to realize the synthesis, is very close to a sufficient condition for optimality.

51. Synthesis of Optimal Controls in Non-Oscillatory Systems of the Second Order.

In § 13.47 we considered the problem of reaching the equilibrium position ($x = 0$, $\dot{x} = 0$) in minimum time for an object governed by (4.23) and (4.24) and have given conditions which must be satisfied in order that the optimal controls take on only the values $u = \pm 1$ and have at most one switching. In the phase coordinates $x^1 = x$, $x^2 = \dot{x}$ these conditions, that is, (4.26) and (4.27), are reproduced here for convenience as follows:

$$f(x^1, x^2, +1) > 0, \qquad f(x^1, x^2, -1) < 0 \qquad \text{for all} \qquad x^1, x^2; \qquad \textbf{(4.39)}$$

$$\frac{\partial f\,(x^1,\, x^2,\, u)}{\partial u} > 0 \qquad \text{for all} \qquad x^1,\, x^2,\, u; \qquad \textbf{(4.40)}$$

$$\frac{\partial f\,(x^1,\, x^2,\, u)}{\partial x^1} \geqslant 0 \qquad \text{for all} \qquad x^1,\, x^2,\, u. \qquad \textbf{(4.41)}$$

However, these conditions are exceedingly restrictive. Consider, for example, the linear object

$$\ddot{x} = -2\,\delta\dot{x} - \omega^2 x + u, \qquad |u| \leqslant 1. \tag{4.42}$$

For $\delta^2 - \omega^2 \geqslant 0$, the corresponding characteristic equation has real roots and therefore (see p. 106) the optimal control takes on only the values $u = \pm 1$ and has at most one switching. However, for $\omega \neq 0$ conditions (4.39) and (4.41) are obviously not satisfied.

The excessiveness of the constraints imposed by conditions (4.39)–(4.41) was pointed out to the author by the Hungarian mathematician Sándor Halász in a personal communication. He cited actual nonlinear engineering objects of the form (4.23) which do not satisfy conditions (4.39) and (4.41); nevertheless, in solving the problem of reaching the equilibrium position in minimum time for these objects, it is natural to expect that the optimal control takes on only the values $u = \pm 1$ and has at most one switching.

For convenience, an object of the form (4.23) will be called non-oscillatory if every time-optimal control for this object takes on only the values $u = \pm 1$ and has at most one switching. In this section, conditions substantially more general than (4.39)–(4.41) are provided under which the object (4.23) is non-oscillatory. Moreover, we construct the synthesis of optimal controls for such objects. In addition, the essentially different significance between conditions (4.39) and (4.41) (or generalizations thereof) will be clarified.

In phase coordinates, (4.23) may be written in the normal form as the system (4.15). We shall assume everywhere in the sequel that condition (4.40) is satisfied. However, conditions (4.39) and (4.41) will be replaced by new, more general conditions.

First of all, let us consider the condition which will replace (4.41). We formulate it in the following manner:

(A)   There exists a continuous function $\varphi(x^1, x^2, u)$ having continuous derivatives $\partial\varphi/\partial x^1$, $\partial\varphi/\partial x^2$ such that the inequality

$$x^2\frac{\partial\varphi}{\partial x^1} + f\frac{\partial\varphi}{\partial x^2} + (\varphi)^2 - \varphi\frac{\partial f}{\partial x^2} - \frac{\partial f}{\partial x^1} \leqslant 0 \tag{4.43}$$

is satisfied for $u = \pm 1$ and for any x^1, x^2.

We note that condition (A) is less restrictive than (4.41), that is, (4.41) implies (A). In order to convince oneself that this is true, it suffices to set $\varphi \equiv 0$. As we shall see below (Theorem 4.9), the object (4.23) is non-oscillatory if condition (A) is satisfied.

Let us consider the particular case of condition (A) which is obtained for $\varphi = \frac{1}{2}(\partial f/\partial x^2)$ (it must be assumed in this case that the function f has continuous second derivatives with respect to x^1, x^2). Thus, we obtain the following condition which is more restrictive than (A):

(A′) The inequality

$$2x^2 \frac{\partial^2 f}{\partial x^1 \partial x^2} + 2f \frac{\partial^2 f}{(\partial x^2)^2} - \left(\frac{\partial f}{\partial x^2}\right)^2 - 4 \frac{\partial f}{\partial x^1} \leqslant 0 \qquad (4.44)$$

is satisfied for $u = \pm 1$ and for any x^1, x^2.

A direct calculation shows that any linear object (4.42) for which $\delta^2 - \omega^2 \geqslant 0$ satisfies condition (A′). Furthermore, (A′) is also satisfied by nonlinear objects which differ "little" from a linear object (4.42) for which $\delta^2 - \omega^2 > 0$.

Consider, for example, the controlled object

$$\ddot{x} = -2\,\delta\dot{x} - \omega^2 x + u + \mu g(x, \dot{x}, u), \qquad |u| \leqslant 1, \qquad (4.45)$$

where δ and ω are constants satisfying the condition $\delta^2 - \omega^2 > 0$ and $g(x, \dot{x}, u)$ is a bounded function with bounded first and second derivatives; moreover, let

$$\left| \frac{\partial^2 g}{\partial x^i \partial x^j} \right| \leqslant \frac{k}{\sqrt{(x^i)^2 + (x^j)^2}}.$$

Then, for a sufficiently small μ, the object (4.45) satisfies condition (A′), and consequently, is non-oscillatory.

As will be shown by Theorem 4.9, any time-optimal control $u(t)$ (not necessarily transferring the phase point to the equilibrium position $x^1 = x^2 = 0$) takes on only the values $u = \pm 1$ and has at most one switching if condition (A) [or (A′)] is satisfied. Of course, this does not guarantee the existence of the synthesis of optimal trajectories leading to the point $x^1 = x^2 = 0$ (or even the possibility of reaching this point from any position belonging to a neighborhood thereof).

In order to ensure the existence of the synthesis of optimal trajectories, we shall impose still another condition on the object (4.23):

(B) $f(x^1, 0, 1) > 0$ for $x^1 \leqslant 0$; $f(x^1, 0, -1) < 0$ for $x^1 \geqslant 0$.

It is clear that (B) is less restrictive than (4.39) since here the fulfillment of the inequalities (4.39) is required not for all x^1, x^2 but only on the semiaxes $x^1 \leqslant 0$ and $x^1 \geqslant 0$.

We note that the linear object (4.42) (or any nonlinear object which differs "little" from it) satisfies condition (B).

Theorems 4.9 and 4.10 which we shall prove below, taken together, assert that the synthesis of time-optimal trajectories leading to the point $x^1 = x^2 = 0$ is realized for the object (4.25), (4.24), (4.40) if conditions (A) and (B) [or (A′) and (B)], that is, substantially weaker conditions than

(4.39) and (4.41), are satisfied. This synthesis turns out to be qualitatively the same as that for the linear object (4.42) for which $\delta^2 - \omega^2 \geqslant 0$.

We proceed now to the formulation and proof of the above-mentioned results.

Theorem 4.9.   *If condition* A [*or* (A′)] *is satisfied, then the object* (4.25), (4.24), (4.40), *is non-oscillatory, that is, every time-optimal control (corresponding to the transition from any initial point to any terminal point) takes on only the values* $u = \pm 1$ *and has at most one switching.*

PROOF.   Let $u(t)$, $t_0 \leqslant t \leqslant t_1$, be a time-optimal control (corresponding to the transition from some initial point to some terminal point), and let $x^1(t)$, $x^2(t)$ be the corresponding optimal trajectory. Let us write the function H and the system of equations for the auxiliary variables ψ_1, ψ_2 according to the maximum principle (see pp. 78, 79);

$$H = \psi_1 x^2 + \psi_2 f(x^1, x^2, u); \tag{4.46}$$

$$\dot{\psi}_1 = -\frac{\partial H}{\partial x^1} = -\psi_2 \frac{\partial f}{\partial x^1},$$
$$\dot{\psi}_2 = -\frac{\partial H}{\partial x^2} = -\psi_1 - \psi_2 \frac{\partial f}{\partial x^2}. \tag{4.47}$$

Let us replace x^1, x^2, u in the right-hand sides of (4.47) by the functions $x^1(t)$, $x^2(t)$, $u(t)$ (introduced above). Then, by virtue of the maximum principle, there exists a nontrivial solution $\psi_1(t)$, $\psi_2(t)$, $t_0 \leqslant t \leqslant t_1$, of (4.47) such that for $t_0 \leqslant t \leqslant t_1$ the function

$$\psi_1(t)x^2(t) + \psi_2(t)f(x^1(t), x^2(t), u)$$

of the variable u attains its maximum for $u = u(t)$. Hence, it follows directly from (4.40) (compare p. 200) that

$$u(t) = \text{sign } \psi_2(t).$$

Therefore, to prove the theorem it suffices to establish that the function $\psi_2(t)$, $t_0 \leqslant t \leqslant t_1$, has at most one root.

Let us suppose on the contrary that $\psi_2(t)$ has more than one root on the interval $t_0 \leqslant t \leqslant t_1$, and let α and β be two adjacent roots of this function. The identity $\psi_2(t) \equiv 0$ is not possible since then it would follow from (4.47) that $\psi_1(t) \equiv 0$. Then the function $\psi_2(t)$ maintains a constant sign on the interval $\alpha < t < \beta$. Since, in addition, $\psi_2(\alpha) = \psi_2(\beta) = 0$, it follows that $\psi_1(\alpha) \neq 0$, $\psi_2(\beta) \neq 0$ because $\psi_1(t)$, $\psi_2(t)$ is a nontrivial solution of (4.47). For the sake of definiteness, let $\psi_1(\alpha) > 0$ [the arguments remain the same in case $\psi_1(\alpha) < 0$]. Then, by the second equation in (4.47),

$$\dot{\psi}_2(\alpha) = -\psi_1(\alpha) < 0;$$

hence, it follows that for $t > \alpha$ the function $\psi_2(t)$ is negative near α. Consequently, it is negative also on the entire interval (α, β). But then it follows from analogous arguments that

$$\psi_1(\beta) < 0. \tag{4.48}$$

We note that $u(t) = -1$ for $\alpha < t < \beta$ since the function $\psi_2(t)$ is negative. Now, let us set

$$\eta(t) = \psi_1(t) + \psi_2(t)\varphi(x^1(t), x^2(t), -1), \tag{4.49}$$

where φ is a function whose existence is asserted in condition (A). We have

$$\dot{\eta} = \dot{\psi}_1 + \dot{\psi}_2\varphi + \psi_2\dot{\varphi} = -\psi_2\frac{\partial f}{\partial x^1} - \varphi\left(\psi_1 + \psi_2\frac{\partial f}{\partial x^2}\right) + \psi_2\left(\frac{\partial \varphi}{\partial x^1}x^2 + \frac{\partial \varphi}{\partial x^2}f\right)$$

$$= \psi_2\left(x^2\frac{\partial \varphi}{\partial x^1} + f\frac{\partial \varphi}{\partial x^2} + \varphi^2 - \varphi\frac{\partial f}{\partial x^2} - \frac{\partial f}{\partial x^1}\right) - \varphi(\psi_1 + \psi_2\varphi)$$

$$= \psi_2\left(x^2\frac{\partial \varphi}{\partial x^1} + f\frac{\partial \varphi}{\partial x^2} + \varphi^2 - \varphi\frac{\partial f}{\partial x^2} - \frac{\partial f}{\partial x^1}\right) - \varphi\eta.$$

Since the expression in the last parentheses is nonpositive [see condition (A)] and $\psi_2(t) < 0$ on the interval (α, β), we obtain

$$\dot{\eta} \geqslant -\eta\varphi(x^1(t), x^2(t), -1), \qquad \alpha < t < \beta. \tag{4.50}$$

Since $\psi_2(\alpha) = \psi_2(\beta) = 0$, it follows from (4.49) that

$$\eta(\alpha) = \psi_1(\alpha) > 0, \qquad \eta(\beta) = \psi_1(\beta) < 0,$$

and therefore $\eta(t)$ vanishes for some t between α and β. Let γ be the root of $\eta(t)$ closest to α. Then

$$\eta(\gamma) = 0, \qquad \eta(t) > 0 \qquad \text{for } \alpha < t < \gamma. \tag{4.51}$$

It follows from (4.50) and (4.51) that

$$\frac{\dot{\eta}}{\eta} \geqslant -\varphi(x^1(t), x^2(t), -1) \qquad \text{for } \alpha < t < \gamma.$$

Hence, integrating, we find

$$\ln \eta(t) - \ln \eta(\alpha) \geqslant -\int_\alpha^t \varphi(x^1(t), x^2(t), -1)\, dt \qquad \text{for } \alpha < t < \gamma,$$

and therefore

$$\eta(\alpha) \leqslant \eta(t)e^{\int_\alpha^t \varphi(x^1(t), x^2(t), -1)\, dt} \qquad \text{for } \alpha < t < \gamma.$$

Hence, for $t \to \gamma$, we obtain

$$\eta(\alpha) \leqslant \eta(\gamma)e^{\int_\alpha^\gamma \varphi(x^1(t), x^2(t), -1)\, dt},$$

which obviously contradicts the relation $\eta(\gamma) = 0$, $\eta(\alpha) > 0$.

The contradiction proves the theorem.

Now, let the non-oscillatory object (4.25), (4.24), (4.40) satisfy condition (B). We shall prove below (Theorem 4.10) that, in this case, the synthesis of optimal trajectories is realized on an open set containing the origin. First of all, let us describe this synthesis. Let Γ_a^+ denote a phase trajectory of (4.25) which corresponds to $u \equiv 1$ and passes through the point $(a, 0)$, where $a \leqslant 0$. Since, by virtue of condition (B), $f(a, 0, 1) > 0$, this curve is directed upward as the point $(a, 0)$ is approached [see the second equation in (4.25)]. We shall prove that, for an object (4.23) which satisfies condition (A), the curve Γ_a^+ intersects the axis $x^2 = 0$ only once, that is, the part of this trajectory up to the point $(a, 0)$ lies entirely in the lower half-plane $x^2 < 0$ and the remaining part lies in the upper half-plane $x^2 > 0$.

Suppose the contrary, namely, that the trajectory Γ_a^+ intersects the axis $x^2 = 0$ more than once. Let $x^1(t)$, $x^2(t)$ denote the parametric representation of the trajectory Γ_a^+, so that $x^1(t)$, $x^2(t)$ satisfy (4.25) for $u \equiv 1$. Let us consider the point of intersection of the trajectory Γ_a^+ and the axis $x^2 = 0$ which is closest to $(a, 0)$, and let α, β be the instants of time at which the trajectory Γ_a^+ passes through these points. Thus, $x^2(\alpha) = x^2(\beta) = 0$, and for $\alpha < t < \beta$ the trajectory Γ_a^+ does not intersect the axis $x^2 = 0$, that is, $x^2(t)$ has a fixed sign. For the sake of definiteness, let $x^2(t) > 0$ for $\alpha < t < \beta$ [the arguments are analogous in case $x^2(t) < 0$]. Differentiating (4.25), we find

$$\ddot{x}^1(t) = \dot{x}^2(t) = f(x^1(t), x^2(t), 1),$$

$$\ddot{x}^2(t) = \frac{\partial f\,(x^1(t),\,x^2(t),\,1)}{\partial x^1}\,\dot{x}^1(t) + \frac{\partial f\,(x^1(t),\,x^2(t),\,1)}{\partial x^2}\,\dot{x}^2(t) = \frac{\partial f}{\partial x^1}\,x^2 + \frac{\partial f}{\partial x^2}\,f.$$

Now, we set

$$\eta(t) = \dot{x}^2(t) - \varphi(x^1(t), x^2(t), 1)\dot{x}^1(t) = f(x^1(t), x^2(t), 1) - \varphi(x^1(t), x^2(t), 1)x^2(t),$$

where $\varphi(x^1, x^2, u)$ is a function whose existence is asserted by condition (A). A direct calculation shows that the relation

$$\dot{\eta} = -x^2(t)\left(\frac{\partial\varphi}{\partial x^1}\,x^2 + \frac{\partial\varphi}{\partial x^2}\,f + \varphi^2 - \varphi\,\frac{\partial f}{\partial x^2} - \frac{\partial f}{\partial x^1}\right) + \eta\left(\frac{\partial f}{\partial x^2} - \varphi\right)$$

holds.

Considering condition (A) and the relation $x^2 > 0$ for $\alpha < t < \beta$, we obtain

$$\dot{\eta} \geqslant \eta\left(\frac{\partial f}{\partial x^2} - \varphi\right).$$

In addition,

$$\eta(\alpha) = f - \varphi x^2\,\Big|_{t=\alpha} = f(x^1(\alpha), x^2(\alpha), 1) = \dot{x}^2(\alpha),$$

$$\eta(\beta) = f - \varphi x^2\,\Big|_{t=\beta} = f(x^1(\beta), x^2(\beta), 1) = \dot{x}^2(\beta).$$

(4.52)

We note that $\dot{x}^2(\alpha) \neq 0$; otherwise it would follow from the relations $\dot{x}^1(\alpha) = x^2(\alpha) = 0$, $\dot{x}^2(\alpha) = 0$ that $(x^1(\alpha), 0)$ is an equilibrium position of (4.25) for $u \equiv 1$, which is not possible since the trajectory Γ_a^+ is not an equilibrium position and therefore no point of this trajectory can be an equilibrium position. Likewise $\dot{x}^2(\beta) \neq 0$. Thus, both numbers in (4.52) are different from zero. Since, in addition, $x^2(t)$ has a constant sign on the interval (α, β), the derivatives $\dot{x}^2(\alpha)$ and $\dot{x}^2(\beta)$ have opposite signs; that is, the numbers $\eta(\alpha)$ and $\eta(\beta)$ also have opposite signs. Consequently, $\eta(t)$ vanishes for some t between α and β. Hence, as in the proof of Theorem 4.9, we can easily arrive at a contradiction which proves that the trajectory Γ_a^+ cannot intersect the axis $x^2 = 0$ twice.

Similarly, let Γ_b^- denote a phase trajectory of (4.25) corresponding to $u \equiv -1$ and passing through the point $(b, 0)$, where $b \geqslant 0$. This trajectory approaches the point $(b, 0)$ from the upper half-plane. The same way as above, it is proved that the curve Γ_b^- intersects the axis $x^2 = 0$ only once, that is, the part of this trajectory up to the point $(b, 0)$ lies entirely in the upper half-plane $x^2 > 0$, and the remaining part lies in the lower half-plane $x^2 < 0$.

Let us consider also the question of the behavior of the trajectories Γ_a^+ and Γ_b^- as $t \to \pm \infty$. Let $x^1(t)$, $x^2(t)$ be a parametric representation of Γ_a^+ (where $a \leqslant 0$). The part of this trajectory up to the point $(a, 0)$ lies, as we have seen, in the lower half-space $x^2 < 0$, and therefore, by virtue of the first equation in (4.25), $\dot{x}^1 < 0$ on this semitrajectory. Therefore, there are two possibilities: either $\lim\limits_{t \to -\infty} x^1(t) = +\infty$ (that is, moving along the phase trajectory Γ_a^+ in the opposite direction, the phase point departs to the right without bound) or $\lim\limits_{t \to -\infty} x^1(t)$ is finite and equal to a number m_a. In the latter case, as can be easily seen, $\lim\limits_{t \to -\infty} x^2(t) = 0$ and $m_a > 0$, that is, the trajectory Γ_a^+ has a single limiting point lying on the positive x^1 semiaxis as $t \to -\infty$.

In fact, since the improper integral

$$\int_{-\infty}^{t_0} x^2(t) \, dt = \int_{-\infty}^{t_0} \dot{x}^1(t) \, dt = x^1(t) \Big|_{-\infty}^{t_0} = x^1(t_0) - m_a$$

(where t_0 is arbitrary) converges, a sequence of numbers $t_1, t_2, \ldots, t_n,$ $\ldots$ can be found such that $\lim\limits_{n \to +\infty} t_n = -\infty$ and $\lim\limits_{n \to \infty} x^2(t_n) = 0$, that is, $(m_a, 0)$ is an α-limiting point of the trajectory Γ_a^+. If the equality $\lim\limits_{t \to -\infty} x^2(t) = 0$ were not satisfied, then a sequence $\tau_1, \tau_2, \ldots, \tau_n, \ldots$ could be found such that $\lim\limits_{n \to \infty} \tau_n = -\infty$ and $x^2(\tau_n) < -\epsilon$, $n = 1, 2, \ldots$. But then the entire segment joining the points $(m_a, 0)$ and $(m_a, -\epsilon)$ would belong to the α-limiting set of the trajectory Γ_a^+. Hence, it would follow that the phase-velocity vector is vertical (parallel to the x^2-axis) at all points of the above segment, that is, $\dot{x}^1 = 0$. This would mean that $\dot{x}^1 = x^2 = 0$ at all points of the above segment, which is obviously contradictory. This contradiction, in fact, proves that $\lim\limits_{t \to -\infty} x^2(t) = 0$. Thus, the point $(m_a, 0)$ is the single limiting point of the trajectory Γ_a^+ as $t \to -\infty$, and therefore, $(m_a, 0)$ is the equilibrium position of (4.25) for $u \equiv 1$. But then $\dot{x}^2 = f(m_a, 0, +1)$ $= 0$; hence we conclude, by virtue of condition (B), that $m_a > 0$. Thus, if $\lim\limits_{t \to -\infty} x^1(t)$ is finite and equal to m_a, then $m_a > 0$ and $\lim\limits_{t \to -\infty} x^2(t) = 0$.

Completely analogous arguments are applicable to the trajectory Γ_a^+ as $t \to +\infty$. Hence it follows, in particular, that the trajectory Γ_a^+, after its intersection with the x^1-axis at the point $(a, 0)$, passes into the upper half-plane and goes to the right necessarily intersecting the positive x^2 semiaxis.

Thus, at the outset, the trajectory Γ_a^+ traverses the fourth quadrant (moreover, either $\lim\limits_{t \to -\infty} x^1(t) = +\infty$, or the trajectory Γ_a^+ has a single limiting point on the positive x^1 semiaxis as $t \to -\infty$), then traverses the third quadrant, intersects the x^1-axis at the point $(a, 0)$, and finally passes to the second quadrant which it necessarily leaves intersecting the positive x^2 semiaxis. The behavior of the trajectory Γ_b^- (for $b > 0$) is investigated in a completely analogous manner. Namely, at the outset, this trajectory traverses the second quadrant [moreover, either $\lim\limits_{t \to -\infty} x^1(t) = -\infty$ or $\lim\limits_{t \to -\infty} x^1(t) = m_b < 0$ and $\lim\limits_{t \to -\infty} x^2(t) = 0$], then passes into the first quadrant, intersects the x^1-axis at the point $(b, 0)$, and finally passes to the fourth quadrant which it necessarily leaves intersecting the negative x^2 semiaxis.

Let α_a^+ denote the portion of the trajectory Γ_a^+ which lies in the lower half-plane $x^2 < 0$. In particular, α_0^+ is the semitrajectory of (4.25) (for $u \equiv +1$) which reaches the point $(0, 0)$. All the semitrajectories α_a^+ (corresponding to negative values of a) fill out an open set on the phase plane which we shall denote by U^+. The boundary of the region U^+ consists of the following: the negative x^1 semiaxis (or a segment thereof), the semitrajectory α_0^+, and a curve Δ^+ which, however, may recede to infinity (see Figure 167). The portion of the plane contained between α_0^+ and the positive part of the x^1-axis will be denoted by U^-.

Similarly, let α_b^- denote the portion of the trajectory Γ_b^- (for $b > 0$) which lies in the upper half-plane $x^2 > 0$. In particular, α_0^- is the semitrajectory of (4.25) (for $u \equiv -1$) which reaches the point $(0, 0)$. All the semitrajectories α_b^- (corresponding to positive values of b) fill out an open set on the phase plane which we shall denote by U^-. The boundary of U^- consists of the following: the positive x^1 semiaxis (or a segment thereof), the semitrajectory α_0^-, and a curve Δ^- which, however, may recede to infinity (see Figure 167). The part of the plane contained between α_0^- and the negative portion of the x^1-axis will be denoted by V^+.

Now we note that every trajectory Γ_a^+ traversing the region V^+ and passing into the second quadrant necessarily reaches the positive x^2 semiaxis and therefore intersects the curve α_0^- (which lies entirely in the second quadrant). We shall consider the trajectory Γ_a^+ only up to the point at which it meets the curve α_0^-. In the same manner, every trajectory Γ_b^- traversing the region U^- and passing into the fourth quadrant necessarily reaches the negative x^2 semiaxis and therefore intersects the curve α_0^+ (which lies entirely in the fourth quadrant). We shall consider the trajectory Γ_b^- only up to the point at which it meets the curve α_0^+.

As a result, the entire region G contained between the curves Δ^+ and Δ^- (this region may coincide with the entire plane if both curves Δ^+ and Δ^- recede to infinity) is filled out by the trajectories of (4.25). Namely, let Γ denote the curve consisting of the semitrajectories α_0^+ and α_0^-. If the point x_0 lies in the region G above the curve Γ, then it is possible to proceed from this point with $u \equiv -1$ (that is, along one of the trajectories Γ_b^-) to an intersection with the curve Γ (more precisely, with the semitrajectory α_0^+). A switching occurs at the instant at which the semitrajectory α_0^+ is reached, and the motion continues along a semitrajectory with $u \equiv 1$ to the origin. If, however, the point x_0 in G lies below the curve Γ, then it is necessary to move first from x_0 with $u \equiv +1$ up to an intersection with the semitrajectory α_0^-, and then (with $u \equiv -1$) along α_0^- to the origin.

Thus, we obtain a family of trajectories of (4.25) filling out the entire region G (see Figure 167). Only one trajectory of the above family emanates from every point of this region. We shall show that all these trajectories are optimal, by which, in fact, the synthesis of optimal trajectories in the region G is realized.

First of all, it must be established that each of the above trajectories satisfies the maximum principle. Let us consider, for example, a phase trajectory which first passes above Γ (with $u \equiv -1$), then at some instant τ intersects the semitrajectory α_0^+, then (for $t > \tau$) proceeds along the curve α_0^+ (for $u \equiv +1$), and finally approaches the origin. The time interval corresponding to this trajectory (from the initial point to the origin) will be denoted by $t_0 \leqslant t \leqslant t_1$. Finally, let $\psi_1(t)$, $\psi_2(t)$ denote the solution of the adjoint system (4.47) corresponding to the trajectory under consideration which satisfies the initial condition

$$\psi_1(\tau) = -1, \qquad \psi_2(\tau) = 0.$$

Then, $\psi_2(\tau) = -\psi_1(\tau) = 1$, and since, according to Theorem 4.9, the function $\psi_2(t)$ has only one root, we have

$$\psi_2(t) < 0 \qquad \text{for } t < \tau, \qquad \psi_2(t) > 0 \qquad \text{for } t > \tau.$$

Consequently, $u = \text{sign } \psi_2(t)$, that is, the trajectory under consideration satisfies the maximum principle. It is similarly established that a trajectory originating in the region G below the curve Γ also satisfies the maximum principle.

The trajectories obtained (filling out the region G and satisfying the maximum principle) realize the regular synthesis in G (compare p. 203). Therefore, all these trajectories are optimal in G. Finally, it is altogether not possible to reach the origin from points not belonging to G (if such points exist) (compare p. 203). This proves the following theorem.

Theorem 4.10. For the controlled object (4.25), (4.24), (4.40) *satisfying conditions* (A) *and* (B) [*or* (A′) *and* (B)] *it is possible to realize the time-optimal motion to the origin from every point belonging to the region G. From points not lying in this region* (*such points exist if at least one of the curves* Δ^+, Δ^- *does not recede to infinity*), *it is altogether not possible to reach the origin. The synthesis of optimal trajectories in the region G is realized as follows: at the points lying above the curve Γ and on the semitrajectory α_0^- we set $u = -1$; at the points lying below the curve Γ and on the semitrajectory α_0^+ we set $u = +1$. Moreover, it turns out that there is at most one switching on each optimal trajectory.*

We note in conclusion that if conditions (A) and (B) are satisfied in some neighborhood of the origin, rather than in the entire plane, then Theorems 4.9 and 4.10 remain valid within this neighborhood of the origin.

Let, for example, (4.25), (4.24), (4.40) be an arbitrary object satisfying the single condition

$$f(0, 0, 1) > 0, \qquad f(0, 0, -1) < 0. \tag{4.53}$$

Then, setting $\varphi(x^1, x^2, u) = -\lambda x^2 u$, where

$$\lambda > \left| \frac{\partial f (0, 0, \pm 1)}{\partial x^1} \right|,$$

we easily find that inequalities (4.45) and condition (B) are satisfied in a sufficiently small neighborhood of the origin. Consequently, the synthesis of optimal controls can be realized in a sufficiently small neighborhood of the origin for any object (4.25), (4.24), (4.40) which satisfies condition (4.53); moreover, each optimal control has at most one switching in this neighborhood.

52. Synthesis of Optimal Controls in Nonlinear Oscillatory Systems of the Second Order.* Let us consider a controlled object whose motion is described by the second-order differential equation (4.23) with the one-dimensional control region U given by (4.24). The introduction of the

---

* This section was written in collaboration with G. Nasrigdinov.

variables $x^1 = x$, $x^2 = \dot{x}$ reduces (4.23) to the normal system (4.25) of the second order. We shall assume that the function $f(x^1, x^2, u)$ in (4.25) is continuous, continuously differentiable with respect to u, twice continuously differentiable with respect to x^1, x^2 and, in addition, satisfies the following conditions:

$$f(0, 0, 1) > 0, \qquad f(0, 0, -1) < 0; \tag{4.54}$$

$$\frac{\partial f(x^1, x^2, u)}{\partial u} > 0 \qquad \text{for all} \qquad x^1, x^2, u; \tag{4.55}$$

$$\frac{\partial f}{\partial x^1} < -\frac{1}{4}\left(\frac{\partial f}{\partial x^2}\right)^2 \qquad \text{for all} \qquad x^1, x^2, u; \tag{4.56}$$

$$\left(\frac{\partial^2 f}{\partial x^1 \, \partial x^2}\right)^2 \leqslant \frac{\partial^2 f}{(\partial x^1)^2} \cdot \frac{\partial^2 f}{(\partial x^2)^2}, \qquad \frac{\partial^2 f}{(\partial x^1)^2} + \frac{\partial^2 f}{(\partial x^2)^2} \leqslant 0 \tag{4.57}$$
$$\text{for} \qquad u = \pm 1 \qquad \text{and all} \qquad x^1, x^2.$$

Condition (4.57) means that for $u = \pm 1$, the quadratic form

$$\frac{\partial^2 f}{(\partial x^1)^2} \xi_1^2 + 2 \frac{\partial^2 f}{\partial x^1 \, \partial x^2} \xi_1 \xi_2 + \frac{\partial^2 f}{(\partial x^2)^2} \xi_2^2$$

takes on only nonpositive values.

The above conditions are satisfied, for example, by the linear object $\ddot{x} + x = u$ (see [9], pp. 27–35) and also by nonlinear objects which differ from this object by a "small" convex addition. Namely, let $\varphi(x^1, x^2, u)$ be an arbitrary convex function of x^1, x^2 [that is, satisfying the condition (4.57)] which has bounded first derivatives

$$\left|\frac{\partial \varphi(x^1, x^2, u)}{\partial x^1}\right| \leqslant M, \qquad \left|\frac{\partial \varphi(x^1, x^2, u)}{\partial x^2}\right| \leqslant M, \qquad \left|\frac{\partial \varphi(x^1, x^2, u)}{\partial u}\right| \leqslant M.$$

Then, setting

$$f(x^1, x^2, u) = -x^1 + u + \mu\varphi(x^1, x^2, u),$$

we obtain, for sufficiently small μ, an object (4.25) which satisfies all the conditions (4.54)–(4.57).

In addition, we note that if the function f has the form

$$f(x^1, x^2, u) = -x^1 + g(x^2, u),$$

where g is a function satisfying the conditions

$$g(0, 1) > 0, \qquad g(0, -1) < 0,$$

$$\left|\frac{\partial g(x^2, u)}{\partial x^2}\right| < 2, \qquad \frac{\partial g(x^2, u)}{\partial u} > 0, \qquad \frac{\partial^2 g}{(\partial x^2)^2} \leqslant 0 \qquad \text{for all} \qquad x^2, u,$$

then the conditions (4.54)–(4.57) are also satisfied (see [1], p. 356).

We shall consider the problem of reaching the origin in the shortest time for the object described by (4.24), (4.25), (4.54)–(4.57). As we shall see, the synthesis of optimal controls for the nonlinear object under consideration will be qualitatively the same as for the linear object $\ddot{x} + x = u$, that is, the optimal trajectories will approach the origin as spirals (see [9], Figure 13, and Figure 182). It is interesting that the "oscillatory" character of the optimal trajectories is associated with the fulfillment of condition (4.56) which may be termed the condition of "strong negativity" of the derivative. As shown in § 13.49, replacing this condition of *non-negativity* of the derivative $\partial f/\partial x^1$ implies that every optimal control has only one switching, and the portrait of the synthesis resembles that of the object $\ddot{x} = u$ (see pp. 24–28, and in particular, Figure 26).

We proceed to the solution of the time-optimal problem for the object (4.24), (4.25), (4.52)–(4.57). First of all, let us find all the optimal trajectories which satisfy the maximum principle (see p. 78). The function H for the object under consideration has the following form:

$$H = \psi_1 x^2 + \psi_2 f(x^1, x^2, u). \tag{4.58}$$

Furthermore, let us write the system of equations for the auxiliary variables ψ_1, ψ_2:

$$\dot{\psi}_1 = -\frac{\partial H}{\partial x^1} = -\psi_2 \frac{\partial f}{\partial x^1},$$
$$\dot{\psi}_2 = -\frac{\partial H}{\partial x^2} = -\psi_1 - \psi_2 \frac{\partial f}{\partial x^2}. \tag{4.59}$$

The maximum condition (along an optimal trajectory, the function H attains its maximum with respect to u) means, in the given case, that the expression $\psi_2 f(x^1, x^2, u)$ attains its maximum [see (4.58)], and therefore, since, by (4.55), the function f increases monotonically with respect to u,

$$u = +1 \quad \text{for} \quad \psi_2 > 0; \quad u = -1 \quad \text{for} \quad \psi_2 < 0.$$

In other words,

$$u = \text{sign } \psi_2 \tag{4.60}$$

(for $\psi_2 = 0$, the value of u is not defined by the maximum condition). In this connection, we will be interested in the law governing the change of ψ_2. According to (4.59), we have

$$\frac{d}{dt} \arctan \frac{\psi_2}{\psi_1} = \frac{\psi_1 \dot{\psi}_2 - \psi_2 \dot{\psi}_1}{\psi_1^2 + \psi_2^2} = \frac{-\psi_1^2 - \dfrac{\partial f}{\partial x^2} \psi_1 \psi_2 + \dfrac{\partial f}{\partial x^1} \psi_2^2}{\psi_1^2 + \psi_2^2}. \tag{4.61}$$

The discriminant of the quadratic form

$$-\psi_1^2 - \frac{\partial f}{\partial x^2} \psi_1 \psi_2 + \frac{\partial f}{\partial x^1} \psi_2^2$$

is

$$\frac{1}{4} \left(-\frac{\partial f}{\partial x^2} \right)^2 + \frac{\partial f}{\partial x^1}.$$

According to (4.56), this discriminant is negative [this, in fact, is the reason for imposing condition (4.56)], and therefore the numerator of (4.61) has a constant sign. It is obvious that it is *negative* (it suffices to set $\psi_1 = 1$, $\psi_2 = 0$), that is,

$$\frac{d}{dt} \arctan \frac{\psi_2}{\psi_1} < 0 \qquad \text{for} \qquad \psi_1^2 + \psi_2^2 \neq 0.$$

In other words, the angle $\arctan \psi_2/\psi_1$ decreases monotonically, that is, the *vector* $\{\psi_1(t), \psi_2(t)\}$, which yields a nonzero solution of (4.59), rotates continuously (possibly changing its length) in the clockwise direction.

Furthermore, by virtue of (4.25), we have for $u = $ const,

$$\ddot{x}^1 = \dot{x}^2,$$

$$\ddot{x}^2 = \frac{d}{dt} f(x^1, x^2, u) = \frac{\partial f}{\partial x^1} \dot{x}^1 + \frac{\partial f}{\partial x^2} \dot{x}^2,$$

and therefore

$$\frac{d}{dt} \arctan \frac{\dot{x}^2}{\dot{x}^1} = \frac{\dot{x}^1 \ddot{x}^2 - \dot{x}^2 \ddot{x}^1}{(\dot{x}^1)^2 + (\dot{x}^2)^2} = \frac{\frac{\partial f}{\partial x^1} (\dot{x}^1)^2 + \frac{\partial f}{\partial x^2} \dot{x}^1 \dot{x}^2 - (\dot{x}^2)^2}{(\dot{x}^1)^2 + (\dot{x}^2)^2}. \qquad \textbf{(4.62)}$$

According to (4.56), the quadratic form in the numerator is negative (for $u = $ const), that is,

$$\frac{d}{dt} \arctan \frac{\dot{x}^2}{\dot{x}^1} < 0.$$

In other words, *on each piece of the solution* $\{x^1(t), x^2(t)\}$ *of system* (4.25) *on which* $u = $ const, *the phase-velocity vector* $\{\dot{x}^1(t), \dot{x}^2(t)\}$ *rotates continuously in the clockwise direction.*

Now, let $x_0 = (x_0^1, x_0^2)$ be an arbitrary point of the phase plane for which $\{x_0^2, f(x_0^1, x_0^2, 1)\} \neq 0$. Let us consider the solution $x(t), \psi(t)$ of systems (4.25) and (4.59) for $u = 1$ which satisfies the initial conditions

$$x^1(0) = x_0^1, \qquad x^2(0) = x_0^2, \qquad \psi_1(0) = 1, \qquad \psi_2(0) = 0. \qquad \textbf{(4.63)}$$

By virtue of (4.63), we find from the second equation in (4.59) that $\dot{\psi}_2(0) = -1$. Consequently, $\psi_2(t) > 0$ for sufficiently small negative t. Let $T_+(x_0)$

denote the negative root of $\psi_2(t)$ closest to zero; if $\psi_2(t)$ does not vanish for $t < 0$, then we set $T_+(x_0) = m_1$, where (m_1, m_2) is the maximum interval of existence of the solution $x(t)$, $\psi(t)$; in particular, it may turn out that $T_+(x_0) = -\infty$. It is clear that $\psi_2(t)$ is positive on the interval $T_+(x_0) < t < 0$. Let $K_+(x_0)$ denote the portion of the trajectory $x(t)$ corresponding to the time interval $T_+(x_0) \leqslant t \leqslant 0$. This portion terminates at x_0; its initial point [which exists when $T_+(x_0) \neq m_1$] will be denoted by $\xi_+(x_0)$. We note that along $K_+(x_0)$ we have $\psi_2(t) > 0$, $u \equiv +1$, that is, condition (4.60) is satisfied. In other words, the *piece* $K_+(x_0)$ satisfies the maximum principle. We also note that, by virtue of the relation $\{x_0^2, f(x_0^1, x_0^2, 1)\} \neq 0$, the point x_0 is not an equilibrium position of system (4.25) for $u = 1$, and therefore the phase-velocity vector $\{x^2, f(x^1, x^2, 1)\}$ is different from zero along the entire arc $K_+(x_0)$. In particular, the relation $f(x^1, x^2, 1) \neq 0$ is satisfied at the points of the arc $K_+(x_0)$ at which $x^2 = 0$.

Similarly, making use of the initial conditions

$$x^1(0) = x_0^1, \qquad x^2(0) = x_0^2, \qquad \psi_1(0) = -1, \qquad \psi_2(0) = 0, \qquad (4.64)$$

we construct for $\{x_0^2, f(x_0^1, x_0^2, -1)\} \neq 0$ the solution of systems (4.25) and (4.59) for $u \equiv -1$; the negative root of the function $\psi_2(t)$ closest to zero will be denoted by $T_-(x_0)$ [if $\psi_2(t) \neq 0$ for $t < 0$, then $T_-(x_0) = m_1$, where (m_1, m_2) is the maximum interval of existence of the solution under consideration]. The piece of the trajectory $x(t)$ corresponding to the time interval $T_-(x_0) \leqslant t \leqslant 0$ will be denoted by $K_-(x_0)$, and the initial point of this piece will be denoted by $\xi_-(x_0)$. This piece $K_-(x_0)$ also satisfies the maximum principle.

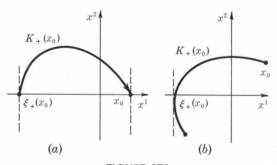

(a)   (b)

FIGURE 173

We now prove that if the point x_0 lies on the x^1-axis, then both points $\xi_+(x_0)$ and $\xi_-(x_0)$ also lie on the x^1-axis; moreover, $K_+(x_0)$ [and also $K_-(x_0)$] is a convex arc lying entirely on one side of the x^1-axis, and the tangents to this arc at its endpoints are parallel to the x^2-axis [Figure 173(a)]; if, however, the point x_0 does not lie on the x^1-axis, then the points x_0 and $\xi_+(x_0)$

[and also x_0 and $\xi_-(x_0)$] lie on opposite sides of the x^1-axis; moreover, $K_+(x_0)$ [respectively, $K_-(x_0)$] is a convex arc having one intersection with the x^1-axis [Figure 173(b)].

As a matter of fact, along the arc $K_+(x_0)$, we have

$$H = \psi_1 x^2 + \psi_2 f(x^1, x^2, +1) = \text{const.} \tag{4.65}$$

Since, in addition, $\psi_2(0) = \psi_2(T_+(x_0)) = 0$, it follows that

$$\psi_1(0)x^2(0) = \psi_1(T_+(x_0))x^2(T_+(x_0)).$$

But the numbers $\psi_1(0)$ and $\psi_1(T_+(x_0))$ are of opposite sign since the vector $\psi(t)$ rotates in the clockwise direction, and 0 and $T_+(x_0)$ are two adjacent zeros of the function $\psi_2(t)$ (Figure 174). Consequently, the numbers $x^2(0) = x_0^2$ and $x^2(T_+(x_0))$ also have *opposite* signs, that is, either both points x_0,

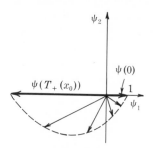

FIGURE 174

$\xi_+(x_0)$ lie on the x^1-axis or they lie on opposite sides of the x^1-axis. In the first case [both points x_0, $\xi_+(x_0)$ lie on the x^1-axis], the constant (4.65) is equal to zero since $x^2(0) = \psi_2(0) = 0$. If an intermediate point $x(\tau)$, $T_+(x_0) < \tau < 0$, of the arc $K_+(x_0)$ were to lie on the x^1-axis, then it would follow from the relations

$$H = \psi_1(\tau)x^2(\tau) + \psi_2(\tau)f(x^1(\tau), x^2(\tau), 1) = 0,$$

$$x^2(\tau) = 0, \qquad f(x^1(\tau), x^2(\tau), 1) \neq 0$$

that $\psi_2(\tau) = 0$. But this contradicts the fact that $T_+(x_0)$ is the negative root of $\psi_2(t)$ *closest* to zero. Consequently, the arc $K_+(x_0)$ does not have common points with the x^1-axis except at the endpoints, that is, it lies entirely on one side of the x^1-axis. The convexity of this arc follows from the fact that the vector $\{\dot{x}^1, \dot{x}^2\}$ tangent to this arc rotates continuously in one direction (clockwise). Finally, since $\dot{x}^1 = x^2 = 0$ at the endpoints of the arc $K_+(x_0)$ and $\dot{x}^2 = f(x^1, x^2, 1) \neq 0$, it follows that the tangents at the endpoints of the arc $K_+(x_0)$ are parallel to the x^2-axis. Now let the points x_0 and $\xi_+(x_0)$ lie on opposite sides of the x^1-axis. We shall show that the arc intersects the

x^1-axis only once. In fact, we have $f(x^1, x^2, 1) \neq 0$ at each point at which the arc $K_+(x_0)$ intersects the x^1-axis, that is, the vector $\{\dot{x}^1, \dot{x}^2\} = \{x^2,$ $f(x^1, x^2, 1)\}$ tangent to the arc $K_+(x_0)$ is different from zero and is parallel to the x^2-axis [Figure 173(b)]. Therefore it suffices to establish that the vector $\{\dot{x}^1, \dot{x}^2\}$ can be parallel to the x^2-axis *at only one* point of the arc $K_+(x_0)$. Let us suppose the contrary: there exist at least two points of the arc $K_+(x_0)$ at which the vector $\{\dot{x}^1, \dot{x}^2\}$ is parallel to the x^2-axis. Let τ_1 and τ_2 be two adjacent instants of time on the interval $T_+(x_0) \leqslant t \leqslant 0$ for which the vector $\{\dot{x}^1, \dot{x}^2\}$ is parallel to the x^2-axis, that is, the angle

$$y_1 = - \arctan \frac{\dot{x}^1}{\dot{x}^2} = \frac{\pi}{2} + \arctan \frac{\dot{x}^2}{\dot{x}^1} + K\pi$$

is a multiple of π. Since the angle y_1 decreases monotonically (because the vector $\{\dot{x}^1, \dot{x}^2\}$ rotates continuously in the clockwise direction), it follows that setting $y_1(\tau_1) = \pi$ we obtain $y_1(\tau_2) = 0$. In other words, the graph of the quantity $y_1(t)$ for t on the interval $[\tau_1, \tau_2]$ is a continuous curve passing from the value π to the value 0. According to (4.62), $y_1(t)$ is a solution of the differential equation

$$\dot{y} = \frac{-1 + \dfrac{\partial f}{\partial x^2} \tan y + \dfrac{\partial f}{\partial x^1} \tan^2 y}{1 + \tan^2 y}. \tag{4.66}$$

The function

$$y_2 = \arctan \frac{\psi_2}{\psi_1},$$

by virtue of (4.61), is also a solution of (4.66). We note, in addition, that the inequalities $T_+(x_0) < \tau_1 < \tau_2 < 0$ also hold [since we have $x^2 \neq 0$ at the endpoints of the arc $K_+(x_0)$, that is, the vector $\{\dot{x}^1, \dot{x}^2\}$ is not parallel to the x^2-axis]. The function $y_2(t)$ is defined [as is $y_1(t)$] to within a constant addend which is a multiple of π; we shall make the definition of the function $y_2(t)$ more precise by assuming that $0 < y_2(\tau_1) \leqslant \pi$. The graphs of the functions $y_1(t)$, $y_2(t)$, as integral curves of the same equation (4.66), either coincide or do not have common points. In the first case (the graphs coincide) we have $y_2(\tau_2) = y_1(\tau_2) = 0$. In the second case we have $y_2(\tau_1) > 0$, $y_2(\tau_2) < 0$ (Figure 175), and therefore $y_2(\tau) = 0$ at some intermediate point τ. Thus, in either case, there exists a point τ of the interval $T_+(x_0) < t < 0$ such that $y_2(\tau) = 0$, that is, $\psi_2(\tau) = 0$. But this contradicts the fact that $T_+(x_0)$ is the negative root of $\psi_2(t)$ closest to zero. This contradiction proves that the vector $\{\dot{x}^1, \dot{x}^2\}$ can be parallel to the x^2-axis at only one point of the arc $K_+(x_0)$.

By the same token, all the properties of the arc $K_+(x_0)$ have been established. An analogous proof can be carried out for the arc $K_-(x_0)$.

It should be noted, as can be seen from the above proof, that the properties of the arcs $K_+(x_0)$ and $K_-(x_0)$ hold when the points $\xi_+(x_0)$ and $\xi_-(x_0)$ do exist. However, if for example the point $T_+(x_0)$ does not exist, then it can only be asserted that $K_+(x_0)$ is a convex arc which intersects the x^1-axis at most once [similarly for the arc $K_-(x_0)$]; moreover, the proof carried out above remains valid (with obvious modifications).

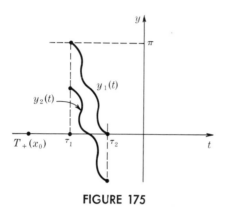

FIGURE 175

We note also that the points $\xi_+(x_0)$ and $\xi_-(x_0)$, as well as the functions $T_+(x_0)$ and $T_-(x_0)$, depend differentiably on x_0. This follows easily from the theorem on the differentiability of solutions with respect to parameters [since, by assumption, the right-hand sides of (4.25) and (4.59) have continuous first derivatives with respect to x^1, x^2, ψ_1, ψ_2].

Let us now proceed to the construction of trajectories which reach the origin and satisfy the maximum principle. Taking the origin O as x_0, let us construct the arc $\sigma_1^+ = K_+(0)$ [this is possible since $\{0, f(0, 0, 1)\} \neq 0$, see (4.54)]. By virtue of the first inequality in (4.54), the constructed arc σ_1^+ lies entirely below the x^1-axis and its initial point a_1^+ lies on the x^1-axis (Figure 176). Now, let z_1 be a point of the arc σ_1^+ distinct from a_1^+. If z_1 does not coincide with O, then it lies *below* the x^1-axis, so that $z_1^2 \neq 0$. If, however, $z_1 = 0$, then $f(z_1^1, z_1^2, -1) < 0$ [see (4.54)]. Thus, in either case, the vector $\{x^2, f(x^1, x^2, -1)\}$ is different from zero at the point z_1. Consequently, it is possible to determine the arc $K_-(z_1)$ (Figure 176). Let z_2 denote the initial point of this arc [this initial point does not exist for $T_-(z_1) = m_1$]. For $z_1 \neq 0$, the point z_2 lies above the x^1-axis, so that $z_2^2 \neq 0$. If, however, $z_1 = 0$, then the point z_2 lies on the x^1-axis, the arc $K_-(z_1)$ lies above the x^1-axis, and therefore, $f(z_2^1, z_2^2, -1) > 0$. Consequently, $f(z_2^1, z_2^2, +1) > 0$ [see (4.55)]. Thus, in either case, the vector $\{x^2, f(x^1, x^2, 1)\}$ is different from zero at the point

z_2, and therefore it is possible to determine the arc $K_+(z_2)$. The initial point of this arc, defined for $T_-(z_2) \neq m_1$, will be denoted by z_3. Then let us construct the arc $K_-(z_3)$ with the initial point z_4, then the arc $K_+(z_4)$, and so on. The resulting trajectory consisting of the arcs

$$\ldots, K_+(z_{2n}), K_-(z_{2n-1}), \ldots, K_+(z_4), K_-(z_3), K_+(z_2), K_-(z_1), z_1$$

(solid line in Figure 176) will be denoted by $\eta_+(z_1)$.

Similarly, if z_1 is an arbitrary point of the arc $\sigma_1^- = K_-(0)$, distinct from the endpoint a_1^- of this arc, then we construct the arc $K_+(z_1)$ whose initial point will be denoted by z_2, then we construct $K_-(z_2)$ with the initial point z_3, then $K_+(z_3)$, and so on (Figure 177). The resulting trajectory will be denoted by $\eta_-(z_1)$.

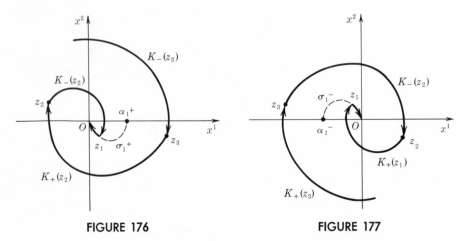

FIGURE 176   FIGURE 177

It is clear that all the trajectories $\eta_+(z_1)$, $\eta_-(z_1)$ are phase trajectories of (4.25) corresponding to the control $u(t)$ which is equal to $+1$ on the arcs of the form $K_+(z_i)$, and equal to -1 on the arcs of the form $K_-(z_i)$.

We shall prove that all of the trajectories $\eta_+(z_1)$, $\eta_-(z_1)$ satisfy the maximum principle. Let us consider the trajectory $\eta_+(z_1)$. Let $x_0(t)$ denote the solution of (4.25) for $u \equiv 1$ describing the arc $\sigma_1^+ = K_+(0)$, $T_+(0) \leqslant t \leqslant 0$, and let $\psi^*(t)$ denote the corresponding solution of (4.59) [satisfying the initial conditions (4.63) at the point $x_0 = 0$]. Furthermore, let $x_{2i-1}(t)$ denote the solution of (4.25) for $u \equiv -1$ describing the arc $K_-(z_{2i-1})$, $T_-(z_{2i-1}) \leqslant t \leqslant 0$, and let $x_{2i}(t)$ denote the solution of (4.25) for $u \equiv 1$ describing the arc $K_+(z_{2i})$, $T_+(z_{2i}) \leqslant t \leqslant 0$ ($i = 1, 2, \ldots$). The corresponding solutions of system (4.59) will be denoted by $\psi^{(2i-1)}(t)$, $\psi^{2i}(t)$. According to the above, the pair of functions $x_0(t)$, $\psi^*(t)$, and also each pair of functions $x_i(t)$, $\psi^{(i)}(t)$, $i = 1, 2, \ldots$, satisfy the maximum principle (for $u \equiv 1$ in the case of even i and for $u \equiv -1$ in the case of odd i). Let θ_i denote the instant of time at which the solution $x_0(t)$ passes through the point z_1 of the arc σ_1^+ [so that $T_+(0) < \theta_1 \leqslant 0$], and let $\psi^{(0)}(t) = (\psi_1^{(0)}(t), \psi_2^{(0)}(t))$ denote the solution of system (4.59) for $x = x_0(t)$, $u \equiv 1$ with the initial condition $\psi_2^{(0)}(\theta_1) = 0$, $\psi_1^{(0)}(\theta_1) = -1$. Both functions

$$y^{(0)}(t) = \arctan \frac{\psi_2^{(0)}(t)}{\psi_1^{(0)}(t)}, \qquad y^*(t) = \arctan \frac{\psi_2^*(t)}{\psi_1^*(t)}$$

are solutions of (4.66) on the interval $T_+(0) < t < 0$, and therefore their graphs either coincide or do not have common points. They cannot coincide for $z_1 \neq 0$ (that is, for $\theta_1 \neq 0$) since $\psi_2^{(0)}(\theta_1) = 0$, $\psi_2^*(\theta_1) \neq 0$, and therefore $y^{(0)}(\theta_1) \neq y^*(\theta_1)$. Thus, these graphs do not have common points. Both functions $y^{(0)}$ and y^* are defined to within an additive constant multiple of π. We shall make their definition more precise by assuming that $y^{(0)}(\theta_1) = \pi$, $0 < y^*(\theta_1) < \pi$. Both functions decrease monotonically [since the vectors $\psi^{(0)}(t)$, $\psi^*(t)$ rotate in the clockwise direction]; moreover, $y^*(t)$ vanishes at $t = 0$. Thus, the graph of $y^{(0)}(t)$ lies *above* the graph of $y^*(t)$,

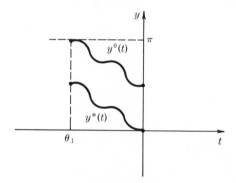

FIGURE 178

which, in turn, lies above the t-axis on the interval $\theta_1 < t < 0$. Consequently, the inequalities $0 < y^{(0)}(t) < \pi$ are satisfied for $\theta_1 < t < 0$, that is, the function $\psi_2^{(0)}(t)$ does not vanish on the interval $\theta_1 < t < 0$. In other words, this function has a constant sign on the above interval. But it follows from (4.59), by virtue of the relations $\psi_2^{(0)}(\theta_2) = 0$, $\psi_1^{(0)}(\theta_1) = -1$, that $\psi_2^{(0)}(\theta_1) = +1$. Consequently, the function $\psi_2^{(0)}(t)$ is *positive* on the interval $\theta_1 < t < 0$. From (4.60), this means that $x_0(t)$, $\psi^{(0)}(t)$ satisfy the maximum principle on the interval $\theta_1 < t < 0$.

Let us now determine the numbers θ_2, θ_3, . . . and the functions $x(t) = (x^1(t), x^2(t))$ and $\psi(t) = (\psi_1(t), \psi_2(t))$ as follows (the number θ_1 was determined above):

$$\theta_{2i} = T_-(z_{2i-1}) + \theta_{2i-1}, \qquad \theta_{2i+1} = T_+(z_{2i}) + \theta_{2i}, \qquad i = 1, 2, \ldots ;$$

$$x(t) = \begin{cases} x_0(t) & \text{for} \quad \theta_1 < t \leqslant 0, \\ x_1(t - \theta_i) & \text{for} \quad \theta_{i+1} < t \leqslant \theta_i, \quad i = 1, 2, \ldots ; \end{cases}$$

$$\psi(t) = \begin{cases} \psi^{(0)}(t) & \text{for} \quad \theta_1 < t \leqslant 0, \\ |\psi^{(1)}(\theta_2 - \theta_1)| \cdot |\psi^{(2)}(\theta_3 - \theta_2)| \cdots |\psi^{(i-1)}(\theta_i - \theta_{i-1})| \psi^{(i)}(t - \theta_i) \\ \qquad \qquad \text{for} \quad \theta_{i+1} < t \leqslant \theta_i, \quad i = 1, 2, \ldots . \end{cases}$$

It can be directly verified that the vector functions $x(t)$, $\psi(t)$ are continuous (that is, have one and the same limit from the right and left for $t = \theta_i$, $i = 1, 2, \ldots$). Furthermore, since the functions $x(t)$, $\psi(t)$ on the interval $\theta_{i+1} \leqslant t \leqslant \theta_i$ are obtained from the functions $x_i(t)$, $\psi^{(i)}(t)$ by a displacement (with respect to t) and by multiplying the function $\psi^{(i)}(t)$ by a positive constant, it follows that the functions $x(t)$, $\psi(t)$ satisfy the maximum principle on this interval. Here, i may take on the values $i = 1, 2, \ldots$, and therefore the functions $x(t)$, $\psi(t)$ satisfy the maximum principle on the entire interval on which they are defined. In addition, by definition, the vector function $x(t)$, $t \leqslant 0$, describes the trajectory $\eta_+(z_1)$. Thus, all the trajectories $\eta_+(z_1)$ satisfy the maximum principle. The arguments are analogous for the trajectory $\eta_-(z_1)$.

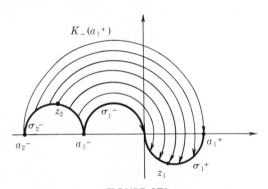

FIGURE 179

Let us now study the topological structure of the trajectories $\eta_+(z_1)$ and $\eta_-(z_1)$ on the phase plane. Let z_1 be an arbitrary interior point of the arc σ_1^+. It lies below the x^1-axis, and therefore the corresponding point $\xi_-(z_1) = z_2$ lies above the x^1-axis. As the point z_1 describes the entire arc σ_1^+, the corresponding point $z_2 = \xi_-(z_1)$ describes an arc σ_2^- which lies above the x^1-axis (Figure 179). Obviously, the points a_1^- and $a_2^- = \xi_-(a_1^+)$ lying on the x^1-axis serve as the endpoints of the arc σ_2^-. The curvilinear rectangle formed by the arcs σ_1^+, σ_1^-, σ_2^-, and $K_-(a_1^+)$ will be denoted by Q_1^- (Figure 180). We note that the tangent to the arc σ_2^- varies continuously since $\xi_-(z_1)$ depends differentiably on z_1, and z_1 describes σ_1^+ with continuously varying tangent.

Furthermore, when the point z_2 describes the arc σ_2^-, the corresponding point $z_3 = \xi_+(z_2)$ describes an arc σ_3^+ with continuously varying tangent which lies below the x^1-axis. The points $a_2 = \xi_+(a_1^-)$ and $a_3^+ = \xi_+(a_2^-)$ which lie on the x^1-axis serve as the endpoints of the arc σ_3^+. The curvilinear rectangle formed by the arcs σ_2^-, $K_+(a_1^-)$, $K_+(a_2^-)$, and σ_3^+ will be denoted by Q_2^+ (Figure 180). Continuing in the same manner, we construct the points a_i^+, a_i^- on the x^1-axis, the arcs σ_1^+, σ_3^+, σ_5^+, . . . lying below the x^1-axis, and the curvilinear rectangles Q_1^-, Q_2^+, Q_3^-, Q_4^+, . . . (Figure 180).

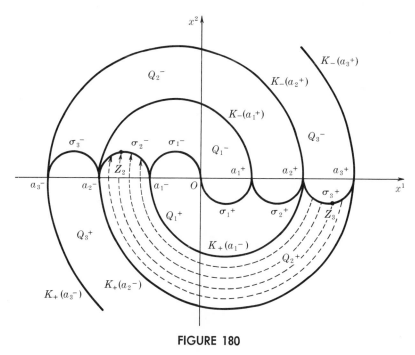

FIGURE 180

Similarly, if the point z_1 describes the arc σ_1^-, then the corresponding point $z_2 = \xi_+(z_1)$ describes an arc σ_2^+ lying below the x^1-axis with endpoints at a_1^+ and a_2^+. The point $z_3 = \xi_-(z_2)$ describes an arc σ_3^+ lying above the x^1-axis with endpoints a_2^- and a_3^-, and so on. The arcs thus constructed form curvilinear rectangles Q_1^+, Q_2^-, Q_3^+, . . . also shown in Figure 180. All curves which are the sides of curvilinear rectangles (solid lines in Figure 180) have continuously varying tangents.

We note that some of these rectangles will not be closed if $\xi_+(z_i)$ or $\xi_-(z_i)$ do not exist. An example of "rectangles" in this case is shown in Figure 181.

We shall now establish that at all interior points of the arc σ_i^+ (or σ_i^-) the tangent to this arc is not parallel to the x^2-axis, so that each of the arcs σ_i^+, σ_i^- can be uniquely projected onto the x^1-axis. This statement is obvious with respect to the arcs σ_1^+ and σ_1^- since these arcs are convex and lie on one side of the x^1-axis, and the tangents to these arcs at the endpoints, which lie on the x^1-axis, are parallel to the x^2-axis.

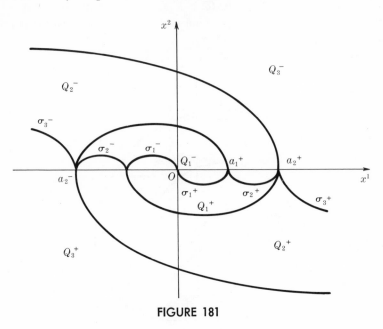

FIGURE 181

Let us adopt the inductive hypothesis that the above property holds for the arcs σ_1^+, σ_2^+, . . . , σ_k^+ and σ_1^-, σ_2^-, . . . , σ_k^- and let us prove its validity for the arc σ_{k+1}^+ (the arguments for the arc σ_{k+1}^- are analogous). Let us suppose, contrary to the assertion being proved, that the tangent to σ_{k+1}^+ at some interior point z_{k+1} is parallel to the x^2-axis.

In the following, let $o(\epsilon)$ denote various infinitesimals of order higher than ϵ, that is, functions which satisfy the condition

$$\lim_{\epsilon \to 0} \frac{o(\epsilon)}{\epsilon} = 0.$$

The point $z(\epsilon)$, lying on the arc σ_{k+1}^+ at distance $\epsilon > 0$ below the point z_{k+1}, has the form

$$z(\epsilon) = z_{k+1} + \epsilon e_2 + o(\epsilon), \qquad (4.67)$$

where $e_2 = \{0, -1\}$. Let $x^*(t)$ and $x^\epsilon(t)$ denote the solutions of system (4.25) for $u \equiv 1$ with the initial conditions

$$x^*(0) = z_{k+1}, \qquad x^\epsilon(0) = z(\epsilon),$$

and let $\psi^*(t)$ and $\psi^\epsilon(t)$ denote the corresponding solutions of (4.59) (also for $u \equiv 1$) with the initial conditions

$$\psi^*(0) = \psi^\epsilon(0) = \{-1, 0\}. \tag{4.68}$$

The positive root of $\psi_2^*(t)$ closest to zero will be denoted by θ. Then it is clear that for $0 \leqslant t \leqslant \theta$ the vector $x^*(t)$ describes the arc $K_+(z_{k+1})$ and $\psi^*(t)$ is the corresponding solution of (4.59). Furthermore, we can write, on the entire interval $0 \leqslant t \leqslant \theta$,

$$x^\epsilon(t) = x^*(t) + \epsilon \delta x(t) + o(\epsilon), \qquad \psi^\epsilon(t) = \psi^*(t) + \epsilon \delta \psi(t) + o(\epsilon),$$

where $\delta x = \{\delta x^1, \delta x^2\}$ and $\delta \psi = \{\delta \psi_1, \delta \psi_2\}$ satisfy the following variational equations [see (4.25), (4.59)]:

$$\delta \dot{x}^1 = \delta x^2,$$

$$\delta \dot{x}^2 = \frac{\partial f}{\partial x^1} \delta x^1 + \frac{\partial f}{\partial x^2} \delta x^2,$$

$$\delta \dot{\psi}_1 = -\frac{\partial f}{\partial x^1} \delta \psi_2 - \psi_2^* \frac{\partial^2 f}{\partial x^1 \partial x^1} \delta x^1 - \psi_2^* \frac{\partial^2 f}{\partial x^1 \partial x^2} \delta x^2, \tag{4.69}$$

$$\delta \dot{\psi}_2 = -\delta \psi_1 - \frac{\partial f}{\partial x^2} \delta \psi_2 - \psi_2^* \frac{\partial^2 f}{\partial x^1 \partial x^2} \delta x^1 - \psi_2^* \frac{\partial^2 f}{\partial x^2 \partial x^2} \delta x^2,$$

in which $x = x^*(t)$ and $u \equiv 1$ must be substituted in f. Moreover, it follows from relations (4.67) and (4.68) that the following initial conditions are satisfied:

$$\delta x^1(0) = 0, \qquad \delta x^2(0) = -1, \qquad \delta \psi_1(0) = \delta \psi_2(0) = 0. \tag{4.70}$$

Furthermore, by definition of the number θ, we have $\psi^*(\theta) = 0$.

From the first two equations in (4.69), we find

$$\frac{d}{dt} \arctan \frac{-\delta x^1}{\delta x^2} = \frac{\delta \dot{x}^2 \, \delta x^1 - \delta \dot{x}^1 \, \delta x^2}{(\delta x^1)^2 + (\delta x^2)^2} = \frac{-(\delta x^2)^2 + \dfrac{\partial f}{\partial x^2} \delta x^1 \, \delta x^2 + \dfrac{\partial f}{\partial x^1} (\delta x^1)^2}{(\delta x^1)^2 + (\delta x^2)^2}.$$

Thus the quantity $\eta = \arctan \dfrac{-\delta x^1}{\delta x^2}$ satisfies the differential equation

$$\dot{\eta} = \frac{-1 - \dfrac{\partial f}{\partial x^2} \tan \eta + \dfrac{\partial f}{\partial x^1} \tan^2 \eta}{1 + \tan^2 \eta}$$

with the initial condition $\eta(0) = \pi$ [see (4.70)]. The quantity $\arctan \psi_2^*/\psi_1^*$ is also a solution of the same differential equation [see (4.66), (4.61)] with the same initial condition [see (4.63)]. Consequently,

$$\arctan \frac{-\delta x^1}{\delta x^2} \equiv \arctan \frac{\psi_2^*}{\psi_1^*}$$

on the entire interval $0 \leqslant t \leqslant \theta$. Since the vector $\psi^*(t)$ rotates in the clockwise direction and θ is a positive root of the function $\psi_2^*(t)$ closest to zero, it follows that $\arctan \psi_2^*(\theta)/\psi_1^*(\theta) = 0$, that is, $\arctan \dfrac{-\delta x^1(\theta)}{\delta x^2(\theta)} = 0$. In other words, we have

$$\delta x^1(\theta) = 0, \qquad \delta x^2(\theta) > 0. \tag{4.71}$$

An obvious calculation shows that

$$\frac{d}{dt}(\delta\psi_1\,\delta x^1 + \delta\psi_2\,\delta x^2) = \delta\dot\psi_1\,\delta x^1 + \delta\psi_1\,\delta\dot x^1 + \delta\dot\psi_2\,\delta x^2 + \delta\psi_2\,\delta\dot x^2$$

$$= -\psi_2^*\left(\frac{\partial^2 f}{\partial x^1\,\partial x^1}(\delta x^1)^2 + 2\frac{\partial^2 f}{\partial x^1\,\partial x^2}\,\delta x^1\,\delta x^2 + \frac{\partial^2 f}{\partial x^2\,\partial x^2}(\delta x^2)^2\right). \tag{4.72}$$

The right-hand side in (4.72) is non-negative on the entire interval $0 \leqslant t \leqslant \theta$ since $\psi_2^*(t) > 0$ on this interval and the quadratic form in parenthesis is non-positive by virtue of (4.57). Therefore, integrating (4.72) from 0 to θ, we obtain

$$[\delta\psi_1(\theta)\,\delta x^1(\theta) + \delta\psi_2(\theta)\,\delta x^2(\theta)] - [\delta\psi_1(0)\,\delta x^1(0) + \delta\psi_2(0)\,\delta x^2(0)] \geqslant 0,$$

which, by (4.70) and (4.71), leads to the relation

$$\delta\psi_2(\theta) \geqslant 0. \tag{4.73}$$

On the other hand, since the point $x^*(\theta) = z_k$ lies on the curve σ_k^- whose tangent at the point z_k is, by inductive assumption, not parallel to the x^2-axis, it follows from (4.71) that the trajectory $x^\epsilon(t) = x^*(t) + \epsilon\delta x(t) + o(\epsilon)$ passes from one side of the curve σ_k^- to the other, and therefore intersects the curve σ_k^- at some time $t^\epsilon < \theta$. But the trajectory $x^\epsilon(t)$ describes the arc $K_+(z^\epsilon)$ for $0 \leqslant t \leqslant t^\epsilon$, so that the time t^ϵ, at which the point $x^\epsilon(t)$ reaches the arc σ_k^-, coincides with the positive root of the equation $\psi_2^\epsilon(t) = 0$ closest to zero. In other words, the vector $\psi^\epsilon(t)$, rotating (for $t > 0$) in the clockwise direction from the position $\psi^\epsilon(0) = \{-1, 0\}$, takes the form $\{q, 0\}$ at the time $t = t^\epsilon$, where $q > 0$. Consequently, at the time $\theta > t^\epsilon$ the vector $\psi^\epsilon(t)$ lies (for a sufficiently small ϵ) in the fourth quadrant, that is, $\psi_2^\epsilon(\theta) < 0$. Finally, recalling that

$$\psi^\epsilon(\theta) = \psi^*(\theta) + \epsilon\delta\psi(\theta) + o(\epsilon) \qquad \text{and} \qquad \psi_2^*(\theta) = 0,$$

we find that $\epsilon\delta\psi_2(\theta) + o(\epsilon) < 0$. Since this relation holds for any sufficiently small ϵ, we obtain $\delta\psi_2(\theta) \leqslant 0$, and therefore [see (4.73)]

$$\delta\psi_2(\theta) = 0.$$

Thus, $\psi_2^\epsilon(\theta) = \psi_2^*(\theta) + \epsilon\delta\psi(\theta) + o(\epsilon) = o(\epsilon)$ and therefore the time t^ϵ at which the function $\psi_2^\epsilon(t)$ vanishes has the form $t^\epsilon = \theta + o(\epsilon)$. Consequently,

$$x^\epsilon(t^\epsilon) = x^\epsilon(\theta) + o(\epsilon) = x^*(\theta) + \epsilon\delta x(\theta) + o(\epsilon);$$

hence, it is seen that $\delta x(\theta)$ is a vector tangent to the arc σ_k^- at the point z_k [since both the points $x^\epsilon(t^\epsilon)$ and $x^*(\theta) = z_k$ lie on the arc σ_k^-]. Now, (4.71) shows that the tangent to the arc σ_k^- at the point z_k is parallel to the x^2-axis, which contradicts the inductive hypothesis.

Thus, there is not a single interior point of the arc σ_i^+ (or σ_i^-) at which the tangent to this arc is parallel to the x^2-axis.

We shall prove that every trajectory $\eta_+(z_1)$, $\eta_-(z_1)$ (except for the two trajectories passing along the sides of rectangles, Figure 182) passing through some rectangle enters it only once, and upon leaving it, approaches the corresponding side of the rectangle at zero angle. Thus, a general portrait of trajectories filling out the plane has the form shown in Figure 182.

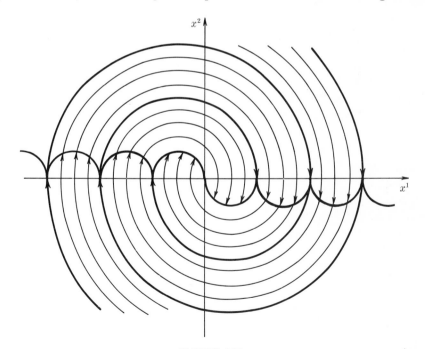

FIGURE 182

Let z_1 be an arbitrary interior point of the arc σ_1^+. Since σ_1^+ is a trajectory of system (4.25) for $u = +1$, the tangent to σ_1^+ at the point $z_1 = (z_1^1, z_1^2)$ is defined by the vector $(z_1^2, f(z_1^1, z_1^2, +1))$ which lies in the second or third quadrant (since $z_1^2 < 0$, compare Figure 179). Furthermore, since $f(z_1^1, z_1^2, +1) > f(z_1^1, z_1^2, -1)$ [see (4.55)], it follows that the vector $(z_1^2, f(z_1^1, z_1^2, -1))$, also lying in the second or third quadrant, is below the vector $(z_1^2, f(z_1^1, z_1^2, +1))$, that is, below the tangent to σ_1^+ at the point z_1. But $(z_1^2, f(z_1^1, z_1^2, -1))$ is the phase velocity vector of the trajectory $\eta_+(z_1)$ at the time at which this trajectory reaches the point z_1. Consequently, the trajectory $\eta_+(z_1)$ [more accurately, its arc $K_-(z_1)$] approaches σ_1^+ *from above* and, moreover, at nonzero angle (equal to the angle between the two vectors we have considered). Thus, traversing the arcs $K_-(z_1)$ in the opposite direction, these arcs move upward from points of σ_1^+, cross the x^1-axis, and terminate at points of σ_2^-. [None of the arcs $K_-(z_1)$ can intersect σ_1^+, since it intersects the x^1-axis only once.] Moreover, the arcs emanating from different points z_1 of σ_1^+ do not intersect one another and do not intersect the sides σ_1^- and $K_-(a_1^+)$ of the rectangle Q_1^-, as they are phase trajectories of the same system (4.25) (for $u \equiv -1$). In other words, the arcs $K_-(z_1)(z_1 \in \sigma_1^+)$ emanating from the points of σ_2^- fill out the rectangle Q_1^-, and approach σ_1^+ at nonzero angles. Therefore, it is clear that these arcs depart from σ_2^- in an upward direction (Figure 179).

We shall now prove that the trajectories $\eta_+(z_1)$ approach σ_2^- from below and, moreover, at nonzero angles. Let $z_2 = (z_2^1, z_2^2)$ be an arbitrary interior point of σ_2^-. The trajectory $\eta_+(z_1)$ passing through the point z_2 approaches this point with a phase velocity $v_+ = \{z_2^2, f(z_2^1, z_2^2, +1)\}$ and departs from the point z_2 with a phase velocity $v_- = \{z_2^2, f(z_2^1, z_2^2, -1)\}$. The vectors v_+ and v_- lie, respectively, in the first and fourth quadrants (since $z_2^2 > 0$). Moreover, $f(z_2^1, z_2^2, +1) > f(z_2^1, z_2^2, -1)$, that is, the vector v_+ lies

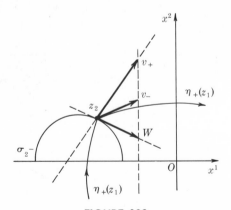

FIGURE 183

above the vector v_- (Figure 183). Furthermore, let h be a number such that the vector $w = \{z_2^2, h\}$ is directed along the tangent to the curve σ_2^- at z_2 (such an h exists, since $z_2^2 \neq 0$ and the tangent to σ_2^-, according to the above proof, is not parallel to the x^2-axis). As we already know, the arc $K_-(z_1)$, being a portion of the trajectory $\eta^+(z_1)$, lies above the arc σ_2^- and therefore the vector v_- lies above* the vector w, that is, $f(z_2^1, z_2^2, -1) > h$. Consequently, $f(z_2^1, z_2^2, +1) > h$, that is, the vector v_+ lies above the vector w. But this means that the trajectory $\eta_+(z_1)$ approaches the arc σ_2^- from below and, moreover, at a nonzero angle (which is equal to the angle between the vectors v_+ and w). The arcs $K_+(z_2)$ [being portions of the trajectories $\eta_+(z_1)$] approaching various points z_2 of the arc σ_2^- do not intersect each other and do not intersect the sides $K_+(a_1^-)$, $K_+(a_2^-)$ of the rectangle Q_2^+ [as they are phase trajectories of the same system (4.25) for $u \equiv 1$]. Thus, all possible arcs $K_+(z_2)(z_2 \in \sigma_2^-)$ emanating from points of σ_3^+ fill out the rectangle Q_2^+ and approach σ_2^- at nonzero angles. These arcs depart from σ_3^+ in a downward direction (Figure 180).

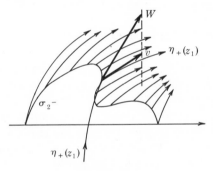

FIGURE 184

Furthermore, we could prove in an analogous manner that the trajectories $\eta_+(z_1)$ approach the arc σ_3^+ from above and, moreover, at nonzero angles. Therefore, analogous arguments may be applied to the rectangle Q_3^- and so on. This, in fact, yields a phase portrait for the trajectories $\eta_+(z_1)$ as depicted in Figure 180. The arguments are also analogous for the trajectories $\eta_-(z_1)$.

Now, it is not difficult to prove that all the trajectories $\eta_+(z_1)$, $\eta_-(z_1)$ are optimal. Let P^2 denote the union of all curvilinear rectangles Q_i^-, Q_i^+, $i = 1$, $2, \ldots$. Let P^1 denote the union of all the sides of these rectangles [that is,

---

* This conclusion (the vector v_- lies above the vector w) essentially makes use of the fact that the arc σ_2^- has no vertical tangents (compare Figure 184). In fact, condition (4.57) was only used in proving that the arcs σ_1^+ and σ_1^- do not have vertical tangents. The author does not know whether the facts being proved in this section will remain valid if condition (4.57) is discarded.

the union of arcs of the form σ_i^-, σ_i^+, $K_+(a_i^-)$, $K_-(a_i^+)$] and let P^0 denote the origin. It is clear that P^0, P^1, P^2 are piecewise smooth sets (see Chapter 4 and also [1], p. 329); moreover, $P^0 \subset P^1 \subset P^2$. The union of arcs of the form $K_+(a_i^-)$, $K_i(a_i^+)$, $i = 1, 2, \ldots$ will be denoted by N, and the union of all the arcs σ_i^-, σ_i^+ will be denoted by Γ. Furthermore, the interiors of the rectangles Q_i^-, Q_i^+ will be considered as two-dimensional cells of the first kind, the arcs σ_i^-, σ_i^+ as one-dimensional cells of the first kind, the remaining arcs σ_i^-, σ_i^+ as one-dimensional cells of the second kind, and the origin as a zero-dimensional cell of the first kind. Finally, let $v(x^1, x^2)$ denote the function given in the region P^2 and taking on the value -1 above the curve Γ and on the arc σ_1^-, and taking on the value $+1$ below the curve Γ and on the arc σ_1^+ (the values on the remaining arcs σ_i^+, σ_i^- are of no importance). It is clear that what has been said above defines the regular synthesis (see p. 184 and also [1], p. 341) in the region P^2; moreover, it is the trajectories $\eta_+(z_1)$, $\eta_-(z_1)$ which are, in fact, the distinguished trajectories. Consequently, the function $v(x^1, x^2)$ and the trajectories $\eta_+(z_1)$, $\eta_-(z_1)$ define the synthesis of optimal trajectories in the region P^2 (Theorem 5 and [2], see Theorem 4.7 and also [1], Theorem 5).

Thus, we obtain the following final result concerning the time-optimality of the object (4.24), (4.25), (4.54)–(4.57). Optimal trajectories which reach the origin have the form of spirals and consist of alternating sections corresponding to the values of the control $u = +1$ and $u = -1$. The switching curve Γ consists of the arcs σ_i^-, σ_i^+ (see Figure 180 and the pertaining text), and the synthesis of optimal controls is realized by the function $v(x^1, x^2)$, which is equal to $+1$ below the curve Γ and on the arc σ_1^+, and to -1 above the curve Γ and on the arc σ_1^-. The qualitative portrait of the optimal trajectories is shown in Figure 182.

chapter 5 Formulation of Other Optimal Control Problems

§ 14. THE PROBLEM WITH VARIABLE ENDPOINTS

53. Preliminary Discussion. In order to approach the formulation of the problem with variable endpoints, we shall consider two control problems. Let us consider the motion of a guided missile which is launched from a given point A and which must hit a fixed target located at another given point B. Moreover, it is natural to pose this problem as that of reaching the point B from the point A in the shortest time since a minimum flight duration minimizes the probability of destruction of the missile by enemy interceptors. We have here a time-optimal problem which, however, upon close scrutiny, is seen to differ from the problems considered previously. In fact, the motion of a missile in space is described, under gross idealization, by six phase coordinates x^i, $i = 1, 2, \ldots, 6$ and by several control parameters. If, for example, x^1, x^2, x^3 are taken as spatial coordinates and x^4, x^5, x^6 as their derivatives (that is, the velocity components), then the motion of the missile can be described by equations of the form

$$\dot{x}^i = x^{i+3}, \quad i = 1, 2, 3; \quad \dot{x}^j = f^j(x, u), \quad j = 4, 5, 6.$$

The initial values (that is, the coordinates of the point A and the initial velocity of the missile) can be assumed to be known. On the other hand, at the terminal instant of the motion we know the values of only the first three coordinates: $x^1 = a$, $x^2 = b$, $x^3 = c$ (that is, the "geographical coordinates" a, b, c of the point B) whereas nothing is known about the velocity components x^4, x^5, x^6. In other words, the velocity of the missile upon reaching the point B is immaterial. This means that any of the infinite number of points of the phase space defined by the relations $x^1 = a$, $x^2 = b$,

$x^3 = c$ may serve as the prescribed terminal state. These equations define a "three-dimensional plane" M in the six-dimensional phase space. Thus we arrive at a time-optimal problem in which the initial state x_0 is given and the terminal state x_1 must be appropriately selected from a set M (planes, lines, and so on, Figure 185) of the phase space.

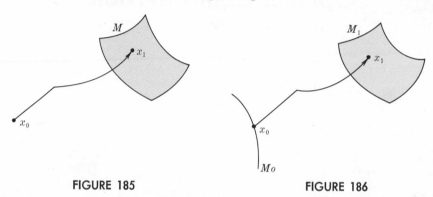

FIGURE 185   FIGURE 186

Here is another problem of a similar kind. An engine is equipped with an actuating mechanism which makes it possible to bring the engine to the required operating state and is then turned off. Let the state of the engine be defined by the phase coordinates $x^1, \ldots, x^k$, and let the state of the actuating mechanism be defined by the phase coordinates $x^{k+1}, \ldots, x^n$. If we assume that the initial states of the engine and actuating mechanism are defined by zero phase coordinates, and that the terminal (operating) state of the engine corresponds to the coordinate values $x^1 = a^1, \ldots, x^k = a^k$, then we arrive at the following problem. An initial point x_0 and a plane M, defined by the equations $x^1 = a^1, \ldots, x^k = a^k$, are given in the phase space of the variables $x^1, \ldots, x^n$; it is required to transfer an object (engine plus actuating mechanism) from the state x_0 to a point of the set M in the shortest time. (The point of the plane M at which the phase point arrives is immaterial since the actuating mechanism is turned off anyway and therefore its state at the terminal time is of no interest.)

In both cases considered above we have a time-optimal problem with a *variable right-hand endpoint* (Figure 185). One can also consider a problem with both endpoints being variable (Figure 186). In this case, two sets M_0 and M_1 are given in the phase space and it is required to go from a point x_0 of M_0 to a point x_1 of M_1 in the shortest time; moreover, the points x_0 and x_1 are not specified in advance.

The sets M_0 and M_1 are usually assumed to be *manifolds*, that is, curves or surfaces of some dimensionality. Therefore, before proceeding to the precise formulation of the problem with variable endpoints, we shall discuss some geometrical concepts related to manifolds.

54. Manifolds and Their Tangent Planes.   Let $f(x) = f(x^1, \ldots, x^n)$
be a real scalar-valued function defined on a region G of the Euclidean space
X with (orthogonal) coordinates $x^1, \ldots, x^n$. If f has first partial derivatives with respect to $x^1, \ldots, x^n$ in G, then the vector

$$\left(\frac{\partial f}{\partial x^1}, \frac{\partial f}{\partial x^2}, \ldots, \frac{\partial f}{\partial x^n} \right),$$

called the *gradient* of the function $f(x)$ and denoted by the symbol grad $f(x)$,
is defined at each point x in G.

The set S of all points $x = (x^1, \ldots, x^n)$ which satisfy the relation

$$f(x^1, x^2, \ldots, x^n) = 0 \tag{5.1}$$

is called a *hypersurface* in X, and (5.1) is the *equation* of this hypersurface.
Let us now assume that the left-hand side of (5.1) has continuous partial
derivatives with respect to $x^1, x^2, \ldots, x^n$. A point $x \in S$ at which the
relations

$$\frac{\partial f(x)}{\partial x^1} = \frac{\partial f(x)}{\partial x^2} = \cdots = \frac{\partial f(x)}{\partial x^n} = 0$$

are satisfied [that is, a point at which grad $f(x)$ vanishes], is called a *singular
point* of the hypersurface S. The points of S at which grad $f(x) \neq 0$ are
called *nonsingular points*. A hypersurface defined by (5.1) with a continuously differentiable left-hand side which contains no singular points is called
a *smooth hypersurface*. All the hypersurfaces considered below are assumed
to be smooth.

For $n = 2$, (5.1) takes the form

$$f(x^1, x^2) = 0,$$

and in this case the concept of a smooth hypersurface reduces to the concept
of a *smooth curve* (in the x^1, x^2-plane). For $n = 3$, (5.1) takes the form

$$f(x^1, x^2, x^3) = 0,$$

and in this case the concept of a smooth hypersurface reduces to the concept
of a *smooth surface* (in x^1, x^2, x^3-space).

If (5.1) is linear, that is, has the form (2.3), then the absence of singular
points signifies that at least one of the coefficients α_i is different from zero.
In this case the hypersurface is called a *hyperplane* (compare p. 40).

Let x_0 be an arbitrary point of the smooth hypersurface S defined by
(5.1). The vector grad $f(x_0)$ (or any vector parallel to it) is called a *normal
vector* (or simply a *normal*) of S at the point x_0. The hyperplane passing
through x_0 and having the vector grad $f(x_0)$ as its normal is called the *tangent*

hyperplane of the hypersurface S at x_0. Each vector emanating from x_0 which lies on the tangent hyperplane is called a *tangent vector* of the hypersurface S at x_0. In other words, a vector which emanates from x_0 is a tangent vector of S if and only if it is orthogonal to grad $f(x_0)$.

Now, let

$$S_1, S_2, \ldots, S_k$$

be smooth hypersurfaces in the space X defined by the equations

$$
\begin{aligned}
f_1(x^1, x^2, \ldots, x^n) &= 0, \\
f_2(x^1, x^2, \ldots, x^n) &= 0, \\
&\cdots \\
f_k(x^1, x^2, \ldots, x^n) &= 0,
\end{aligned}
\tag{5.2}
$$

respectively. The *intersection* M of all these hypersurfaces, that is, the set of all points $x \in X$ which simultaneously satisfy all the equations (5.2), is called an $(n - k)$-dimensional *(smooth) manifold* in X if the following condition is satisfied: at each point $x \in M$ the vectors

$$\text{grad } f_1(x), \text{grad } f_2(x), \ldots, \text{grad } f_k(x) \tag{5.3}$$

are linearly independent. Thus, by definition, an r-dimensional manifold in X is defined by a system of $n - r$ equations. In particular, an $(n - 1)$-dimensional manifold is defined by one equation. Thus, $(n - 1)$-dimensional manifolds of X coincide with hypersurfaces. One-dimensional manifolds are also called curves. Let us note that requiring the vectors (5.3) to be independent is equivalent to requiring the rank of the functional matrix

$$
\begin{pmatrix}
\dfrac{\partial f_1(x)}{\partial x^1} & \dfrac{\partial f_1(x)}{\partial x^2} & \cdots & \dfrac{\partial f_1(x)}{\partial x^n} \\[2mm]
\dfrac{\partial f_2(x)}{\partial x^1} & \dfrac{\partial f_2(x)}{\partial x^2} & \cdots & \dfrac{\partial f_2(x)}{\partial x^n} \\[2mm]
\cdot \quad \cdot & \cdot \quad \cdot & \cdots & \cdot \quad \cdot \\[2mm]
\dfrac{\partial f_k(x)}{\partial x^1} & \dfrac{\partial f_k(x)}{\partial x^2} & \cdots & \dfrac{\partial f_k(x)}{\partial x^n}
\end{pmatrix}
\tag{5.4}
$$

to be maximal (that is, equal to k).

If the equations in (5.2), which define the $(n - k)$-dimensional manifold M, are linear, then M is called an $(n - k)$-dimensional *plane* in X. In other words, an $(n - k)$-dimensional plane is the intersection of k hyperplanes whose normal vectors are linearly independent. One-dimensional planes are also called *straight lines*.

Let M be a smooth $(n - k)$-dimensional manifold in X defined by (5.2) and let x be a point of M. Let L_i denote the tangent hyperplane of the hypersurface $f_i(x^1, x^2, \ldots, x^n) = 0$ at x $(i = 1, 2, \ldots, k)$. The intersection of the hyperplanes $L_1, L_2, \ldots, L_k$ is an $(n - k)$-dimensional plane which is called the *tangent plane* of M at the point x. A vector emanating from x lies on the tangent plane (that is, is a *tangent vector* of M at x) if it is orthogonal to all the vectors (5.3).

Finally, let us note another simple fact which we shall use later. Let

$$x^i = \varphi^i(\xi), \qquad i = 1, 2, \ldots, n, \tag{5.5}$$

be a parametric representation of a curve in X. In vector form, $x = \varphi(\xi)$. The tangent vector of this curve at the point corresponding to $\xi = \xi_0$ has the form

$$\left(\frac{d\varphi^1(\xi_0)}{d\xi}, \frac{d\varphi^2(\xi_0)}{d\xi}, \ldots, \frac{d\varphi^n(\xi_0)}{d\xi} \right) = \frac{d\varphi(\xi_0)}{d\xi}. \tag{5.6}$$

If the curve (5.5) lies entirely on the smooth manifold M (of arbitrary dimension), then the tangent vector (5.6) of this curve is also a tangent vector of M at the point $\varphi(\xi_0)$. Conversely, given a tangent vector of M at the point $x_0 \in M$, there exists a curve on M passing through x_0 whose tangent vector is the given vector. In other words, a vector emanating from an arbitrary point $x_0 \in M$ is a tangent vector of M if and only if it is tangent to some curve which lies in M.

55. The Transversality Conditions and the Formulation of the Theorem.
Let us now pass to the formulation of the optimal control problem with variable endpoints. Let S_0 and S_1 be smooth manifolds in X of arbitrary (but less than n) dimensions r_0 and r_1. Let us pose the problem: *find an admissible control $u(t)$ which transfers the phase point from some position $x_0 \in S_0$ (not specified in advance) to some position $x_1 \in S_1$ in minimum time* (Figure 186). We shall call this problem the time-optimal problem with variable endpoints. If both S_0 and S_1 degenerate into points, the problem with variable endpoints becomes the previous problem which we have already considered (the problem with *fixed endpoints*).

It is clear that if the points x_0 and x_1 on the manifolds S_0 and S_1 were known, we would have a problem with fixed endpoints. Therefore, the control $u(t)$, optimal in the sense of the problem with variable endpoints, is also optimal in the previous sense, that is, the maximum principle remains valid also for the problem with free endpoints.

However, in this case it is also necessary to have additional relations from which one can determine the positions of x_0 and x_1 on S_0 and S_1. As a matter of fact, the *transversality conditions* formulated in this section are such relations. These conditions allow us to write $r_0 + r_1$ relations involving the coordinates of the endpoints x_0 and x_1. On the other hand, since the number of unknown parameters (in comparison with the problem with fixed endpoints) is also increased by $r_0 + r_1$ (since the position of x_0 on the r_0-dimensional manifold S_0 is characterized by r_0 parameters, and the position of x_1 on the r_1-dimensional manifold S_1 is characterized by r_1 parameters), the maximum principle together with the transversality conditions forms a "sufficient" system of relations for solving the given optimal problem with variable endpoints.

It is not difficult to get an idea of the character of the transversality conditions by continuing the arguments of § 2.5 and § 2.6. First, let us consider the case in which only the left-hand endpoint of the trajectory is variable. In other words, let us consider the time-optimal problem for the case in which the point x_1 is given and any point of the given manifold M_0 may be chosen as x_0 (Figure 187). Let $u(t)$, $t_0 \leqslant t \leqslant t_1$, be an optimal control

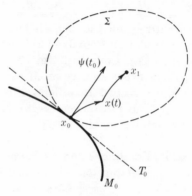

FIGURE 187

which provides a solution of this problem, and let $x(t)$ be the corresponding optimal trajectory. Let Σ denote the set of all points of the phase space from which it is possible to reach the point x_1 exactly in the optimal time $t_1 - t_0$; in particular, the point $x_0 = x(t_0)$ belongs to Σ (this point lies on the manifold M_0). The set Σ is defined by the equation $\omega(x) = -(t_1 - t_0)$, where $-\omega(x)$ is the duration of optimal motion from the point x to the point x_1. If we assume, as in § 2.5 and § 2.6, that the function $\omega(x)$ has continuous derivatives, then Σ is an $(n - 1)$-dimensional *smooth manifold*, that is, a hypersurface in X [we note that by virtue of Bellman's equation (1.18) we have grad $\omega(x) \neq 0$ at every point $x \neq x_1$]. Let T_0 denote the tangent plane to the hypersurface Σ at the point x_0.

It can be readily seen that *any vector tangent to the manifold M_0 at the point x_0 lies in the hyperplane T_0.* In fact, let there exist a tangent vector v of the manifold M_0 at x_0 which does not lie in the hyperplane T_0. Then the derivative $\partial \omega(x)/\partial x$, taken in the direction of the vector v, is different from zero. It can be assumed (if necessary, by reversing the direction of the vector v) that this derivative is positive. Now, if we take a curve lying on the manifold M_0 which emanates from x_0 and is tangent to the vector v, then we find (by virtue of the positiveness of the above derivative), that $\omega(x)$ increases in moving along this curve. In other words, there is a point x_0' on the manifold M_0 such that $\omega(x_0') > \omega(x_0)$. But this means that the duration of the optimal motion from the point x_0' (to the point x_1) is less than that from the point x_0 in spite of the assumed optimality of the trajectory $x(t)$. The contradiction obtained proves our assertion.

Now, let $\psi(t)$ denote a vector which corresponds to the optimal control by virtue of the maximum principle, that is, $\psi(t) = \mathrm{grad}\ \omega(x(t))$ [see (1.22)]. Since the vector $\psi(t_0) = \mathrm{grad}\ \omega(x_0)$ is normal to the hypersurface Σ defined by the equation $\omega(x) = \mathrm{const}$, it is orthogonal to all the vectors which lie in the tangent plane T_0 and, in particular, to all the tangent vectors of the manifold M_0. In fact, this is the

Transversality condition (at the left-hand endpoint): *the vector $\psi(t_0)$ is orthogonal to all the tangent vectors of the manifold M_0 at the point x_0.*

If the right-hand endpoint x_1 is not fixed but may vary on a manifold M_1, then we have an analogous

Transversality condition (at the right-hand endpoint): *the vector $\psi(t_1)$ is orthogonal to all the tangent vectors of the manifold M_1 at the point x_1.*

The easiest way to obtain this condition is to reverse the course of time in (1.3); the optimal process will be preserved and the roles of the left- and right-hand endpoints will be interchanged.

Thus, we arrive at the following proposition:

Theorem 5.1.   Let $u(t)$, $t_0 \leqslant t \leqslant t_1$, be an admissible control transferring the phase point from the state $x_0 \in M_0$ to the state $x_1 \in M_1$ and let $x(t)$ be the corresponding trajectory. In order that $u(t)$, $x(t)$ yield the solution of the optimal problem with variable endpoints, it is necessary that there exist a nonzero continuous vector-function $\psi(t)$ which satisfies the maximum principle (p. 78) and, in addition, satisfies the transversality condition at both endpoints of the trajectory $x(t)$.

[If either of the manifolds M_0, M_1 degenerates to a point, then the transversality condition at the corresponding endpoint of the trajectory $x(t)$ is replaced by the condition that the trajectory pass through this point.]

Using the previous arguments, this theorem may be considered as established only in the case in which the function $\omega(x)$ has continuous second derivatives. The deficiencies of such a proof have already been discussed (pp. 17, 18). Therefore, we shall present here an alternate proof which does not utilize properties of the function $\omega(x)$.

56. Proof (The Case of a Variable Right-Hand Endpoint).

First, let us consider the case in which only the right-hand endpoint of the trajectory is variable, that is, let us consider the time-optimal problem in which the point x_0 is prescribed, but any point of a manifold M_1 may be chosen as x_1. Let $u(t)$, $x(t)$, $t_0 \leqslant t \leqslant t_1$, be the optimal process yielding the solution of the posed problem. Let us consider the cone K constructed in the proof of the maximum principle (p. 63). We assert that *it is possible to draw through the point* $Q = x_1$ *a support hyperplane* Γ_1 *of* K *which contains all the tangent vectors of the manifold* M_1 *at the point* x_1.

Let us assume just for the time being that the above assertion has been proved, and let us show how to derive the needed theorem from this assertion. Let n denote the normal vector of the above support hyperplane Γ_1 and let its direction be chosen in such a way that the cone K lies in the negative half-space (that is, the inequality $n \cdot \overrightarrow{QP} \leqslant 0$ is satisfied for any point $P \in K$, compare p. 75). Then we can carry out the arguments of § 5.21 which complete the proof of the maximum principle and obtain the vector-function $\psi(t)$ which satisfies the conditions of Theorem 2.2. Moreover, $\psi(t_1) = n$ [see (2.68)]. By definition, the vector $n = \psi(t_1)$ is orthogonal to all the vectors lying in the hyperplane Γ_1 and, in particular, to all the tangent vectors of the manifold M_1. Thus, the transversality conditions are satisfied at the right-hand endpoint.

Thus, it remains to prove the validity of the above statement concerning the existence of the support hyperplane Γ_1. Let us assume the opposite: there does not exist such a hyperplane. Let L_1 denote the tangent plane of the manifold M_1 at the point x_1, and let N_1 denote its orthogonal complement (that is, the plane formed by all the vectors emanating from x_1 orthogonal to the plane L_1). Let π denote the orthogonal projection of the phase space X on the plane N_1. Under this projection the entire plane L_1 maps into one point x_1 and the cone K is projected into a convex cone $\pi(K)$ with vertex at the same point x_1. If the cone $\pi(K)$ were not to fill out the entire plane N_1, then it would be possible to draw in N_1 a support hyperplane Γ' to the cone $\pi(K)$ through the point x_1; but then the set of all the points of the space X which are projected into Γ' would be a hyperplane of the space X which is a support hyperplane of the cone K containing the plane L_1, that is, the desired hyperplane Γ_1. Since we have assumed that there does not exist such a hyperplane, the *cone* $\pi(K)$ *fills out the entire plane* N_1.

Let us select a simplex in N_1 (of the same dimension as N_1) containing the point $Q = x_1$ in its interior, and let $B^{(0)}$, $B^{(1)}$, . . . , $B^{(q)}$ be its vertices. Since $B^{(0)}$, $B^{(1)}$, . . . , $B^{(q)}$ belong to the cone $\pi(K) = N_1$, there exist points $A^{(0)}$, $A^{(1)}$, . . . , $A^{(q)}$ in K which project under π into $B^{(0)}$, $B^{(1)}$, . . . , $B^{(q)}$. Let us choose the quantities $\delta t^{(\alpha)}$, $\tau_i^{(\alpha)}$, $l_i^{(\alpha)}$, $v_i^{(\alpha)}$ ($\alpha = 0, 1, \ldots, q$; $i = 1, 2,$ . . . , $s^{(\alpha)}$) such that the corresponding displacement vectors coincide with $\overrightarrow{QA^{(0)}}$, $\overrightarrow{QA^{(1)}}$, . . . , $\overrightarrow{QA^{(q)}}$ [compare formulas (2.45)–(2.49)]. As on p. 68, we can assume that all the points $\tau_i^{(\alpha)}$ are mutually distinct.

We now set

$$\delta t = k^{(0)} \delta t^{(0)} + k^{(1)} \delta t^{(1)} + \cdots + k^{(q)} \delta t^{(q)}, \qquad \bar{l}_i^{(\alpha)} = k^{(\alpha)} l_i^{(\alpha)}, \quad (5.7)$$

where the $k^{(\alpha)}$ are non-negative numbers. Then the displacement vector $\overrightarrow{QB}$, corresponding to the quantities $\delta t, \tau_i^{(\alpha)}, \bar{l}_i^{(\alpha)}, v_i^{(\alpha)}$ ($\alpha = 0, 1, \ldots, q; i = 1, 2, \ldots, s^{(\alpha)}$), equals [compare (2.52)]

$$\overrightarrow{QB} = k^{(0)} \overrightarrow{QA}^{(0)} + k^{(1)} \overrightarrow{QA}^{(1)} + \cdots + k^{(q)} \overrightarrow{QA}^{(q)}.$$

Since all the points $\tau_i^{(\alpha)}$ are mutually distinct, we can consider the perturbations of the control $u(t)$ and the trajectory $x(t)$ which correspond to the quantities $\delta t, \tau_i^{(\alpha)}, \bar{l}_i^{(\alpha)}, v_i^{(\alpha)}$; moreover, for the corresponding perturbed trajectory we have [compare (2.54)]

$$x^*(t_1 - \epsilon \delta t) = x(t_1) + \epsilon \overrightarrow{QB} + o(\epsilon).$$

The point $x^*(t_1 - \epsilon \delta t)$ depends continuously on the quantities $k^{(0)}, k^{(1)}, \ldots, k^{(q)}$.

The (q-dimensional) plane passing through the points $A^{(0)}, A^{(1)}, \ldots, A^{(q)}$ will be denoted by N. As does N_1, the plane N intersects L_1 at the single point $Q = x_1$. Using Lemma 2.10 (p. 73) we can construct continuous non-negative functions $k^{(\alpha)}(C), \alpha = 0, 1, \ldots, q$, in N such that the relation

$$\overrightarrow{QC} = k^{(0)}(C) \cdot \overrightarrow{QA}^{(0)} + k^{(1)}(C) \cdot \overrightarrow{QA}^{(1)} + \cdots + k^{(q)}(C) \cdot \overrightarrow{QA}^{(q)}$$

is satisfied for any point C on N [compare (2.54)].

Now, making use of the quantities $k^{(\alpha)} = k^{(\alpha)}(C), \alpha = 0, 1, \ldots, q$, and formulas (5.7), let us determine the quantities $\delta t, \bar{l}_i^{(\alpha)}$ (leaving the quantities $\tau_i^{(\alpha)}, v_i^{(\alpha)}$ unchanged). The corresponding perturbed trajectory $x^*(t)$ will now be denoted by $x_C^*(t)$, and the quantity δt by δt_C. The following relation holds for the perturbed trajectory $x_C^*(t)$:

$$x_C^*(t_1 - \epsilon \delta t_C) = C_\epsilon + o(\epsilon),$$

where C_ϵ is a point such that $\overrightarrow{QC_\epsilon} = \epsilon \overrightarrow{QC}$ [compare (2.57)]. Moreover, the point $x_C^*(t_1 - \epsilon \delta t_C)$ depends continuously on C.

Furthermore, let S denote a sphere of radius one with center lying in the plane N (Figure 188). Since N is a (non-orthogonal) complement of L_1, it follows that the sphere S is linked to L_1, that is, it cannot be contracted to a point by a continuous deformation without coming into contact with the plane L_1. (The exact definition of the concept "linkage" lies in the domain of topology; we shall confine ourselves here to the intuitive "comprehension" of the above assertion.)

As C traverses the sphere S, the point C_ϵ traverses the sphere S_ϵ of radius ϵ with center at Q, and the point $x_C^*(t_1 - \epsilon \delta t_C)$ describes a "closed surface" L_ϵ which is close to S_ϵ. For a sufficiently small ϵ, both the surface L_ϵ and the sphere S_ϵ are linked with the plane L_1, and hence also with the manifold M_1 to which the plane L_1 is tangent; let us consider such a value of ϵ as fixed.

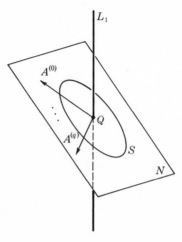

FIGURE 188

Finally, let us consider the point [compare (2.59)]

$$x_C^*((1 - \sigma)t_0 + \sigma(t_1 - \epsilon \delta t_C)), \qquad 0 \leqslant \sigma \leqslant 1. \tag{5.8}$$

As C traverses the sphere S, the point (5.8) describes a closed surface $L^{(\sigma)}$ (for any fixed value of the parameter σ). For $\sigma = 1$, we obtain the surface $L^{(1)}$ which coincides with L_ϵ, and for $\sigma = 0$, the point (5.8) coincides with the point $x_C^*(t_0) = x_0$, that is, the surface $L^{(0)}$ degenerates into the single point x_0. Thus, as σ varies from 1 to 0, we obtain a continuous deformation of the surface $L^{(1)} = L_\epsilon$ into a single point. But the surface L_ϵ is linked with the manifold M_1, and therefore the deforming surface must come in contact with the manifold M_1. In other words, there exists a number $\sigma_0, 0 < \sigma_0 < 1$, and a point $C \in N$ such that the point (5.8) belongs to the manifold M_1:

$$x_C^*((1 - \sigma_0)t_0 + \sigma_0(t_1 - \epsilon \delta t_C)) \in M_1.$$

But we have

$$(1 - \sigma_0)t_0 + \sigma_0(t_1 - \epsilon \delta t_C) = t_1 - (1 - \sigma_0)(t_1 - t_0) - \sigma_0 \epsilon \delta t_C < t_1.$$

Thus, the perturbed trajectory $x_C^*(t)$ reaches the manifold M_1 before the time t_1, which means that the original process $x(t), u(t)$ was not optimal. The contradiction obtained proves the existence of the desired hyperplane Γ_1, which completes the proof.

57. Proof (General Case).

We have proved Theorem 5.1 for the case in which only the right-hand endpoint of the trajectory is variable. We shall indicate the changes which must be made in this proof if both endpoints are variable. Let $u(t)$, $x(t)$, $t_0 \leqslant t \leqslant t_1$, be an optimal process solving the problem with variable endpoints, and let $x_0 \in M_0$, $x_1 \in M_1$ be the endpoints of the trajectory $x(t)$.

We choose a curve G on the manifold M_0 emanating from the point x_0, and let $x = \xi(\epsilon)$ be a parametric representation of this curve [$\epsilon \geqslant 0$, $\xi(0) = x_0$], and let

$$w = \frac{d\xi}{d\epsilon}\bigg|_{\epsilon=0}$$

denote the vector tangent to G at x_0. Thus, for points of the curve G, we have

$$x = x_0 + \epsilon w + o(\epsilon). \tag{5.9}$$

Let us consider the trajectory $x^*(t)$ emanating at time t_0 from the point (5.9) (lying on the curve G and consequently on the manifold M_0) under the action of the perturbed control $u^*(t)$ (corresponding to the quantities τ_i, v_i, l_i; $i = 1, \ldots, s$, compare p. 58). Then the following relation holds for the trajectory $x^*(t)$:

$$x^*(t_1 - \epsilon\delta t) = x(t_1) - \epsilon\delta t f(x(t_1), u(t_1)) + \epsilon \sum_{i=1}^{s} \Delta(\tau_i, h_i) \\ + \epsilon\Delta(t_0, w) + o(\epsilon). \tag{5.10}$$

This relation differs from (2.36) only by the presence of the additional term $\epsilon\Delta(t_0, w)$ in the right-hand side and may be explained as follows. If the initial point x_0 is not shifted to the position (5.9), then the relation (2.36) holds. If, however, the control is not perturbed, that is, if $u^*(t) = u(t)$, but the initial point shifts to the position (5.9), then, by Corollary 2.6 (p. 52) and the relation (2.35), we have

$$x^*(t_1) = x(t_1) + \epsilon\Delta(t_0, w) + o(\epsilon).$$

Performing simultaneously the shifting of the initial point x_0 and the perturbing of the control $u(t)$, we obtain the combined effect which is expressed by formula (5.10). The accurate derivation of (5.10) is carried out precisely like that of the relation (2.36) (pp. 60, 62) and therefore is omitted here.

Formula (5.10) can be written in the form

$$x^*(t_1 - \epsilon\delta t) - x(t_1) = \epsilon\overrightarrow{QA} + \epsilon\Delta(t_0, w) + o(\epsilon).$$

where $\overrightarrow{QA}$ is a displacement vector [see (2.42)], that is, a vector which belongs to the cone K, and w is a vector tangent to the manifold M_0. Let the vector

$$\overrightarrow{QD} = \overrightarrow{QA} + \Delta(t_0, w)$$

emanate from the point Q. As the point A describes the cone K and w describes all possible tangent vectors of the manifold M_0 at the point x_0, the vectors $\overrightarrow{QD}$ obviously fill out a convex cone $\tilde{K}$ with vertex Q. The cone $\tilde{K}$ contains the entire cone K and all vectors of the form $\Delta(t_0, w)$.

Thus, if the initial point x_0 is permitted to vary over the manifold M_0, then (2.36) is replaced by (5.10), and the cone K is replaced by the cone $\tilde{K}$. Repeating now almost verbatim the previous arguments (pp. 240–242) we conclude that if the process $x(t)$, $u(t)$ is optimal, then *there exists a support hyperplane Γ_1 of $\tilde{K}$ passing through the point x_1 and containing all the tangent vectors of the manifold M_1 at the point x_1.* Proceeding from this hyperplane we construct, as before (p. 240), the vector-function $\psi(t)$ satisfying the maximum principle and the transversality condition at the right-hand endpoint (we note that the cone K is contained in $\tilde{K}$ and therefore Γ_1 is a support hyperplane not only of $\tilde{K}$ but also of K). It remains to prove that the transversality condition is satisfied also at the left-hand endpoint.

Let w be an arbitrary tangent vector of the manifold M_0 at the point x_0. Then, both the vectors $\Delta(t_0, w)$ and $\Delta(t_0, -w) = -\Delta(t_0, w)$ belong to the cone $\tilde{K}$ and therefore the inequalities

$$\psi(t_1)\Delta(t_0, w) \leqslant 0, \qquad \psi(t_1)(-\Delta(t_0, w)) \leqslant 0$$

are satisfied. It follows from these inequalities that $\psi(t_1)\Delta(t_0, w) = 0$. But by virtue of Corollary 2.9 (p. 56), this relation can be rewritten in the form $\psi(t_0)w = 0$. Since this holds for any vector w tangent to the manifold M_0 at the point x_0, the transversality condition at the left-hand endpoint is satisfied.

Thus, Theorem 5.1 has been completely proved.

58. An Oscillation Theorem. As an example illustrating the application of transversality conditions, let us consider the following problem.

Given the differential equation

$$\ddot{x} + u(t)\dot{x} + v(t)x = 0 \tag{5.11}$$

with variable (piecewise continuous) coefficients $u(t)$, $v(t)$ subject to the conditions

$$\alpha \leqslant u \leqslant \alpha', \qquad \beta \leqslant v \leqslant \beta', \tag{5.12}$$

where α, α', β, β' are positive constants for which the roots of the equation $\lambda^2 + \alpha'\lambda + \beta' = 0$ are complex, it is required to determine the shortest possible distance between two adjacent zeros of the nontrivial solutions of (5.11).

Let $x(t)$ be a nontrivial solution of (5.11) [for a certain choice of the coefficients $u(t)$ and $v(t)$], and let t_0 and t_1 ($t_0 < t_1$) be two adjacent zeros of this solution. Since $x(t_0) = x(t_1) = 0$, we have $\dot{x}(t_0) \neq 0$ and $\dot{x}(t_1) \neq 0$ [since the solution $x(t)$ is nontrivial]. We note that, by multiplying $x(t)$ by a non-zero number, this function remains a solution of (5.11) and that its zeros are unaltered. Therefore, without loss of generality, we can assume that $\dot{x}(t_0) = 1$. It is obvious that $\dot{x}(t_1) < 0$ since $x(t)$ maintains a fixed sign on the interval (t_0, t_1). Thus, we need only consider the solution $x(t)$ of (5.11) which satisfies the initial conditions $x(t_0) = 0$ and $\dot{x}(t_0) = 1$; by "controlling" the coefficients $u(t)$ and $v(t)$ it is required to satisfy the conditions $x(t_1) = 0$, $\dot{x}(t_1) < 0$ in the shortest time.

Introducing the quantities $x^1 = x$, $x^2 = \dot{x}$, we can rewrite (5.11) in the form of the following system of equations:

$$\dot{x}^1 = x^2,$$
$$\dot{x}^2 = -vx^1 - ux^2 \tag{5.13}$$

[under the same constraints (5.12) imposed on the coefficients]. Now, our problem consists in transferring the phase point from the state $x^1 = 0$, $x^2 = 1$ to a state satisfying the conditions $x^1 = 0$, $x^2 < 0$ in the shortest time. In other words, we obtain a time-optimal problem with variable right-hand endpoint, in which the initial point x_0 has the coordinates $(0, 1)$, and the terminal point x_1 must lie on the manifold M_1, which is the negative portion of the x^2-axis (Figure 189).

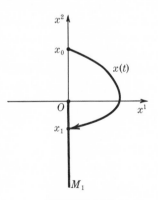

FIGURE 189

Let us write the function H for the problem under consideration:

$$H = \psi_1 x^2 - \psi_2(vx^1 + ux^2);$$

the equations for the auxiliary variables are

$$\dot{\psi}_1 = v\psi_2,$$
$$\dot{\psi}_2 = -\psi_1 + u\psi_2. \tag{5.14}$$

In addition, let us write the transversality condition at the right-hand endpoint. Since every tangent vector of the manifold M_1 is parallel to the x^2-axis, that is, proportional to the vector $w = (0, 1)$, the transversality condition takes the form $\psi(t_1)w = 0$, that is, $\psi_1(t_1) \cdot 0 + \psi_2(t_1) \cdot 1 = 0$. Thus, the transversality condition at the right-hand endpoint yields the relation

$$\psi_2(t_1) = 0. \tag{5.15}$$

Assume that we know the optimal process which gives the solution of the problem we have posed, and let $\psi_1(t)$, $\psi_2(t)$ be the corresponding solution of (5.14) satisfying the maximum condition and the transversality condition (5.15). If $\psi_2(t)$ were to have infinitely many zeros on the interval $[t_0, t_1]$, then there would exist a limit point θ of these zeros. At the point θ we would have $\psi_2(\theta) = 0$, $\dot{\psi}_2(\theta) = 0$, and therefore, by virtue of the second equation in (5.14), $\psi_1(\theta) = 0$. But the relations $\psi_1(\theta) = \psi_2(\theta) = 0$ contradict the fact that $\psi_1(t)$, $\psi_2(t)$ is a nontrivial solution. Consequently, $\psi_2(t)$ can have only a finite number of zeros on the interval $[t_0, t_1]$.

Since the function H attains a maximum with respect to u and v, the term $-\psi_2 v x^1$ also attains a maximum with respect to v, and thus $-\psi_2 v$ attains a maximum with respect to v since $x^1 > 0$ on the entire interval (t_0, t_1). Consequently, $v = \beta'$ if ψ_2 is negative, and $v = \beta$ if ψ_2 is positive. Since $\psi_2(t)$ has only a finite number of zeros, it follows that the function $v(t)$ is uniquely defined by the maximum condition and is piecewise continuous.

Similarly, if $x^2(t)$ were to have infinitely many zeros, at a limit point τ of these zeros we would have $x^2(\tau) = 0$, $\dot{x}^2(\tau) = 0$ and, therefore, by virtue of the second equation in (5.13), we would have $-v(\tau)x^1(\tau) = 0$. But $v(\tau) \neq 0$ since both constants β and β' in (5.12) are positive. Therefore, $x^1(\tau) = 0$. It follows from the relations $x^1(\tau) = x^2(\tau) = 0$ that the solution $x^1(t)$, $x^2(t)$ is trivial and therefore cannot satisfy the required initial condition $x(t_0) = x_0$. Consequently, the function $x^2(t)$ has only a finite number of zeros on the interval $[t_0, t_1]$.

Now, it follows from the maximum condition that at any time t the function $-\psi_2 u x^2$ must assume its maximum (with respect to u) at the point $u = u(t)$, and therefore $u(t)$ is a uniquely defined piecewise continuous function (taking on the values α, α').

Thus, both functions $u(t)$, $v(t)$ which realize the optimal condition are piecewise continuous so that by (5.14), $\psi_1(t)$ and $\psi_2(t)$ satisfy the relations

$$\ddot{\psi}_2 = -\dot{\psi}_1 + u\dot{\psi}_2 = -v\psi_2 + u\dot{\psi}_2.$$

Reversing the course of time, that is, considering the function $\varphi(t) = \psi_2(t_0 + t_1 - t)$ instead of $\psi_2(t)$, we find

$$\ddot{\varphi} = -v\varphi - u\dot{\varphi},$$

that is, $\varphi(t)$ satisfies (5.11). Moreover, from the transversality condition (5.15), we obtain $\varphi(t_0) = 0$. But as a matter of fact, t_1 is the least possible root of a function which satisfies (5.11) and vanishes at the point $t = t_0$. Consequently, $\varphi(t)$ maintains the same sign over the entire interval (t_0, t_1). Returning to $\psi_2(t)$, we conclude that $\psi_2(t)$ *maintains the same sign over the entire interval* (t_0, t_1).

Let us determine the sign of ψ_2 on this interval. We note for this purpose that, by virtue of (5.15), inequality (E) (p. 79) takes the form $\psi_1(t_1)x^2(t_1) \geqslant 0$, and since $x^2(t_1) < 0$, we have $\psi_1(t_1) \leqslant 0$. The equality $\psi_1(t_1) = 0$ cannot hold since $\psi_2(t_1) = 0$, and the solution $\psi_1(t)$, $\psi_2(t)$ is nontrivial. Consequently, $\psi_1(t_1) < 0$. By virtue of (5.15), the second relation in (5.14) now yields $\psi_2(t_1) = -\psi_1(t_1) > 0$. Finally, it follows from $\psi_2(t_1) = 0$, $\psi_2(t_1) > 0$ that $\psi_2(t)$ is negative on the interval (t_0, t_1).

We conclude from the inequality $\psi_2 < 0$ (see above) that $v = \beta'$ on the entire interval (t_0, t_1). In addition, the condition $-\psi_2 u x^2 = \max$ takes the form $u x^2 = \max$, and therefore,

$$u = \begin{cases} \alpha' \text{ for } x^2 > 0 \\ \alpha \text{ for } x^2 < 0. \end{cases}$$

Thus, in the upper half-plane $x^2 > 0$, the optimal trajectory satisfies the system

$$\begin{aligned} \dot{x}^1 &= x^2, \\ \dot{x}^2 &= -\beta'x^1 - \alpha'x^2, \end{aligned} \tag{5.16}$$

and in the lower half-plane $(x^2 < 0)$ it satisfies the system

$$\begin{aligned} \dot{x}^1 &= x^2, \\ \dot{x}^2 &= -\beta'x^1 - \alpha x^2. \end{aligned} \tag{5.17}$$

Both systems (5.16), (5.17) have complex eigenvalues (since we have assumed that the roots of $\lambda^2 + \alpha'\lambda + \beta' = 0$ are complex) and therefore the phase trajectories of these systems are spirals moving about the origin in the clockwise direction. Hence, it is clear that the desired optimal trajectory can only be a curve consisting of two pieces: a piece of a trajectory of (5.16) located in the first quadrant and a piece of a trajectory of (5.17) located in the fourth quadrant (Figure 189).

The time $T = t_1 - t_0$ of motion along this trajectory (from x_0 to x_1) is, in fact, the desired minimum distance between two adjacent zeros of the nontrivial solutions of (5.11). It is not difficult to calculate this time by directly solving (5.16) and (5.17). Namely, let $-\mu_1 \pm i\nu_1$ denote the roots of the characteristic equation of (5.16), and let $-\mu_2 \pm i\nu_2$ denote the roots of the characteristic equation of (5.17) with μ_1, μ_2, ν_1, ν_2 positive. Then

$$T = T_1 + T_2, \quad T_1 = \frac{1}{\nu_1} \arctan \frac{\nu_1}{\mu_1}, \quad T_2 = \frac{1}{\nu_2}\left(\pi - \arctan \frac{\nu_2}{\mu_2}\right) \quad (5.18)$$

(we do not indicate the corresponding elementary calculations). Thus, we obtain the following theorem:

Theorem 5.2. For the equation under consideration [see (5.11), (5.12)], *the minimum possible distance T between two adjacent zeros t_0, t_1 of a nontrivial solution is determined from formula* (5.18). *This minimum distance is attained if $v \equiv \beta'$ and the coefficient u takes on the value α' on the interval $[t_0, t_0 + T_1]$ and the value α on the remaining portion of the interval $[t_0, t_1]$.*

Of course, the previous arguments do not provide a complete proof of this theorem. In fact, we have proved only that no trajectory other than the one we have found could be optimal; however, there is no a priori proof that an optimal trajectory actually exists. In order to complete the proof, one can use the sufficient conditions for optimality.

First of all, it is clear by virtue of the linearity of (5.13) that the maximum principle is satisfied not only by the trajectory we have found (Figure 189), but also by any trajectory obtained from the latter by a dilation with center at the origin. Thus, the entire phase plane with punctured origin is filled out by similar trajectories which satisfy the maximum principle (Figure 190). These trajectories start on one of the x^2 semiaxes and terminate

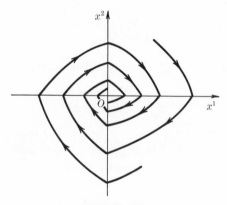

FIGURE 190

on the other. The control parameters take on the values $u = \alpha'$, $v = \beta'$ in the first and third quadrants, and the values $u = \alpha$, $v = \beta'$ in the second and fourth quadrants. We can regard the above trajectories (satisfying the maximum principle) as not starting or terminating on the x^2-axis but as spreading without bound to both sides in the form of spirals (as shown in Figure 190).

Let us now determine the function $\omega(x)$, considering it to be equal to zero on the positive x^2 semiaxis, assuming the value $-t$ at the point x, where t is the duration of the motion along the spiral from the point lying on the positive x^2 semiaxis to the point x. The function $\omega(x)$ is multivalued: in addition to the value $\omega(x)$, it also takes on, at the point x, all the values $\omega(x) + 2kT$ $(k = 0, 1, 2, \ldots)$, since after the time $2T$, the spiral starting on the positive x^2 semiaxis once again returns to this semiaxis. If, however, the above spirals and the function $\omega(x)$ are considered on an infinite-sheeted "Riemann surface" branching about the origin, then $\omega(x)$ becomes a single-valued continuous function. The arguments of § 12.46 (pp. 187–194) are applicable to this function, which makes it possible to assert that $\omega(x)$ is continuously differentiable and satisfies Bellman's equation everywhere except at the points of a one-dimensional piecewise smooth set. The fact that all the trajectories terminated at the point x_1 was not used in § 12.46 in proving these properties of the function $\omega(x)$, but the remaining assumptions were imposed here. Therefore, by virtue of the fundamental lemma (p. 179), the estimate (4.2) is valid for the transition time. In particular, if the point x_0 lies on the positive x^2 semiaxis and the point x_1 lies on the negative x^2 semiaxis, then

$$\omega(x_0) = 2kT, \qquad \omega(x_1) = T + 2lT,$$

and therefore, for the transition time from the point x_0 to the point x along any trajectory, we have

$$t_1 - t_0 \geqslant \omega(x_1) - \omega(x_0) = T + 2(l - k)T.$$

The number $l - k$ depends on the sheets on which the points x_0 and x_1 are located (that is, on how many revolutions about the origin the trajectory makes). In any case (by virtue of the non-negativity of the number $t_1 - t_0$), the inequality $t_1 - t_0 \geqslant T$ is valid; in fact, this is the inequality which proves that the transition from the point x_0 to the negative portion of the x^2-axis during a time less than T is not possible.

Thus, Theorem 5.2 has been completely proved.

§ 15. THE GENERAL MAXIMUM PRINCIPLE

59. Formulation of the Problem.

In this section we shall formulate the problem of optimal control in a sense different from that of time-optimality and give its solution. The idea of the general problem of optimal control consists in evaluating the "quality" of the transition process from the point x_0 to x_1 not in terms of the time spent, but in terms of some other quantity (work, amount of heat, fuel consumption, and so on). Most frequently, the quality of the transition process is evaluated by means of an integral functional

$$I = \int_{t_0}^{t_1} f^0(x, u) \, dt. \tag{5.19}$$

The function $f^0(x, u)$, which is chosen according to the meaning of the problem, is assumed to satisfy the same conditions as the functions $f^i(x, u)$ on the left-hand sides of the equations of motion (1.2).

The problem which we shall consider in this section can now be formulated as follows. *Given two points x_0 and x_1 in the phase space X; from among all admissible controls $u(t)$, $t_0 \leqslant t \leqslant t_1$, which transfer the phase point from x_0 to x_1, it is required to pick one for which the functional (5.19) takes on the least possible value.* (The time t_0 of departure from x_0 and the time t_1 of arrival at x_1 are not given beforehand.) The control $u(t)$ which solves the formulated problem as well as the corresponding trajectory will be called optimal, as before.

We note that in case $f^0(x, u) \equiv 1$, the functional (5.19) coincides with the transition time: $I = t_1 - t_0$. Thus, the time-optimal problem is a special case of the general problem of optimal control.

The solution of the problem formulated above will be presented in this section under the additional assumption that the function $f^0(x, u)$ is positive (for all x, u). In this case, the solution, as we shall see, follows directly from the maximum principle formulated on p. 78. Theorem 5.3 thus obtained (see below) is also called the maximum principle. It should be noted that Theorem 5.3 is valid also without the additional assumption of positivity of the function f^0; the general proof of this theorem may be found in the monograph [9] cited in the preface.

60. Fundamental Theorem.   Thus, let us assume that $f^0(x, u) > 0$. The idea of the solution of the problem we have formulated consists in the following. On every trajectory we introduce a new time τ related to the old time by the differential relation $d\tau = f^0(x, u) \, dt$. In the new time, the functional (5.19) takes the form $I = \int_{\tau_0}^{\tau_1} d\tau = \tau_1 - \tau_0$, that is, the problem formulated above becomes a time-optimal problem. Let us proceed to develop this idea.

Let $u(t)$, $t_0 \leqslant t \leqslant t_1$, be an admissible control transferring the phase point from x_0 to x_1, and let $x(t)$ be the corresponding trajectory. We set

$$\tau(t) = \int_{t_0}^{t} f^0(x(t), u(t)) \, dt, \qquad t_0 \leqslant t \leqslant t_1.$$

The function $\tau(t)$ is continuous and monotonically increasing (since $f^0 > 0$) and therefore the inverse function $t(\tau)$ exists. It is clear that [see (5.19)]

$$\tau(t_0) = 0, \qquad \tau(t_1) = I, \qquad t(0) = t_0, \qquad t(I) = t_1;$$

moreover,

$$\frac{d\tau(t)}{dt} = f^0(x(t), u(t)), \qquad \frac{dt(\tau)}{d\tau} = \frac{1}{f^0(x(t(\tau)), u(t(\tau)))}.$$

Let us now define vector functions $v(\tau)$ and $y(\tau)$ by setting

$$v(\tau) = u(t(\tau)), \qquad y(\tau) = x(t(\tau)), \qquad 0 \leqslant \tau \leqslant I,$$

where $u(t)$, $x(t)$ is the process under consideration. It is clear that $v(\tau)$ is a piecewise continuous function taking on values in the control region U. Furthermore, we have

$$\frac{dy^i(\tau)}{d\tau} = \frac{dx^i(t(\tau))}{d\tau} = \frac{dx^i(t(\tau))}{dt} \cdot \frac{dt(\tau)}{d\tau}$$

$$= f^i(x(t(\tau)), u(t(\tau))) \frac{1}{f^0(x(t(\tau)), u(t(\tau)))} = \frac{f^i(y(\tau), v(\tau))}{f^0(y(\tau), v(\tau))}.$$

Thus, $y(\tau)$ is a solution of the system

$$\frac{dy^i}{d\tau} = \frac{f^i(y, v)}{f^0(y, v)}, \qquad i = 1, \ldots, n, \tag{5.20}$$

which corresponds to the admissible control $v(\tau)$. This trajectory joins the points x_0 and x_1, and the time spent on the transfer from x_0 to x_1 is I:

$$y(0) = x(t(0)) = x(t_0) = x_0, \qquad y(I) = x(t(I)) = x(t_1) = x_1.$$

Taking different processes $u(t)$, $x(t)$ for the system (1.2) we will have correspondingly different processes $v(t)$, $y(t)$ for the system (5.20), each having its own transition time I determined by (5.19). Hence it is clear that if the process $u(t)$, $x(t)$ is optimal in the sense of the above problem, that is, ascribes the least possible value to the functional (5.19), then the corresponding process $v(\tau)$, $y(\tau)$ is optimal in the sense of time-optimality for the system (5.20).

In fact, let us consider such optimal processes $u(t)$, $x(t)$, $t_0 \leqslant t \leqslant t_1$, and accordingly $v(\tau)$, $y(\tau)$, $0 \leqslant \tau \leqslant I$. Since the process $v(\tau)$, $y(\tau)$ is time-optimal, the assertions of Theorem 2.12 are valid for this process. We denote the auxiliary variables by $\varphi_1, \ldots, \varphi_n$ and write the function H for the system (5.20):

$$H(\varphi, y, v) = \sum_{i=1}^{n} \varphi_i \frac{f^i(y, v)}{f^0(y, v)}.$$

We also write the system of equations for the auxiliary variables:

$$\frac{d\varphi_i}{d\tau} = -\frac{\partial H}{\partial y^i} = -\sum_{\alpha=1}^{n} \varphi_\alpha \frac{\partial}{\partial y^i}\left(\frac{f^\alpha(y, v)}{f^0(y, v)}\right)$$

$$= -\frac{1}{f^0(y, v)}\sum_{\alpha=1}^{n} \varphi_\alpha\left(\frac{\partial f^\alpha(y, v)}{\partial y^i} - \frac{f^\alpha(y,v)}{f^0(y, v)}\cdot\frac{\partial f^0(y, v)}{\partial y^i}\right) \qquad (5.21)$$

$$= -\frac{1}{f^0(y, v)}\left[\left(\sum_{\alpha=1}^{n} \varphi_\alpha \frac{\partial f^\alpha(y, v)}{\partial y^i}\right) - H(\varphi, y, v)\frac{\partial f^0(y, v)}{\partial y^i}\right].$$

According to the maximum principle (Theorem 2.12) corresponding to the process $v(\tau)$, $y(\tau)$ there exists a nontrivial solution $\varphi(\tau)$ of (5.21) which satisfies conditions (D) and (E) of this theorem:

$$H(\varphi(\tau), y(\tau), v(\tau)) = \max_{v \in U} H(\varphi(\tau), y(\tau), v), \qquad (5.22)$$

$$H(\varphi(I), y(I), v(I)) \geqslant 0. \qquad (5.23)$$

Now, let us set

$$\psi_i(t) = \varphi_i(\tau(t)), \qquad i = 1, \ldots, n, \qquad t_0 \leqslant t \leqslant t_1,$$

$$\psi_0 = -H(\varphi(\tau), y(\tau), v(\tau)).$$

According to Theorem 2.13, $\psi_0 = $ const, and (5.23) can be rewritten in the form

$$\psi_0 \leqslant 0.$$

The derivatives of the functions $\psi_i(t)$ can be easily found from (5.21):

$$\frac{d\psi_i(t)}{dt} = \frac{d\varphi_i(\tau(t))}{dt} = \frac{d\varphi_i(\tau(t))}{d\tau}\cdot\frac{d\tau(t)}{dt}$$

$$= -\frac{1}{f^0(y(\tau(t)), v(\tau(t)))}\left[\sum_{\alpha=1}^{n} \varphi_\alpha(\tau(t))\frac{\partial f^\alpha(y(\tau(t)), v(\tau(t)))}{\partial y^i}\right.$$

$$\left. - H(\varphi(\tau(t)), y(\tau(t)), v(\tau(t)))\frac{\partial f^0(y(\tau(t)), v(\tau(t)))}{\partial y^i}\right] \times f^0(x(t), u(t)).$$

Since $t(\tau(t)) = t$ [the functions $t(\tau)$ and $\tau(t)$ are inverses] and therefore $y(\tau(t)) = x(t(\tau(t))) = x(t)$, $v(\tau(t)) = u(t(\tau(t))) = u(t)$, these expressions can be rewritten as follows:

$$\frac{d\psi_i(t)}{dt} = -\sum_{\alpha=1}^{n} \psi_\alpha(t)\frac{\partial f^\alpha(x(t), u(t))}{\partial x^i} - \psi_0 \frac{\partial f^0(x(t), u(t))}{\partial x^i}. \qquad (5.24)$$

Finally, the maximum condition (5.22) can be written as

$$\sum_{i=1}^{n} \varphi_i(\tau) \frac{f^i(y(\tau), v(\tau))}{f^0(y(\tau), v(\tau))} = -\psi_0, \qquad \sum_{i=1}^{n} \varphi_i(\tau) \frac{f^i(y(\tau), v)}{f^0(y(\tau), v)} \leqslant -\psi_0$$

or, upon removing the denominator (recall that $f^0 > 0$),

$$\sum_{i=1}^{n} \varphi_i(\tau) f^i(y(\tau), v(\tau)) + \psi_0 f^0(y(\tau), v(\tau)) = 0,$$

$$\sum_{i=1}^{n} \varphi_i(\tau) f^i(y(\tau), v) + \psi_0 f^0(y(\tau), v) \leqslant 0.$$

Substituting $\tau(t)$ for τ in these relations, we obtain

$$\sum_{i=1}^{n} \psi_i(t) f^i(x(t), u(t)) + \psi_0 f^0(x(t), u(t)) = 0,$$

$$\sum_{i=1}^{n} \psi_i(t) f^i(x(t), u) + \psi_0 f^0(x(t), u) \leqslant 0 \qquad (5.25)$$

(for $t_0 \leqslant t \leqslant t_1$ and $u \in U$).

The form of (5.24) and (5.25) suggests the introduction of the vector

$$\psi = (\psi_0, \psi_1, \ldots, \psi_n) = (\psi_0, \psi)$$

and the function

$$\mathcal{H}(\psi, x, u) = \sum_{i=0}^{n} \psi_i f^i(x, u) \qquad (5.26)$$

(the summation is from 0 to n). Now, formulas (5.24) mean that the functions $\psi_1(t), \ldots, \psi_n(t)$ satisfy the system of equations

$$\frac{d\psi_i}{dt} = -\frac{\partial \mathcal{H}(\psi, x(t), u(t))}{\partial x^i}, \qquad i = 1, \ldots, n, \qquad (5.27)$$

and the maximum relation (5.25) can be rewritten in the form

$$\max_{u \in U} \mathcal{H}(\psi(t), x(t), u) = \mathcal{H}(\psi(t), x(t), u(t)) \equiv 0, \qquad t_0 \leqslant t \leqslant t_1. \qquad (5.28)$$

Accordingly, we have proved (under the assumption that $f^0 > 0$) the following theorem for the above optimal problem [see (1.2) and (5.19)].

Theorem 5.3 (The Maximum Principle). Let $u(t)$, $t_0 \leqslant t \leqslant t_1$, be an admissible control transferring the phase point from x_0 to x_1 and let $x(t)$ be the corresponding trajectory. In order that the process $u(t)$, $x(t)$, $t_0 \leqslant t \leqslant t_1$, be optimal [in the sense of minimizing the functional (5.19)], it is necessary that *there exist a constant $\psi_0 \leqslant 0$ and a nontrivial solution $\psi(t)$, $t_0 \leqslant t \leqslant t_1$, of the system (5.27) such that the maximum condition (5.28) is satisfied for any time t which is a point of continuity of the control $u(t)$.*

61. The Problem with Variable Endpoints. Let us now consider the problem with variable endpoints. In other words, given smooth manifolds M_0 and M_1 of arbitrary dimensions (less than n), find an admissible control $u(t)$ which transfers the phase point from some (unspecified) position $x_0 \in M_0$ to another position $x_1 \in M_1$ which ascribes the least possible value to the functional (5.19). In this case also the introduction of the new time τ used in deriving the proof of Theorem 5.3 makes it possible (provided the condition $f^0 > 0$ is satisfied) to obtain the following proposition directly from Theorem 5.1.

Theorem 5.4.   Let $u(t)$, $t_0 \leqslant t \leqslant t_1$, be an admissible control which transfers the phase point from a position $x_0 \in M_0$ to a position $x_1 \in M_1$, and let $x(t)$ be the corresponding trajectory. In order that $u(t)$, $x(t)$ provide the solution of the optimal problem [in the sense of minimizing the functional (5.19)] with variable endpoints, it is necessary that there exist a constant ψ_0 and a nonzero continuous vector function $\psi(t)$ which satisfy the maximum principle (Theorem 5.3) and, in addition, the transversality condition at both endpoints of the trajectory $x(t)$.

Moreover, the formulation of the transversality conditions remains the same as before (p. 239).

62. Bellman's Equation and Sufficient Conditions for Optimality. Let us consider once again the optimality of the process (1.3) in the sense of minimizing the functional (5.19), and let us study the problem of finding optimal trajectories with fixed endpoints; moreover, the terminal phase point x_1 will be considered fixed, and various points of the phase space will be taken as the initial point x. We shall make assumptions analogous to Hypotheses 1 and 2 in § 2.5.

Hypothesis 1'.   For any point x distinct from x_1 there exists an optimal process transferring x to x_1.

Let $T(x)$ denote the value of the functional (5.19) corresponding to the optimal process transferring x to x_1, and set $\omega(x) = -T(x)$.

Hypothesis 2'.   The function $T(x)$ or $\omega(x)$ is continuous and has continuous partial derivatives with respect to x^1, x^2, . . . , x^n everywhere except at the point x_1.

Now, introducing (as in § 15.58) the new time $\tau = \int_{t_0}^{t} f^0 \, d\tau$, the optimal [in the sense of minimizing the functional (5.19)] processes become time-optimal processes for the system (5.20). Moreover, it turns out that $T(x)$

is the time $\tau_1 - \tau_0$ of the optimal transfer from x to x_1 and Hypotheses 1′ and 2′ become Hypotheses 1 and 2 (§ 2.5) for the system (5.20). Consequently [see (1.18)], the relation

$$\max_{u \in U} \sum_{i=1}^{n} \frac{\partial \omega(x)}{\partial x^i} \cdot \frac{f^i(x, u)}{f^0(x, u)} = 1 \qquad (x \neq x_1)$$

holds for (5.20), or, since $f^0 > 0$,

$$\max_{u \in U} \sum_{i=1}^{n} \frac{\partial \omega(x)}{\partial x^i} f^i(x, u) = f^0(x, u) \qquad (x \neq x_1). \tag{5.29}$$

Thus, *under Hypotheses 1′ and 2′, the function ψ satisfies* (5.29); *moreover, the maximum in* (5.29) *is attained for optimal processes.*

Equation (5.29) is called Bellman's equation for the process (1.2) and for optimality in the sense of minimizing functional (5.19). Everything said about Hypotheses 1 and 2 at the end of § 2.5 can be repeated for Hypotheses 1′ and 2′.

Similarly (by introducing a new time), the basic results of Chapter 4 can be extended to the case of optimality in the sense of minimizing the functional (5.19). Let us give the formulation of the theorems thus obtained.

Theorem 5.5.   Let M be a piecewise smooth set of dimension less than n lying in the phase space X and let $\omega(x)$ be a continuous function defined on X having continuous derivatives with respect to x^1, x^2, . . . , x^n at points not belonging to M. Furthermore, let $\omega(x_1) = 0$ for a point $x_1 \in X$. Let us assume that for every point $x_0 \in X$ distinct from x_1 there exists an admissible control $u(t) = u_{x_0}(t)$, $t_0 \leqslant t \leqslant t_1$, transferring the phase point [moving according to (1.3)] from x_0 to x_1 which satisfies the relation

$$\int_{t_0}^{t_1} f^0(x(t), u(t)) \, dt = -\omega(x_0).$$

In order that all the controls $u_{x_0}(t)$ be optimal it is necessary and sufficient that the function $\omega(x)$ satisfy (5.29) *at all points x_1 which do not belong to M.*

In considering optimality in the sense of minimizing the functional (5.19), the definition of regular synthesis (§ 12.45) remains the same, with only condition (F) to be reformulated so as to apply to the functional:

(F) The value of the functional (5.19) calculated along the distinguished trajectories (terminating at the point x_1) is a continuous function of the initial point x_0.

Theorem 5.6.   If a regular synthesis is realized in the set G for (1.3) [assuming the existence of continuous derivatives $\partial f^i / \partial x^j$, $\partial f^i / \partial u^k$ and that $f^0(x, u) > 0$], then all the distinguishing trajectories are optimal (in G).

We note that (5.29) (if Hypotheses 1' and 2' are satisfied) and Theorem 5.2 may be proved by direct generalization of arguments given in § 2.5 and § 11.42 (and not by reducing the proof to the case of time-optimality by introducing the new time τ). In such a method of proof, these propositions can be established without the assumption of the validity of the inequality $f^0(x, u) > 0$.

§ 16. VARIOUS GENERALIZATIONS

63. The Maximum Principle for Nonautonomous Systems. Let us consider briefly the changes which take place in the formulation and proof of the maximum principle in the case in which the law of motion (1.3) becomes nonautonomous, that is, the right-hand side of this equation explicitly contains the variable t in addition to x and u:

$$\dot{x} = f(x, t, u) \tag{5.30}$$

(the control region U is assumed to be independent of time). The time t_0 is assumed to be given, whereas the time t_1 at which the trajectory passes through the point x_1 is to be determined.

Let us first consider the time-optimal problem with fixed endpoints x_0 and x_1 for the object (5.30). It turns out that *if the right-hand sides* $f^i(x, t, u)$ *of* (5.30) *and their derivatives* $\partial f^i (x, t, u)/\partial x^j$ *are continuous* (*jointly in the variables* x, t, u), *then all the arguments* in § 4.14–§ 5.21 *pertain without change, and therefore, for the nonautonomous system* (5.30), *the formulation of the maximum principle is preserved verbatim as on p.* 78. Of course, the right-hand sides of the conditions (A), (B), (C), (D), and (E) also explicitly contain the variable t; for example, the function H [see (B)] is to be written in the form

$$H(\psi, x, t, u) = \psi f(x, t, u) = \sum_{\alpha = 1}^{n} \psi_\alpha f^\alpha(x, t, u).$$

What is more, the maximum principle is preserved also in the case in which the right-hand side of (5.30) is piecewise continuous in t. In a more detailed and precise formulation this means the following. Let

$$\cdots < \theta_{k-1} < \theta_k < \theta_{k+1} < \cdots$$

be a (finite or infinite) number of fixed points on the t-axis and let each finite interval contain only a finite number of these points. Furthermore, for each k, let the interval Δ_k be somewhat greater than the interval $\theta_{k-1} < t < \theta_k$, and let the functions $\varphi_k^i(x, t, u)$, $i = 1, \ldots, n$, be defined and continuous together with their partial derivatives $\partial \varphi_k^i/\partial x^j$ for $x \in X$, $t \in \Delta_k$, and $u \in U$.

We define the function $f^i(x, t, u)$ by setting

$$f^i(x, t, u) = \varphi_k^i(x, t, u) \qquad \text{for} \quad \theta_{k-1} < t < \theta_k$$

(the values of the function f^i at the points $t = t_k$ play no role). In other words, the function f^i is "sewn together" from the various functions φ_k^i; the functions f^i, $i = 1, \ldots, n$, may have discontinuities at the points $t = t_k$. It turns out that the maximum principle is completely preserved also for the form of the right-hand sides $f^i(x, t, u)$ indicated here. The proof carried out in § 4.14–§ 5.24 also applies without change.

However, Theorem 2.13 is invalid for the nonautonomous system (5.30): the function $M(t) = H(\psi(t), x(t), t, u(t))$ is now not constant. The character of the dependence of this function on t can be defined in an entirely different way by assuming additionally that the *function f is continuous and has a continuous derivative $\partial f(x, t, u)/\partial t$*.

Let us introduce another auxiliary variable x^{n+1} defined by

$$\dot{x}^{n+1} = 1, \qquad x^{n+1}(t_0) = t_0.$$

Evidently, we then have $x^{n+1} \equiv t$. Let X^* denote the $x^1, x^2, \ldots, x^n$, x^{n+1}-space. The system (5.30) can be rewritten, with the aid of the variable x^{n+1}, in the following autonomous form

$$\begin{aligned} \dot{x}^i &= f^i(x, x^{n+1}, u), \qquad i = 1, \ldots, n, \\ \dot{x}^{n+1} &= 1. \end{aligned} \tag{5.31}$$

Moreover, we must find an optimal trajectory in X^* which joins the point $(x_0^1, x_0^2, \ldots, x_0^n, t_0)$ with a point on the straight line M_1 passing through the point $(x_1^1, x_1^2, \ldots, x_1^n, 0)$ parallel to the x^{n+1}-axis (since the final value of the variable x^{n+1}, that is, the time of arrival of the moving point to the position x_1, is not prescribed). Thus, we arrive at the usual optimal problem with fixed left-hand endpoint and variable right-hand endpoint.

Namely, let us write the maximum principle and the transversality condition for the above problem. According to Theorems 5.2 and 2.12, to solve this problem one constructs the function

$$\psi_1 f^1(x, x^{n+1}, u) + \psi_2 f^2(x, x^{n+1}, u) + \cdots + \psi_n f^n(x, x^{n+1}, u) + \psi_{n+1} \cdot 1.$$

This function will be denoted by H^* (and not by H as in Theorem 2.12) preserving the notation H for the function

$$H(\psi, x, t, u) = \psi_1 f^1(x, t, u) + \psi_2 f^2(x, t, u) + \cdots + \psi_n f^n(x, t, u).$$

Thus, considering that $x^{n+1} \equiv t$, we can write

$$H^* = H + \psi_{n+1}. \tag{5.32}$$

The system of equations for the auxiliary variables ψ_1, ψ_2, $\ldots$, ψ_{n+1} corresponding to the object (5.31) has the form

$$\dot{\psi}_i = -\frac{\partial H^*}{\partial x^i} = -\frac{\partial H}{\partial x^i}, \qquad i = 1, \ldots, n, \tag{5.33}$$

$$\dot{\psi}_{n+1} = -\frac{\partial H^*}{\partial x^{n+1}} = -\frac{\partial H}{\partial t} = -\sum_{\alpha=1}^{n} \frac{\partial f^\alpha(x, t, u)}{\partial t} \psi_\alpha. \tag{5.34}$$

Finally, the transversality condition at the right-hand endpoint of the trajectory means that the straight line M_1 (parallel to the x^{n+1}-axis) is orthogonal to the vector $\{\psi_1(t_1), \psi_2(t_1), \ldots, \psi_n(t_1), \psi_{n+1}(t_1)\}$. In other words

$$\psi_{n+1}(t_1) = 0. \tag{5.35}$$

Together with (5.34) this yields

$$\psi_{n+1}(t) = \int_t^{t_1} \sum_{\alpha=1}^{n} \frac{\partial f^\alpha(x, t, u)}{\partial t} \psi_\alpha \, dt. \tag{5.36}$$

According to Theorems 2.12 and 5.1, corresponding to the optimal process $x(t)$, $u(t)$ there exists a nontrivial solution $\{\psi_1(t), \ldots, \psi_n(t), \psi_{n+1}(t)\}$ of (5.33), (5.34) such that during the entire motion the function H^* takes on its maximum value (with respect to u) and the relation

$$H^*\Big|_{t=t_1} \geqslant 0 \tag{5.37}$$

is satisfied. In addition, according to Theorem 2.13,

$$H^* = \text{const} \tag{5.38}$$

is valid for the functions $x^i(t)$, $u^k(t)$, $\psi_i(t)$. But by (5.32), the maximality of H^* means that H also attains its maximum (with respect to u), that is,

$$H(\psi(\tau), x(\tau), \tau, u(\tau)) = \max_{u \in U} H(\psi(\tau), x(\tau), \tau, u) \tag{5.39}$$

for any time τ which is a point of continuity of the control $u(t)$. Furthermore, it follows from (5.32) and (5.35) that (5.37) can be rewritten in the form

$$H\Big|_{t=t_1} \geqslant 0$$

that is,

$$H(\psi(t_1), x(t_1), t_1, u(t_1)) \geqslant 0. \tag{5.40}$$

By the same token, a new proof is obtained for the existence of functions $\psi_1(t)$, $\ldots$, $\psi_n(t)$ satisfying (5.33), (5.39), and (5.40), that is, a proof of the maximum principle for the nonautonomous system (5.30). Furthermore, comparing (5.32), (5.36), and (5.38), we obtain

$$H(\psi(t), x(t), t, u(t)) + \int_t^{t_1} \sum_{\alpha=1}^{n} \frac{\partial f^\alpha(x(t), t, u(t))}{\partial t} \psi_\alpha(t) \, dt = \text{const} \geqslant 0,$$

that is,

$$H(\psi(t), x(t), t, u(t)) = \int_{t_1}^{t} \sum_{\alpha=1}^{n} \frac{\partial f^\alpha(x(t), t, u(t))}{\partial t} \psi_\alpha(t) \, dt - \psi_0, \quad (5.41)$$

where ψ_0 is a nonpositive constant.

Thus, *if $x(t)$, $u(t)$ is a time-optimal process of the nonautonomous system* (5.30), *then there exists a nontrivial vector function* $\psi(t) = \{\psi_1(t), \psi_2(t), \ldots, \psi_n(t)\}$ *which satisfies the maximum principle and relation* (5.41).

The case in which optimality for the object (5.30) is understood in the sense of minimizing the functional

$$I = \int_{t_0}^{t_1} f^0(x, t, u) \, dt \quad (5.42)$$

[compare (5.19)], where f^0 is continuous and continuously differentiable in $x^1, \ldots, x^n, t$, is treated in a completely analogous manner. Going over from the space X to the space X^* with the aid of the new variable $x^{n+1} \equiv t$ [see (5.31)], and repeating the arguments on pp. 250–253 for the function $\mathcal{H}$ (see Theorem 5.3), we obtain

$$\mathcal{H} = \psi_0 f^0(x, t, u) + \psi_1 f^1(x, t, u) + \cdots + \psi_n f^n(x, t, u);$$

$$\mathcal{H}^* = \mathcal{H} + \psi_{n+1}; \quad (5.32')$$

$$\psi_i = -\frac{\partial \mathcal{H}^*}{\partial x^i} = -\frac{\partial \mathcal{H}}{\partial x^i}, \quad i = 1, \ldots, n; \quad (5.33')$$

$$\psi_{n+1} = -\frac{\partial \mathcal{H}^*}{\partial x^{n+1}} = -\frac{\partial \mathcal{H}}{\partial t} = -\sum_{\alpha=0}^{n} \frac{\partial f^\alpha(x, t, u)}{\partial t} \psi_\alpha; \quad (5.34')$$

$$\psi_{n+1}(t_1) = 0; \quad (5.35')$$

$$\psi_{n+1}(t) = \int_t^{t_1} \sum_{\alpha=0}^{n} \frac{\partial f^\alpha(x, t, u)}{\partial t} \psi_\alpha \, dt. \quad (5.36')$$

The relation

$$\max_{u \in U} \mathcal{H}^*(\psi(t), x(t), t, u) = \mathcal{H}^*(\psi(t), x(t), t, u(t)) \equiv 0, \qquad t_0 \leqslant t \leqslant t_1$$

[see (5.28)] now takes the form

$$\max_{u \in U} \mathcal{K}(\psi(t), x(t), t, u) = \mathcal{K}(\psi(t), x(t), t, u(t))$$

$$= \int_{t_1}^{t} \sum_{\alpha=0}^{n} \frac{\partial f^\alpha(x(t), t, u(t))}{\partial t} \, \psi_\alpha(t) \, dt. \tag{5.41'}$$

Thus, *if* $x(t)$, $u(t)$, $t_0 \leqslant t \leqslant t_1$, *is an optimal process* [*in the sense of minimizing the functional* (5.42)] *of the nonautonomous system* (5.30), *then there exists a constant* $\psi_0 \leqslant 0$ *and a nontrivial solution* $\psi(t)$, $t_0 \leqslant t \leqslant t_1$, *of* (5.33') *such that the maximum condition* (5.41') *is satisfied for any time t which is a point of continuity of the control* $u(t)$.

The transversality conditions are formulated in the case of fixed manifolds M_0 and M_1 in the same manner as in the autonomous case. In the case of a moving manifold M_1, for example, if from a given point x_0 we must arrive at a point $x_1(t)$, moving according to a certain law, the transversality conditions can be readily obtained by the same method, that is, by a transition from the nonautonomous system in the space X to the autonomous system in the space X^*.

64. Optimal Processes with Parameters.

Let us consider the following optimal problem. Functions $f^i(x, u, w)$, $i = 0, 1, \ldots, n$ are given, where the arguments x, u, w are points of spaces of dimensions n, r, s, respectively,

$$x = (x^1, x^2, \ldots, x^n), \qquad u = (u^1, \ldots, u^r), \qquad w = (w^1, \ldots, w^s).$$

In addition, a set U (the control region) is given in $u^1, \ldots, u^r$-space. The functions f^i and their partial derivatives $\partial f^i / \partial x^j$, $\partial f^i / \partial w^k$ ($i = 0, \ldots, n$; $j = 1, \ldots, n$; $k = 1, \ldots, s$) are assumed to be continuous for $u \in U$ and for any x, w.

The law of motion of the object (in the phase space X of the variables $x^1, \ldots, x^n$) is given by the equation

$$\frac{dx^i}{dt} = f^i(x, u, w), \qquad i = 1, \ldots, n. \tag{5.43}$$

The optimal problem that we want to consider now consists in the following. Given two points x_0 and x_1 in X, it is required to choose a fixed point w (that is, at the outset, a value of the parameter w is to be selected which remains constant during the entire motion) and an admissible control $u(t)$ such that the corresponding trajectory $x(t)$, emanating from x_0 at time t_0, passes at time t_1 through x_1 and such that the integral

$$J = \int_{t_0}^{t_1} f^0(x(t), u(t), w) \, dt \tag{5.44}$$

takes on the least possible value. In solving this problem we shall additionally assume that $f^0 > 0$ (for any values of the arguments). We note that
for $f^0 \equiv 1$ we obtain the time-optimal problem for the object (5.43).

The problem considered here has some special features as compared
with the optimal problem considered in § 1.4 (or in § 15.57). In the optimal
problem of § 1.4 and § 15.57, each piece of the optimal trajectory was also
an optimal trajectory (since the "improvement" of a piece of trajectory
resulted in the improvement of the entire trajectory). Here, in the problem
with parameters, this is not so. As a matter of fact, the optimal values of the
parameter w for the entire trajectory and for its pieces may not coincide,
that is, if $u(t)$, w yield the solution of the optimal problem posed in this
section, and moreover, if the control $u(t)$ is defined on the interval $t_0 \leqslant t$
$\leqslant t_1$, then on a smaller interval it may be possible to "improve" the control
$u(t)$ at the expense of changing the parameter w.

To solve the problem posed here we shall proceed in the following
manner. Let us add to the equations of motion of the object (5.43) another
s equations

$$\frac{dw^i}{dt} \equiv 0, \qquad i = 1, \ldots, s. \tag{5.45}$$

Equations (5.43) and (5.45) describe the motion of a new object in the
phase space X^* of the variables $x^1, \ldots, x^n, w^1, \ldots, w^s$. Furthermore,
let M_0 denote the set of all points $(x^1, \ldots, x^n, w^1, \ldots, w^s)$ of X^* satisfying the condition

$$(x^1, \ldots, x^n) = x_0,$$

and let M_1 denote the set of all points satisfying the condition

$$(x^1, \ldots, x^n) = x_1$$

(where x_0 and x_1 are the points given in the statement of the problem under
consideration). It is clear that M_0 and M_1 are s-dimensional planes in X^*
parallel to the last s axes (that is, to the axes $w^1, \ldots, w^s$).

Now, let us pose the problem for the object (5.43), (5.45) as the problem
of arriving from a certain point of the plane M_0 to a certain point of the
plane M_1 with the aid of a control $u(t)$, $t_0 \leqslant t \leqslant t_1$, which provides the least
possible value of the functional (5.44).

It can be readily seen that the problem with variable endpoints posed
here is equivalent to the problem with constant parameters w. In fact, if
the functions

$$u(t), \, x^*(t) = (x^1(t), \ldots, x^n(t), w^1(t), \ldots, w^s(t))$$

satisfy (5.43), (5.45) and transfer the phase point from a state $x_0^* \in M_0$

to a state $x_1^* \in M_1$, then, by (5.45), we have

$$w^i = \text{const}, \qquad i = 1, \ldots, s.$$

Furthermore, the quantities

$$u(t),\, x(t) = (x^1(t),\, \ldots,\, x^n(t)),\, w = (w^1,\, \ldots,\, w^s)$$

satisfy (5.43). In addition, $x(t_0) = x_0$ and $x(t_1) = x_1$ (since $x_0^* \in M_0$ and $x_1^* \in M_1$). Thus, any transfer process of (5.43), (5.45) from the plane M_0 to the plane M_1 yields a transfer process from the point x_0 to the point x_1 of (5.43) with the condition $w = \text{const}$ (and vice versa). This, in fact, proves that the problem with variable endpoints is equivalent to the problem with constant parameters w considered at the beginning of this section.

Let us now proceed to the solution of the above problem with variable endpoints. The function $\mathcal{H}$ for this problem has the form

$$\mathcal{H} = \psi_0 f^0(x, u, w) + \psi_1 f^1(x, u, w) + \cdots + \psi_n f^n(x, u, w) + \psi_{n+1} \cdot 0$$

$$+ \cdots + \psi_{n+s} \cdot 0 = \psi_0 f^0 + \psi_1 f^1 + \cdots + \psi_n f^n.$$

The system of differential equations for the auxiliary variables can be written as follows:

$$\dot{\psi}_i = -\frac{\partial \mathcal{H}}{\partial x^i} = -\sum_{\alpha=0}^{n} \psi_\alpha \frac{\partial f^\alpha(x, u, w)}{\partial x^i}, \qquad i = 1, \ldots, n; \qquad (5.46)$$

$$\dot{\psi}_{n+j} = -\frac{\partial \mathcal{H}}{\partial w^j} = -\sum_{\alpha=0}^{n} \psi_\alpha \frac{\partial f^\alpha(x, u, w)}{\partial w^j}, \qquad j = 1, \ldots, s. \qquad (5.47)$$

Further, let us write the maximum condition [see (5.29)]:

$$\max_{u \in U} \mathcal{H}(\psi(t), x(t), u, w) = \mathcal{H}(\psi(t), x(t), u(t), w) \equiv 0,$$

$$(5.48)$$

$$t_0 \leqslant t \leqslant t_1.$$

Finally, since both manifolds M_0, M_1 are planes parallel to the last s axes of the space X^*, the transversality conditions take the following form:

$$\psi_{n+1}(t_0) = \cdots = \psi_{n+s}(t_0) = 0, \qquad (5.49)$$

$$\psi_{n+1}(t_1) = \cdots = \psi_{n+s}(t_1) = 0. \qquad (5.50)$$

Thus, by virtue of Theorems 5.4 and 5.3, we obtain the following result. *In order that* $u(t)$, $x(t)$, w *yield the solution of the posed problem with variable endpoints (or, equivalently, the original problem with constant param-*

*eters w), it is necessary that there exist a constant $\psi_0 \leqslant 0$ and a nonzero con-
tinuous vector function $\psi^*(t) = (\psi_1(t), \ldots, \psi_{n+s}(t))$ which satisfy* (5.46),
(5.47), *the maximum condition* (5.48), *and the transversality conditions* (5.49),
(5.50). This, in fact, is the theorem which yields the solution of the formulated problem. However, we shall modify somewhat relations (5.47), (5.49),
and (5.50) in order to present this theorem in a more convenient form
(namely, we shall eliminate the functions $\psi_{n+1}, \ldots, \psi_{n+s}$). It should be
noted, first of all, that by (5.47), (5.49), we have

$$\psi_{n+j}(t) = -\int_{t_0}^{t} \sum_{\alpha=0}^{n} \psi_\alpha \frac{\partial f^\alpha(x, u, w)}{\partial w^j} \, dt, \qquad j = 1, \ldots, s.$$

Then relation (5.50) takes the following form:

$$\int_{t_0}^{t_1} \sum_{\alpha=0}^{n} \psi_\alpha \frac{\partial f^\alpha(x, u, w)}{\partial w^j} \, dt = 0, \qquad j = 1, \ldots, s. \tag{5.51}$$

Thus, instead of the three systems (5.47), (5.49), (5.50), we obtain
the single system (5.51); moreover, the consideration of the functions
$\psi_{n+1}, \ldots, \psi_{n+s}$ becomes unnecessary. In other words, we obtain the
following result.*

*Theorem 5.7. In order that $u(t), x(t), w$ yield the solution of the problem
with parameters, it is necessary that there exist a constant $\psi_0 \leqslant 0$ and a nonzero continuous vector function $\psi(t) = (\psi_1(t), \ldots, \psi_n(t))$ satisfying the
system* (5.46), *the maximum condition* (5.48), *and the additional relations*
(5.51).

For $f^0 \equiv 1$, that is, in the case of the time-optimal problem for the object
(5.43) (with the parameter $w = $ const), we have

$$\mathcal{H} = \psi_0 + (\psi_1 f^1 + \cdots + \psi_n f^n) = \psi_0 + H,$$

and therefore, the relations (5.46), (5.48), and (5.51) take the form

$$\psi_i = -\frac{\partial H}{\partial x^i} = -\sum_{\alpha=1}^{n} \psi_\alpha \frac{\partial f^\alpha(x, u, w)}{\partial x^i}, \qquad i = 1, \ldots, n, \tag{5.46'}$$

$$\max_{u \in U} H(\psi(t), x(t), u, w) = H(\psi(t), x(t), u(t), w) = \text{const} \geqslant 0, \tag{5.48'}$$

$$\int_{t_0}^{t_1} \sum_{\alpha=1}^{n} \psi_\alpha \frac{\partial f^\alpha(x, u, w)}{\partial x^j} \, dt = 0, \qquad j = 1, \ldots, s. \tag{5.51'}$$

---

* The corresponding theorem in [9] is formulated erroneously.

We note that Theorem 5.7 differs from Theorem 5.3 (or 2.12) by the presence of s additional relations (5.51); this, in fact, makes it possible to solve the problem since s additional unknowns $w^1, w^2, \ldots, w^s$ are introduced into this problem.

It can be proved that if the parameter w is allowed to vary not in the entire space W of variables $w^1, \ldots, w^s$, but only in some closed domain $W_1 \subset W$, having a piecewise smooth boundary, then condition (5.51) is replaced by the relations

$$\int_{t_0}^{t_1} \sum_{\alpha=0}^{n} \psi_\alpha \frac{\partial f^\alpha(x, u, w)}{\partial \tau} \, dt \leqslant 0,$$

where the derivative under the integral sign is taken along any direction τ emanating from the point w and directed into the domain W_1. In other words, for any differentiable curve $w(\theta)$, emanating for $\theta = 0$ from the point w and passing through the region W_1, the relation

$$\int_{t_0}^{t_1} \sum_{\alpha=0}^{n} \psi_\alpha \frac{\partial f^\alpha(x, u, w(\theta))}{\partial \theta} \bigg|_{\theta=0} dt \leqslant 0$$

must be satisfied.

65. The Isoperimetric Problem and the Problem with Fixed Time.

The optimal problem considered in this section will be called *isoperimetric* (in analogy with the terminology used in geometry and the calculus of variations). As in the previous section, we shall confine ourselves to the autonomous case.

Let the functions

$$f^0(x, u), f^1(x, u), \ldots, f^n(x, u), g^1(x, u), \ldots, g^k(x, u)$$

be continuous together with their partial derivatives with respect to x^1, $\ldots$, x^n. As always, it is assumed that the point $u = (u^1, \ldots, u^r)$ may take on values in a prescribed set U (the "control region"), in $u^1, \ldots$, u^r-space, and that the point $x = (x^1, \ldots, x^n)$ may run through the entire phase space X.

The isoperimetric optimal problem is now formulated as follows. Consider an object moving according to the law

$$\dot{x}^i = f^i(x, u), \qquad i = 1, \ldots, n. \tag{5.52}$$

Given two points x_0 and x_1 in X and k real numbers $\eta^1, \eta^2, \ldots, \eta^k$. From among all the admissible controls $u(t)$, $t_0 \leqslant t \leqslant t_1$, transferring the phase point from the state x_0 to the state x_1, it is required to choose one which satisfies the conditions

$$\int_{t_0}^{t_1} g^j(x, u) \, dt = \eta^j, \qquad j = 1, \ldots, k \qquad (5.53)$$

and makes the functional (5.19) take on the least possible value.

In solving this problem it is natural to assume that the functions f^0, $g^1, \ldots, g^k$ are linearly independent.

To solve this problem, let us consider the following system of differential equations:

$$\dot{y}^j = g^j(x, u), \qquad j = 1, \ldots, k, \qquad (5.54)$$

with initial conditions

$$y^1(t_0) = \cdots = y^k(t_0) = 0. \qquad (5.55)$$

Since the right-hand sides of (5.54) do not contain the variables y^j, the solution of this system can be written [considering the initial values (5.55)] in the form

$$y^j(t) = \int_{t_0}^{t} g^j(x, u) \, dt, \qquad j = 1, \ldots, k.$$

Thus, by (5.54), (5.55), the conditions (5.53) are equivalent to the conditions

$$y^j(t_1) = \eta^j, \qquad j = 1, \ldots, k. \qquad (5.56)$$

We see that the isoperimetric problem we have formulated is equivalent to the following problem. We consider an object moving in the phase space X^* of variables $x^1, \ldots, x^n, y^1, \ldots, y^k$ according to the law (5.52), (5.54). It is required to find an admissible control $u(t)$, $t_0 \leqslant t \leqslant t_1$, transferring this object from the state $x = x_0$, $y = 0$ [see (5.55)] to the state $x = x_1$, $y = \eta$ [see (5.56)] and ascribing the least possible value to the functional (5.19). This is the usual optimal problem with fixed endpoints; Theorem 5.3 can be used to solve this problem. According to this theorem, to solve the problem it is necessary to construct the function (we denote this function by $\mathcal{K}^*$ rather than $\mathcal{K}$)

$$\mathcal{K}^* = \psi_0 f^0(x, u) + \psi_1 f^1(x, u) + \cdots + \psi_n f^n(x, u)$$
$$+ \psi_{n+1} g^1(x, u) + \cdots + \psi_{n+k} g^k(x, u). \qquad (5.57)$$

The corresponding system of equations for the auxiliary variables $\psi_1, \ldots, \psi_n, \ldots, \psi_{n+k}$ has the form

$$\dot{\psi}_i = -\frac{\partial \mathcal{K}^*}{\partial x^i} = -\sum_{\alpha=0}^{n} \psi_\alpha \frac{\partial f^\alpha(x, u)}{\partial x^i} - \sum_{\beta=1}^{k} \psi_{n+\beta} \frac{\partial g^\beta(x, u)}{\partial x^i}, \qquad (5.58)$$

$$i = 1, \ldots, n,$$

$$\dot{\psi}_{n+j} = -\frac{\partial \mathcal{K}^*}{\partial u^j} \equiv 0, \qquad j = 1, \ldots, k. \qquad (5.59)$$

We see from (5.59) that $\psi_{n+j} = $ const; the value of this constant will be denoted by λ_j. Thus relations (5.58), (5.59) take the following form:

$$\dot{\psi}_i = -\sum_{\alpha=0}^{n} \psi_\alpha \frac{\partial f^\alpha(x,\,u)}{\partial x^i} - \sum_{\beta=1}^{k} \lambda_\beta \frac{\partial g^\beta(x,\,u)}{\partial x^i},$$

(5.60)

$$i = 1,\,\ldots,\,n,\qquad \lambda_j = \text{const},\qquad j = 1,\,\ldots,\,k,$$

and the function $\mathcal{K}^*$ has the form

$$\mathcal{K}^* = \sum_{\alpha=0}^{n} \psi_\alpha f^\alpha(x,\,u) + \sum_{\beta=1}^{k} \lambda_\beta g^\beta(x,\,u).$$

Finally, let us write the maximum condition:

$$\max_{u\in U} \mathcal{K}^*(\psi(t),\,x(t),\,u) = \mathcal{K}^*(\psi(t),\,x(t),\,u(t)) \equiv 0,$$

$$t_0 \leqslant t \leqslant t_1.$$

(5.61)

If the vector function $(\psi_1(t),\,\ldots,\,\psi_n(t))$ were identically equal to zero, then, by (5.61), the identity

$$\psi_0 f^0 + \lambda_1 g^1 + \cdots + \lambda_k g^k \equiv 0$$

would hold; moreover, at least one of the coefficients $\lambda_1,\,\ldots,\,\lambda_k$ would differ from zero since, according to what has been said above, the vector function

$$(\psi_1(t),\,\ldots,\,\psi_n(t),\,\psi_{n+1}(t),\,\ldots,\,\psi_{n+k}(t))$$
$$= (\psi_1(t),\,\ldots,\,\psi_n(t),\,\lambda_1,\,\ldots,\,\lambda_k)$$

is nontrivial. But this contradicts the linear independence of the functions $f^0, g^1, \ldots, g^k$. Consequently, the vector function $\psi(t) = (\psi_1(t),\,\ldots,\,\psi_n(t))$ is nontrivial.

Thus, *in order that* $u(t)$, $x(t)$ *[transferring the phase point from the state x_0 to x_1 and satisfying the relations (5.53)] yield the solution of the posed isoperimetric problem, it is necessary that there exist constants* $\psi_0 \leqslant 0, \lambda_1,\,\ldots,\,\lambda_k$ *and a nonzero continuous vector function* $\psi(t) = (\psi_1(t),\,\ldots,\,\psi_n(t))$ *satisfying the system (5.60) and the maximum condition (5.61).*

As an example of the application of this theorem, which yields the solution of the isoperimetric problem, let us consider the optimal problem with fixed time. The optimal problem is formulated as in § 15.59, but with the condition that the time t_0 of the start of the motion of a point (from the state x_0) and the time t_1 of the point's arrival at x_1 are prescribed, so that the time $t_1 - t_0$ is fixed. Denoting the given time of motion $t_1 - t_0$ by η^1 and setting

$g^1(x, u) \equiv 1$, we can write the condition imposed in the above problem in the form

$$\int_{t_0}^{t_1} g^1(x, u)\, dt = \eta^1. \tag{5.62}$$

Thus, we arrive at the isoperimetric problem for the object (5.52) with a single relation (5.62). In order to solve this isoperimetric problem, let us construct the function $\mathcal{JC}^*$. Since $g^1 \equiv 1$, this function has the form

$$\mathcal{JC}^* = \sum_{\alpha=0}^{n} \psi_\alpha f^\alpha(x, u) + \lambda_1 = \mathcal{JC} + \lambda_1, \tag{5.63}$$

where $\lambda_1 = \text{const}$. Furthermore, system (5.60) takes the form

$$\dot{\psi}_i = -\frac{\partial \mathcal{JC}^*}{\partial x^i} = -\frac{\partial \mathcal{JC}}{\partial x^i},$$

and the maximum condition (5.61) can be written in the form [see (5.63)]

$$\max_{u \in U} \mathcal{JC}(\psi(t), x(t), u) = \mathcal{JC}(\psi(t), x(t), u(t)) \equiv -\lambda_1.$$

We see that the constant λ_1 (about which nothing is known) is actually not needed; that is, we obtain the following

Theorem 5.8.   In order that $u(t)$, $x(t)$ yield the solution of the above optimal problem with fixed time, it is necessary that there exist a constant $\psi_0 \leqslant 0$ and a nonzero continuous vector function

$$\psi(t) = (\psi_1(t), \ldots, \psi_n(t))$$

satisfying the system

$$\dot{\psi}_i = -\frac{\partial \mathcal{JC}}{\partial x^i}, \qquad i = 1, \ldots, n,$$

and the maximum condition

$$\max_{u \in U} \mathcal{JC}(\psi(t), x(t), u) = \mathcal{JC}(\psi(t), x(t), u(t)),$$
$$t_0 \leqslant t \leqslant t_1,$$

where

$$\mathcal{JC} = \sum_{\alpha=0}^{n} \psi_\alpha f^\alpha(x, u).$$

It is also not difficult to obtain (by the methods in § 16.63) the solution of the isoperimetric problem and the solution of the problem with fixed time for the case in which the functions f^i and g^i depend explicitly on t.

References

[1] Boltyanskii, V. G., "Sufficient Conditions for Optimality and the Justification of the Dynamic Programming Method," Izv. Akad. Nauk SSSR Ser. Mat., *28* (1964), 481–514. English translation in SIAM J. Control, *4* (1966), 326–361.

[2] Boltyanskii, V. G. and E. Ya. Roitenberg, "An Example of Synthesis of a Second-Order Nonlinear System," Kibernetika (Kiev), No. 4 (1966), 52–56. English translation in Cybernetics, *2* (1966), No. 4, 42–45.

[3] Cairns, S. S., "On the Triangulation of Regular Loci," Ann. of Math., *35* (1934), 579–587.

[4] Dubovitskii, A. Ya., "On Differentiable Mappings of an n-Dimensional Cube into a k-Dimensional Cube" (in Russian), Mat. Sb., *32* (74) (1953), 443–464.

[5] Eaton, J. H., "An Iterative Solution to Time-Optimal Control," J. Math. Anal. Appl., *5* (1962), 324–344.

[6] Neustadt, L. W., "Synthesizing Time-Optimal Control Systems," J. Math. Anal. Appl., *1* (1960), 484–493.

[7] Pontryagin, L. S., "Smooth Manifolds and Their Application in the Homotopy Theory" (in Russian), Trudy Mat. Inst. Steklov, No. 45, Izdat. Akad Nauk SSSR, Moscow, 1955.

[8] Pontryagin, L. S., *Ordinary Differential Equations*. Translated from the Russian by Leonas Kacinskas and Walter B. Counts, Addison-Wesley Publishing Company, Inc., Reading, Mass., 1962.

[9] Pontryagin, L. S., V. G. Boltyanskii, R. V. Gamkrelidze, and E. F. Mishchenko, *The Mathematical Theory of Optimal Processes*. Translated from the Russian by K. N. Trirogoff, edited by L. W. Neustadt. Interscience Publishers, New York, 1962.

[10] Shilov, G. E., *An Introduction to the Theory of Linear Spaces*. Translated from the Russian by Richard A. Silverman. Prentice-Hall, Englewood Cliffs, N.J., 1961.

[11] Tricomi, F. G., *Differential Equations*. Translated from the Italian by Elizabeth A. McHarg, Hefner Publishing Company, New York, 1961.

Athans, M. and P. L. Falb, *Optimal Control; an Introduction to the Theory and its Applications*, McGraw-Hill, Inc., New York, 1966.

Bellman, R. E., *Dynamic Programming*, Princeton University Press, Princeton, N.J., 1957.

Lee, E. B., and L. Markus, *Foundations of Optimal Control Theory*, John Wiley & Sons, Inc., New York, 1967.

Index